Rays, Waves and Photons

A compendium of foundations and emerging technologies of pure and applied optics

Rays, Waves and Photons

A compendium of foundations and emerging technologies of pure and applied optics

William L Wolfe

Professor Emeritus, James C. Wyant College of Optical Sciences, University of Arizona, Tucson, Arizona, USA

IOP Publishing, Bristol, UK

ISBN 978-0-7503-2612-4 (ebook)
ISBN 978-0-7503-2610-0 (print)
ISBN 978-0-7503-2613-1 (myPrint)
ISBN 978-0-7503-2611-7 (mobi)

DOI 10.1088/978-0-7503-2612-4

Version: 20200801

IOP ebooks

British Library Cataloguing-in-Publication Data: A catalogue record for this book is available from the British Library.

Published by IOP Publishing, wholly owned by The Institute of Physics, London

IOP Publishing, Temple Circus, Temple Way, Bristol, BS1 6HG, UK

US Office: IOP Publishing, Inc., 190 North Independence Mall West, Suite 601, Philadelphia, PA 19106, USA

I want to dedicate this book to my entire family, especially to the memory of my loving and supportive wife, who has put up with all my time at the computer and library instead of going out to eat.

Contents

Appendices

Prologue

This book started in a very unusual way. My grandson, Garrett, was interested in Thomas Aquinas College in California. It has an unusual curriculum. All students are required to take exactly the same courses. There are only about 350 students. The curriculum consists of studying liberal arts by reading and studying the classical works written by the masters—in the original, or translations of them. It seemed like a good idea to me for philosophy, literature, and theology, but I was not so sure about mathematics and science. I learned that Euclid's *Elements* was part of the math curriculum and that they learned optics by reading Newton's *Opticks* and Huygens' *Treatise of Light.* I was familiar with *Opticks*, and I tracked down and perused the other two. They were interesting reading, but certainly not good, modern texts to learn mathematics and optics. During this same time, I was also asked to review a manuscript for SPIE that traced the history of light. It took me through the works of Planck and the origin of the laser, among other topics. They were all very interesting reading, and I was turned on.

As I started my thinking about the history of optics, making an outline and recalling the significant events, I realized that I knew many of the players, and this became something of a memoir. You will immediately realize that I did not know Galileo or Newton, but I did know many of the more modern people who have contributed to many of our latter advances. I presented an award to Rudolf Kingslake, one of the first stars at the Institute of Optics and a renowned pupil of Conrady. I also presented an award to André Maréchal, a principal at the Institut d'Optique. I attended meetings with and traveled with H H Hopkins, who invented the fiber optic endoscope (with others). (The periods have been omitted after Hopkins' initials according to IOP convention; this convention is used throughout.) Francis Turner was a colleague at Arizona and a pupil of Marianus Czerny, one of the pioneers of infrared technology and the designer of the first radiation slide rule. Francis himself was one of the pioneers in thin film technology. In fact, Francis gave me the Czerny slide rule I show in that chapter; I call it the Czerny–Turner rule. Aden Meinel, one of the inventors of the multiple mirror telescope, was director of the Optical Science Center when I was there. I collaborated with Bob Jones, one of the giants of polarized light. I had the honor of instituting the High-Speed Photography award of SPIE, commemorating Harold Edgerton.

Of course, there were even more in the infrared field, my chosen specialty. I discussed the Dove missile with Edwin Land and David Grey of Polaroid. I worked with Warren Arnquist on the 1958 edition of Proc IRE, in which he published an extensive history of the subject. I consider Stan Ballard, who organized that Proc IRE edition, to be one of my mentors. Luc Biberman, another mentor, was on the original Sidewinder team. John Jamison, who had much to do with our early missile launch detectors, was a colleague. It was Werner Weihe who gave me advice about making our mercury telluride detectors available to others. I have fond memories of discussing infrared with Bill Craven and Larry Nichols, who followed McClean in Sidewinder development. Paul Kruse, my colleague of Honeywell days, made it possible for me to be one of the first to design systems based on mercury cadmium telluride and to continue later with microbolometer arrays.

My colleagues at the Wyant College of Optical Sciences have been invaluable in both making this a better book and educating me in their respective disciplines—Bob Shannon on lens design; Angus Macloud on thin films and filters; Marlon Scully, Pierre Meystre, and Murray Sargent on quantum optics; Roland Shack on all sorts of classical optics; Stacey Dereniak on infrared; Jim Wyant on interferometers and organizations; Harry Barrett on medical optics and tomography; Jose Sasian on lithography; and John Grievencamp for the many pictures from his collection of old optical instruments. John Bruning and SPIE gave me permission to use much from his fine articles on the history of lithography.

Brittani Scarpulla did an extremely thorough job editing my prose and correcting my grammar. In a true optical sense, I ground the prose into shape, and she polished it off.

I could go on, but it is enough to remember and honor many friends and colleagues who have made major contributions to modern optics. If I have not named you here, and we were good friends and colleagues, please forgive me; it would be too much to list everyone who has critiqued, criticized, commented, kibitzed, and contributed to this tome and my career. Thank you.

Permissions and acknowledgements

Figures 4.3 and 4.4 were provided by Professor John Grievencamp from his collection of old optical instruments at the James C. Wyant College of Optical Sciences at the University of Arizona.

The Optical Society graciously provide permission for the use of its logo and helped with the description.

SPIE, the International Society for Optics and Photonics, graciously gave permission for the use of its logo and several of the book images.

The Institute of Optics in the Hajim School of Engineering gave their permission for the use of their logos

CREOL, the College of Optics and Photonics, kindly gave permission for the use of its logo.

The IRIA logo was created by me and we did not need permission.

For figure 13.6, the Boston Police Department gave permission for the use of its iconic infrared image of the Boston Bomber hiding in a boat under a polyethylene cover.

Author biography

William L Wolfe

William L (Bill) Wolfe was born in Yonkers, NY on Easter Sunday in 1931. He earned a BS in physics from Bucknell University, an MS in physics and MSE in electrical engineering from the University of Michigan, where he also worked as a Research Engineer and Lecturer. He left academia for the Honeywell Radiation Center in 1966 where he was Chief Engineer and the Department Manager of the Electro-optics Systems Department. He joined what was then the Optical Sciences Center of the University of Arizona and is now the James C Wyant College of Optical Sciences as Professor of Optical Sciences in 1969. He retired as an Emeritus Professor in 1995. He is still vertical most of the day.

He was president of SPIE in 1989 and received its gold medal, its highest award, in 1999. He received Bucknell's award as Most Successful in a Chosen Career in 1954, the Civilian Service medal from the Army, NASA Pioneer Venus Mission Achievement Award, SPIE President's award and others.

He has consulted for many American companies, been an expert witness many times and served on advisory committees for all the military services, the National Academy of Science, the National Bureau of Standards and the Department of Justice.

He is the author of four technical books on infrared system design, radiometry and spectral imaging, two nontechnical ones on optics for the layman and fly fishing float trips; he is also the editor of two infrared handbooks, one on optics in general and an associate editor of two other optics handbooks. He was the editor of the journals: *Proceedings of the Infrared Information Symposia and Infrared Physics and Technology as well as the book series Optical Physics and Technology.*

He is proud of his three children who all have had professional careers and his six grandchildren all but two of whom are graduated from college.

He enjoys singing, fly fishing and crossword puzzles. He sings a few, catches a few and solves a few.

IOP Publishing

Rays, Waves and Photons

A compendium of foundations and emerging technologies of pure and applied optics

William L Wolfe

Chapter 1

Introduction—pleased to meet you

1.1 In the beginning

The history of light and optics parallels much of the history of science. Light was obvious to the ancients, and it was a curiosity. The blue sky, rainbows, bending in media, its speed, stars, the Sun and Moon. Many advances in science were fostered or made possible by light or by the use of it. Light is at the heart of astronomy, which, in turn, has revealed most of the mysteries of our Universe and its origin. These include our heliocentric world and the Big Bang itself. The nature of blackbody emission, considered by Max Planck, opened the doors to quantum mechanics. Light was fundamental in the conception of both special and general relativity by Einstein. There would be no telescopes without the advent of lenses and mirror design. The construction of the Bohr atom was only possible with the results of spectroscopy, and modern ideas of atomic structure have come from analysis of such things as the structure of spectral lines. The ubiquitous smart phone has its origins in the photoelectric effect, detectors, transistors, integrated circuits, and the charge-coupled device. And, of course, the GoPro is a camera.

1.2 Optics today

If mathematics is the queen of science as some say, then optics must be the handmaiden. In addition to providing the impetus and tools for the realization of our heliocentric world, the origin of the Universe, the basis of quantum mechanics, and the elicitation of atomic and molecular structure, mentioned above, it provides the foundation of many modern conveniences and even essentials. Our blood and urine samples go to the clinical laboratory where about 75% of the tests are optical. A Google search goes to one of several processing sites, each of which has about one million diode lasers and fiber optics. Our television programs come to us via optical cables that provide many more data channels than copper wire. Without optical cables and laser communication, we could not tune in to channel 656!

We are on the verge of an entire revolution in lighting. Incandescent bulbs are being replaced by compact fluorescents and LED's, which are much more efficient and long lasting. The 2014 Nobel Prize in physics was for the efficient, white LED. We are also on the verge of real three-dimensional television based on holographic principles. Our street intersections are monitored by digital cameras; our houses are protected by infrared intrusion detectors; some of our cars have lidar crash avoidance sensors and night vision systems; and we have many more mundane optical devices that serve us. These include spectacles that may be bifocals, trifocals, or multifocals, photochromic sunglasses, electrochromic windows, and even self-adjusting spectacles. Our crops are monitored overhead by remote sensing devices that measure soil moisture, crop conditions, and degree of ripeness. Although we may not like it, radar speed sensors are augmented by laser-based ones.

Optics is playing a major role in manufacturing. Robots have taken over many of the repetitive tasks formerly done by assembly workers. These devices have optical sensors and laser-based distance measurers. Three-dimensional printing has shown its face and promises to be a valuable technique for producing many objects—at least if they are not too big. Infrared cameras are proving valuable in searching for lost persons and for police operations, perhaps most notably in the detection of the Boston Marathon bomber. Spectroscopy is valuable for identifying substances for forensics, including DNA. It is equally useful in a myriad of tasks in chemistry and biology laboratories. Autonomous vehicles have made their debut. We may soon see the day when the local cab shows up at our doorstep, takes us to our destination, and then follows its next instruction—all without a driver who we do not have to tip. It is not unreasonable to imagine semis traveling cross country with no drivers as well, competing with trains. Trains no longer have men in cabooses, and they may not have engineers in the engines in the future.

Where would sports be without binoculars for viewers, camcorders for sports practice, laser rangers for golfers, laser tension measurers for tennis racquets, and TV's and TV remotes for armchair quarterbacks?

Our nightly TV reports on the weather depend upon several satellites in both geosynchronous and polar orbit to image the projected Earth in the infrared and to probe it with lasers. Computer jocks have a great time taking these images and turning them into fancy colored renditions. These satellites would not remain erect if it were not for infrared horizon sensors.

The military depends upon optics both old and new. Periscopes are still valuable to submarine commanders and binoculars to field commanders. But we now also have laser designators and directed-energy weapons. The designators are in the field, but laser weapons are still in the testing stage. For decades we have detected the launch of rockets that could be ICBM's or satellites by infrared detection means—sensing the plume from geosynchronous orbit. We have successfully intercepted test launches over the Pacific with infrared homing devices in rockets. The Sidewinder missile and all its descendants have been the mainstay of our Air Force air-to-air missile program. Infrared cameras enable vehicles to drive in the dark, and similar devices have provided the ability to direct weapons. Telescope technology has enabled us to put reconnaissance satellites in orbit to monitor any suspicious activity. Even our ships have been made lighter and more secure by the use of fiber optics communication, rather than open-air transmissions or copper cables.

Our modern electronics—cell phones, digital cameras, computers, and similar devices—would not be possible without optical lithography, which enables the production of integrated circuits. One of the still-unrealized optical accomplishments is controlled fusion for the generation of sustainable, green energy. It is based on the use of many giant lasers bringing a small pellet to sufficient temperature that it undergoes atomic fusion. Scientists have reported a small success: a little more energy out than in.

It is my purpose in this book to chronicle the small and giant steps humankind has taken in the field of optics to get us to this advanced state of accomplishment. In each area it starts with very crude, and perhaps naive, ideas before advancing through the ages to our present state of knowledge and capability. I have written this mostly on a non-technical basis for the general public, but there are some things that can only be appreciated if more technical concepts are employed. I suggest that the lay-reader simply get the gist of these discussions and move on—or ask a knowledgeable friend. I hope my colleagues will enjoy it. By the way, I do not know what to call my colleagues. Opticians make optics, and optometrists test eyes; so, we may be optikers, opticists, or even opticalists. But our glasses are always half full.

1.3 Organization

Most histories, including histories of science, are strictly chronological in nature. I have chosen a different approach. Each part of optics is covered individually. For instance, the history of the telescope starts with the work of Lippershey, follows it with Galileo and Newton, and ends with the monster multi-mirror, adaptive mirror telescopes of today. Spectroscopy starts with the rainbow, goes through Newton, and continues on with Czerny, Fraunhofer, and Vanasse. I have done this in part because we each have our own favorite parts of optics, and partly because each area has its own history, although there are overlaps. Other histories of one part of optics or another have been published, but there does not seem to be a history of most of the various fields of optics in any one place. That was my challenge, and I hope I have covered enough areas in enough detail to meet it. In addition to the technical topics in optics, I have included descriptions of some of the major optical institutes and the major societies. There are certainly more centers with optical activity than I have included, but listing them all is out of the question. A wonderful reference for these is the annual publication by SPIE (look up SPIE in that chapter), which describes just about every place there is any optical activity. And there are local and national societies that I have omitted, partly by ignorance and partly because of lack of space.

1.4 Information sources

Wherever possible, the original publications, translations, or descriptions by the original authors have been used and cited. Of course, the older descriptions and references are less certain. The details are lost in obscurity. Secondary references have to be relied upon. One source that has been used repeatedly is the internet, especially Wikipedia and the various original works provided by Project Gutenberg and others. Wikipedia is a starting place, but in every case I have checked and supplemented it with the original citations, even though Wikipedia has proven quite reliable. The book by Kingslake on *The History of the Photographic Lens* was a great help, as was *The History of the Telescope* by King and Jay Enoch's *History of Mirrors Dating Back* 8000 *Years*, especially for those early references. The September 1958 issue of the *Proceedings of Radio Engineers*, (now *Proceedings of the Institute of Electrical and Electronic Engineers*) was extremely helpful in describing the history of infrared developments during the 1900s. Thank you Warren Arnquist! Emil Wolf's wonderful book, *Principles of Optics*, has many references to the original articles. The book is actually by Born and Wolf; Max Born was a principal in much of the development of quantum mechanics and optics. Emil expanded greatly on Born's earlier book in German called *Optik.* D J Lovell's *Optical Anecdotes* has provided some very nice background on a number of subjects. Other sources and further reading are in the Bibliography.

Some of the references are used quite frequently, and they are therefore abbreviated. There is a separate section that identifies all these abbreviations and publications.

1.5 Limitations

The subjects covered, the various fields of optics, are certainly those of interest to the author and do not include all of optics. For instance, the field of ophthalmology is not covered, nor are color and vision, nor radioactivity and quantum mechanics, although they have a lot of optics in them. I have excluded the spectral region below the ultraviolet, below about 0.3 μm, largely because the techniques at shorter wavelengths are different. For instance, x-rays use mostly grazing incidence optics; lenses are not available. I stopped at the long wave infrared, almost 1 mm, again because techniques involving longer waves are more radio-like than optics-like. I define the field of optics as those topics I have included in this book!

1.6 Idiosyncrasies

I have my writing style and way of saying things in print—and verbally. I am old fashioned enough to consider terms like *man* and *his* to represent humanity, both sexes. If there were sex inclusive words for these, I would use them, but I chose not to clutter the text with such things as he/she, man or woman, and the like. I refuse to refer to those things in the road as person-hole covers! I have used non-gender specific words where it did not clutter the concepts.

1.7 Terminology and symbols

For the most part, the terms and language used here are in common use, as are the symbols—at least in the technical literature. I have tried to define all others when they are used. There is a separate section on the technical symbols I have used. There is also a glossary of terms that may not be entirely familiar or are used in a special way. Dates are followed by BC if they are before the birth of Christ. Otherwise there is no addition; I do not use AD. I have followed the citation of scientists with their birth and death dates in parentheses, but some are not there because I could not find them. That is usually because they are still living, have common names, or were not considered by history to be important enough. I tried.

Enjoy!

IOP Publishing

Rays, Waves and Photons

A compendium of foundations and emerging technologies of pure and applied optics

William L Wolfe

Chapter 2

Acousto-optics—did you hear what I saw?

This relatively modern field relates to the interaction between light waves and sound waves. The main interactions involve sound waves that change the density of a material, and therefore its refractive index, so that refraction and diffraction of the light waves can be effected. I have found no similar effect of light changing the local density of solid matter. The relatively small energy of a photon makes any significant interaction unlikely. However, light can affect the motion of molecules, which can generate a sound.

The first reference to the interaction of light with sound[1] is that of Léon Brillouin (1889–1969), a French physicist who is renowned for his work on wave theory. He postulated in 1922 that when sound propagates in a medium, it sets up a grating—a diffraction grating (see Diffraction, chapter 6). Since sound waves are variations in pressure in a medium, these variations induce local changes in density and therefore changes in refractive index. If the sound is relatively monochromatic, a single tone, it is a sinusoidal grating, and there are only two diffraction orders. It wasn't until 10 years later that there was experimental evidence[2] concerning these ideas by Peter Joseph William Debye (1884–1966) and Francis Weston Sears (1898–1975), and it proved Brillouin wrong! There were multiple orders. It was then realized that the interaction, and in particular the acoustic waves, was much more complicated than had been supposed. They were definitely not simple sinusoids. A series of theoretical investigations was carried out by Sir Chandrasekhara Venkata Raman (1888–1970) that better described the interactions[3], but it was not until the advent of the laser that this technology blossomed.

These studies of acousto-optics led to the modulation and deflection of light waves for various applications. This is accomplished by the establishment of a dynamic transmission diffraction grating of refractive index variation. One example

[1] Brillouin L 1922 *Ann. Phys.* **17** 88

[2] Debye P and Sears F W 1932 *Proc. Natl. Acad. U.S.A.* **18** 409; Lucas R and Biquard P 1932 *Phys. Rad.* **3** 464

[3] Raman C V and Nath N S N 1935 *Proc. Indian Acad. Sci.* **2** 406; Raman C V and Nath N S N 1936 *Proc. Indian Acad. Sci.* **3** 75

doi:10.1088/978-0-7503-2612-4ch2

is an imaging spectrometer based on an acousto-optical tunable filter[4]. Another is an AO light modulator in which the output angle is a function of the ratio of the light wavelength to the sound wavelength, and the intensity is a function of the sound intensity[5]. The Jeffree cell (a variation and improvement on the Kerr cell, invented by J H Jeffree in 1934) was used in the Scophony system in 1934[6] (figure 2.1). (See Displays, chapter 7, for a discussion of the Scophony system.)

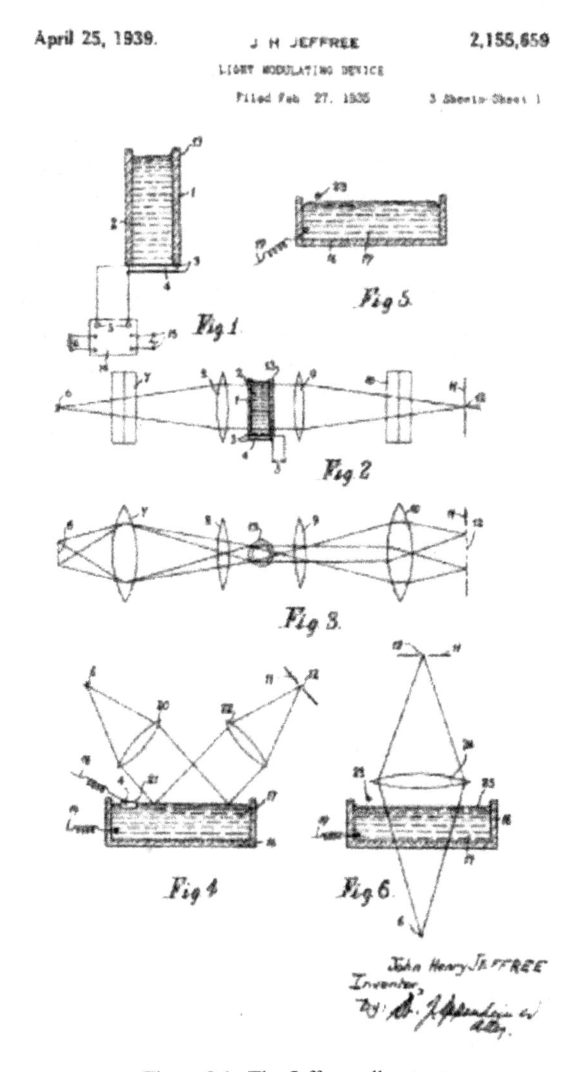

Figure 2.1. The Jeffree cell patent.

[4] Taylor L H *et al* 1995 *Proc. SPIE* **2480** 334; Bass M *et al* 1995 *Handbook of Optics* (New York: McGraw Hill); Wolfe W 1997 *Introduction to Imaging Spectrometers* (Bellingham, WA: SPIE Press).
[5] Eklund H, Roos A and Eng S T 1975 *Opt. Quantum Electron.* **7** 73.
[6] Singleton T 1988 *The Story of Scophony* (London: Royal Television Society).

A fascinating and very modern application is acousto-optical spectroscopy of glucose cells. Some diabetic patients are required to measure their glucose levels every day. This usually involves taking a sample with a hypodermic needle or a pin prick. This modern technique eliminates the needle and takes the sample in much the same way that Bones of *Star Trek* took his samples. A light of the proper wavelength is shone on the skin. It penetrates a reasonable distance and excites the glucose molecules. They vibrate as a result of this excitation and generate sound waves. The level of sound is an indication of the level of glucose[7]. Other non-invasive techniques are discussed in the section Medical Optics (chapter 19).

Photo-acoustic technology, as it is also called, can also examine the nature of blood cells without invasion. When blood cells are illuminated with laser light of appropriate frequencies (in the near infrared) they 'sing.' They emit high frequency sound. Normal cells sing a different song than abnormal ones, i.e. have a different frequency spectrum, so that differentiation can be made to diagnose such things as malaria and sickle-cell anemia[8].

Photo-acoustic microscopy is now used for diagnosing melanoma[9].

Another activity, which might be considered acousto-optics, was sonovision, invented by Lloyd Cross in 1968[10]. It was sort of a way to 'see' sound. A laser beam was shone on the reflective drumhead of a loudspeaker. The reflected beam would then create all sorts of patterns based on the deformation patterns of the drumhead caused by the audio signals.

[7] Christison G and MacKenzie H 1993 *Med. Biol. Eng. Comput.* **31** 284.
[8] Kolios M *et al* 1993 *Biophys. J.* **105** 59–67.
[9] Michaud S 2014 *Opt. Photonics News Oct.*
[10] Cross L 1968 *US Patent* 779510.

IOP Publishing

Rays, Waves and Photons
A compendium of foundations and emerging technologies of pure and applied optics
William L Wolfe

Chapter 3

Binoculars—from both sides now

Binoculars are essentially a pair of telescopes, one for each eye. There have been various designs to make sure that the images are erect and not inverted or reverted—so that right is right and left is left as the object is.

The **first binoculars** are said to have been made by Hans Lippershey (1570–1619), the inventor of the telescope (see Telescopes, chapter 42) Why not just use two of them? When the States General of the Netherlands saw Lippershey's telescope, they asked for one for both eyes. He obliged in 1608[1]. Galileo Galilei (1564–1642) is also said to have proposed binoculars for military usage, and, of course, they would be of the Galilean design (see Telescopes again). Unfortunately, although the image is erect, the telescope is long and therefore not very easy to use. The next obvious step was to use the more compact Keplerian form for the optics, named after Johannes Kepler (1571–1630) (Telescopes one more time). It is significantly shorter, but it generates an inverted image. This image can be re-inverted—that is, made erect—by the use of a prism.

The first such **prism binocular** was patented by Ignazio Porro (1801–1875) in 1854. The Porro prism, shown here, inverts but does not revert the image, turning it upside down; a pair at 90° would do both. Later, other prisms were used, notably roof prisms. One of the earliest designs was by Ernst Karl Abbe (1840–1905) and Albert König (1871–1946) in 1905[2]. Another is the Amici prism of Giovanni Battista Amici (1786–1863). The Porro and roof designs allow the objectives to be further apart than the eyepieces, thus providing improved depth perception; whereas the roof prism systems are generally in line, making for a more compact design. A roof prism replaces the simple apex of the Porro prism with a roof-like structure, as I have tried to show. The Abbe–König is a roof prism, and so is the pentaprism; there are others. Most Porro prism binoculars use a pair of them in each side, thereby making the image upright and with the correct left-to-right orientation (figure 3.1).

---

[1] *Binoculars—History and Design*; Wikipedia Online https://en.wikipedia.org/wiki/Binoculars; Yoder P and Vukobratovich D 2011 *A Field Guide to Binoculars and Scopes* (Bellingham, WA: SPIE Press).
[2] Zeiss C 1905 *German Patent* 180644.

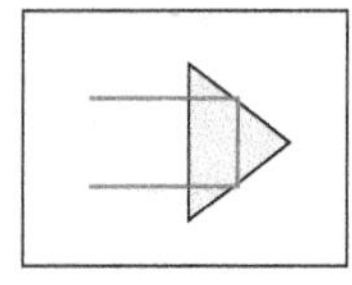 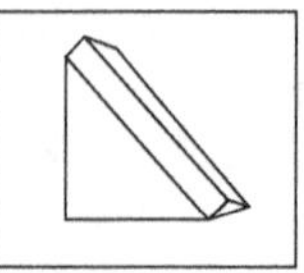 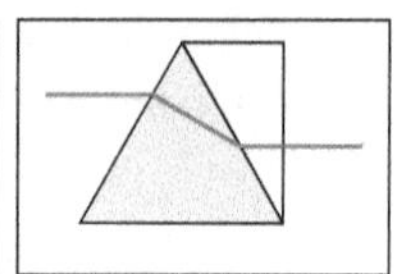 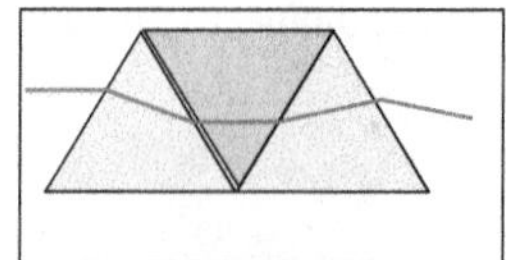

Figure 3.1. Prisms.

Designers have been improving and adding features to these basic binoculars. Some versions now have zoom lenses; some integral cameras; some colored lenses; and the TV now touts very lightweight ones, which, as far as I can tell, are simply the old-fashioned Galilean version with no prisms and just single lens objectives and eyepieces. They are called **Zoomies** and promise 300% magnification[3]. In the usual terms describing binoculars, this is three power, or 3×. That is what Lippershey claimed and Galileo improved by a factor of ten.

While on the topic, it is probably useful to discuss the specifications of binoculars. They are usually cited as, for instance, 12 × 50. This means a **magnification** of twelve times with an **exit pupil** diameter of 50 mm. The size of the exit pupil determines how easy it is to see the image. The human pupil is about 3 mm, so it can range over a 50 mm exit pupil but must be spot on with a 3 mm one. The size of the objective lens, its diameter, determines the collecting power of the binocular. But the larger it is, the heavier it is. The **field of view** may be specified in angular terms or as a linear dimension at a given range, say 100 m at 1000 m. The distance from the rear of the eyepiece to the image is the **eye relief** and is most important for those who wear glasses.

Some small binoculars are called opera glasses. They are just small binoculars with small apertures and limited magnification, but ample eye relief—just enough to let you see the diva clearly! The binoculars shown here are Galilean, not prismatic. The opera glasses are prismatic with the objectives closer together than the eye pieces. Figure 3.2 shows the relative sizes of an 8 × 24 set of binoculars and opera glasses.

Figure 3.2. Binoculars and opera glasses.

There may be incremental improvements in binoculars like more compact lenses, maybe of nano materials, but the significant change will be to a pair of miniature cameras, one in each ocular.

[3] Lots of TV advertising.

Chapter 4

Cameras—got the picture?

The first known cameras are called **camera obscuras**, or dark rooms[1], or obscured vaulted chambers. They consisted of rooms with little holes in the ceiling or a wall. Today we would call them pinhole cameras. The earliest records are those of Aristotle (384–322 BC) and, perhaps, Mozi (470–391 BC) at about the same time in China[2]. Aristotle apparently understood the process by which an image was formed[3], but perhaps not completely. He at least understood how square objects can be projected by round holes. Euclid of Alexandria (about 300 BC) used the camera obscura as proof that light travels in straight lines[4]. Several authors[2] cite and describe the camera obscura in the 16th century, but it was Giambattista della Porta (1535–615) who first described a lens in it in 1558[5]. Although it is still under debate, David Hockney[6] (1937–) and Charles M Falco (1948–) make the case that some Renaissance artists used the camera obscura to form an image that they could trace and then develop into a complete picture[7]. I used it once as a demonstration to a sixth grade class; it works well. The kids inside could watch the images of their classmates outside in the patio appear to be dancing upside down on the back wall! Cameras have evolved over the years with various modifications of the pinhole camera by Kodak, such as its Brownie through the various single lens reflexes, the 'instant' Polaroid camera, and now digital ones. During that time, the rangefinder and its integral version were introduced, as well as modular cameras with interchangeable lenses, backs, rangefinders, viewfinders, etc. The advent of 35 mm film

[1] Gemshein H 1981 *The Origins of Photography* (Milan: Electa).

[2] Needham J 1986 *Science and Civilization in China* (Taipei: Caves Books).

[3] Aristotle, *Problems* (330 BC).

[4] Euclid, *Optica* (300).

[5] Porta G D 1558 *Magia Naturalis* (Naples); available online.

[6] Hockney D 2001 *Secret Knowledge: Rediscovering the Lost Techniques of the Old Masters* (New York: Viking Studio).

[7] Personal communication with Falco.

doi:10.1088/978-0-7503-2612-4ch4

in 1892 by Thomas Alva Edison (1847–1931) and William Dickson (1860–1935), as well as projectors, was a great boon to the industry.

The first 'mass produced' camera is probably the **Daguerreotype** in 1839, named after Louis-Jacques-Mandé Daguerre (1787–1851). It was a wooden box with a simple lens and a daguerreotype photographic plate. This latter formed a positive image on copper coated with silver. Probably the first ones used a pinhole, but later ones used a Chevalier lens and then a Petzval lens (see Lenses, chapter 16).

The **Brownie**, probably the first really popular camera, was preceded by the **Kodak**, which was offered by George Eastman (1854–1932) in 1888 shortly after he developed paper film. The Brownie came in several versions; it was a simple, heavy, cardboard box about six inches on a side with a meniscus lens[8] when it was introduced in 1900[9] and cost $1.00. It was just a step away from the pinhole camera. It took 2 1/4 by 2 1/4 inch (57 mm) images on roll film. It was cheap, easy to use, and gave rise to many people taking snapshots of all sorts of things. This is the origin of what I call the Kodak effect: give the camera away and make money on all the film you sell. Others call it the Gillette effect: give away the razor and make it up on blades. The Brownie used a roll of 127 film (4 × 4 cm format) that, after it was exposed, had to be sent back to the Kodak factory for development. Later versions of the camera included a smaller box, a doublet lens, and a flash attachment (figure 4.1).

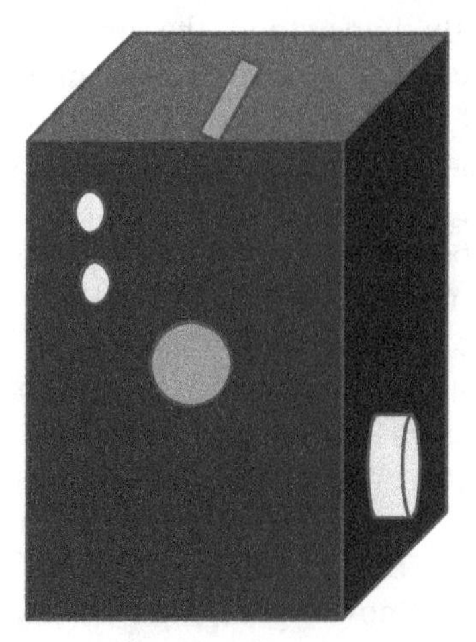

Figure 4.1. Brownie camera representation.

The **first 35 mm** camera was introduced by Ernst Leitz GmbH as the **Leica I** in 1925[10]. Kodak answered with the **Retina I** in 1934[11], made by Kodak AG in

[8] US Patent 725034 (1903).

[9] *The Youth's Companion*, May 17 1900.

[10] Rogliatti G 1985 *Leica, The First Sixty Years* (Hove, UK: Hove Collectors Books).

[11] Nagel H 1977 *Zauber der Camera—Beispielle aus dem Kodak-Nagel-Werk* (Stuttgart: Deutsche Verlags-Anstalt).

Germany. Kodak bought the German firm of Nagel Werk, owned by Helmut Nagel (1914–1996), who developed the camera. It was notable for using a film cartridge that could be loaded in the light. Argus introduced the **Argus A and C3** as much cheaper versions[12]. The Japanese entered the competition with the **Contax II**, which had a rangefinder in the middle of the viewfinder. This entry by the Japanese was greeted enthusiastically by our GI's, who brought many home from the Korean War. I did too. The first camera with an integral rangefinder was the **Kodak Autographic Special** of 1916[13]. Rangefinder cameras were largely displaced by single lens reflex cameras that provided the operator with the capability of real focusing and seeing what you get—the camera version of the computer WYSIWYG.

Probably the first twin lens reflex camera was a Voigtländer, patented in 1886 by Caius C Beagg[14].

The **single lens reflex** camera, SLR, was anticipated by some versions of a camera obscura with flip mirrors, but the first arrangement in a camera is dated 1861 with a British patent and later in the US in 1884. Single lens reflex cameras have been very popular over the years. They provide the photographer with a true representation of what he will take—in focus. The first modern version was patented by John Sutton (1819–1875) in 1861. The first American patent appeared three years later. The **Exacta Vest Pocket camera** was the first SLR to use **127 film**, in 1933[15]. The **Hasselblad 1600F** was the first, and maybe only, SLR that was **fully modular;** it had interchangeable lenses, backs, and viewfinders; it appeared in 1948[16]. The name indicates that the fastest shutter speed is 1/1600 s, and it was a focal plane shutter. The **Olympus Pen F** was the first SLR that used **half frame 35 mm film** in 1963[17]. The **Canon AE-1** was the first SLR with a **microprocessor**, but not digital 'film,' appearing in 1976[18] (figure 4.2).

Figure 4.2. Example of an SLR.

[12] US patent 2051061 (1936).

[13] Kodak No. 1 Autographic Special, Camerapedia online.

[14] Beagg C 1886 *US Patent* 339736.

[15] Aguila C and Rouah M 1989 *Exacta Cameras 1933–1978* (Hove, UK: Hove Photo Books).

[16] Nordin R 1997 *Hasselblad System Compendium* (Hove, UK: Hove Books).

[17] Nakamura K 2005 *Classic Cameras, Olymus Pen F*, online.

[18] Shell B 1994 *Compendium Handbook of the Canon System* (Hove, UK: Hove Books).

The first **automatic exposure camera** was the **Super Kodak Six-20** in 1938[19], but it was extremely expensive and finally replaced by the **Agfa Optima** in 1959 after electronic components became much less expensive and drastically smaller—after the transistor. Many of us used separate exposure meters—forms of photometers—before the advent of these cameras.

The first **'instant' camera** was developed and sold by Edwin 'Din' Land (1909–1991), an exceptional inventor, scientist, and entrepreneur. He formed the Polaroid Corporation based on his Polaroid material, a linear polymer that can polarize light. He expanded into the camera market with his invention of the instant camera, based on instantaneous processing of the film. However, in the process, he also developed the original **folding SLR** camera, the FX 80, in 1972. The Polaroid Model 95 was his first camera in 1948. It was followed by the **Swinger** in 1965 and then the SX-70 in 1972[20] (figure 4.3).

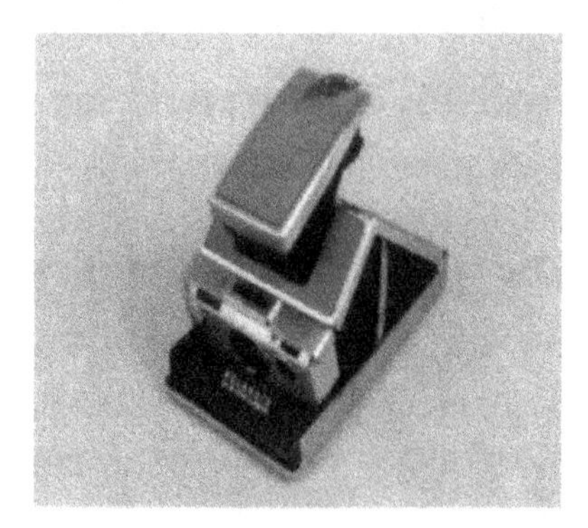

Figure 4.3. Polaroid camera.

The first **color photograph** of a sort was also generated by Thomas Sutton in 1861[21], based on a technique described by James Clerk Maxwell (1831–1979) in 1855[22], who took three black and white images through three different color filters. These were then projected with three projectors using corresponding filters. True panchromatic film was developed forty years later and used in all the cameras listed here after its introduction.

Digital cameras use arrays of detectors rather than film, and they match the sensitivity and resolution of film. The silver grains in film range from about 5 μm to 15 μm, and silicon detectors now can be about as small as that. Digital cameras are now described in terms of the number of pixels (picture elements) they have, the number of individual detectors. And that number now ranges up to about

[19] Coe B 1978 *Cameras from Daguerretypes to Instant Cameras* (Sweden: Nordbok).
[20] Wikipedia, Polaroid SX 70. Online.
[21] Hunt R 2004 *The Reproduction of Colour* (New York: Wiley).
[22] Maxwell J 1861 *The British Journal of Photography*.

45 megapixels. The sensitivities are about the same as high-speed film and the formats almost identical. The cameras themselves come in a variety of types, including SLR's and point and shoot. All have integrated rangefinders and exposure meters. And since the advent of the integrated circuit and charge-coupled devices, every smartphone has an integral point-and-shoot camera. The **first digital camera** was invented and cobbled together by Steve Sasson (1950–) of Kodak in 1975 using an old Super 8 camera and a bunch of computer parts[23]. The output was on a cassette. People had trouble understanding how such a kluge that only put images on a television screen would ever be viable. It was about a cubic foot in size and had only about 0.3 megapixels. But look what happened!

Charge-coupled devices were the invention that allowed these cameras to be viable. These consist of an array of detector elements that gather charge and pass it on from one to the next, thereby avoiding wire connections. They were invented by George E Smith (1930–) and Willard Boyle (1924–2011) in 1969[24]. Just imagine a 45 megapixel device with 45 million wires!

Movie cameras employ a series of individual pictures shown in rapid sequence; each frame lasts for less than the integration time of the eye, about 0.1s. Thus, mechanisms for the rapid advance of these frames must be used in addition to the lenses and other elements of a still camera. And there must be arrangements for proper tensioning, positioning, loops and intermittency—all this inside a light-tight container. The first of these was apparently by Louis Aimé Augustin Le Prince (1841–vanished 1890) in 1888. It was recorded at 12 frames per second and lasted for 2.11 s[25] (probably 24 frames and header and trailer). Soon after, Thomas Edison, the great American inventor, produced his version in 1897,[26] which he called a Kinetographic camera. The **first color cine camera** is the Mitchell Technicolor Camera of 1932, using color separation techniques akin to those of Maxwell. Kodak subsequently introduced a 16 mm version, replacing the earlier 35 mm film in 1923[27]. Most commercial filming is done with 35 mm film cameras, and all personal photography was with 16 mm until the Super 8 came out in 1932[28]. These cameras have been largely replaced with their electronic equivalents; images are stored as bits and bytes on hard drives, thumb drives, and disks. By the way, a byte is eight bits—and a nibble is half a byte, four bits.

In a sense, the first movie camera was a manual one. You flipped individual photographs very quickly (figure 4.4).

[23] Let's Go Digital Online Magazine, September 2007; Prakel D 2010 *The Visual Dictionary of Photography* (Lausanne: AVA).

[24] Boyle W and Smith G 1973 *US Patent* 3792322.

[25] Guinness Book of Records.

[26] The Standard Edition, *Edison and the Kinetographic* Camera, online; US Patent 589168 (1897).

[27] Kodak Super 8 Film History, online.

[28] Ibid.

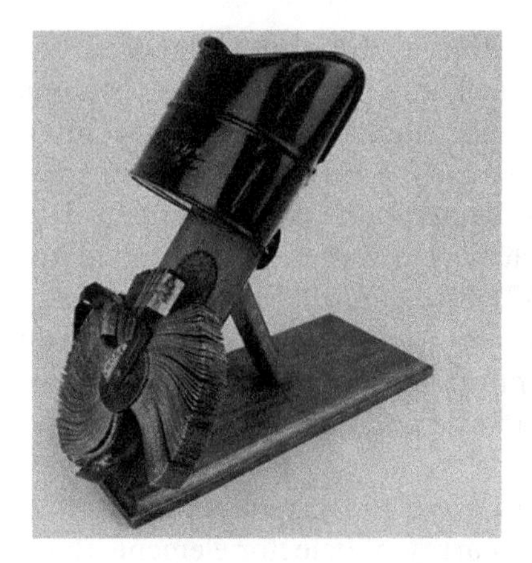

Figure 4.4. Early movie camera.

The most modern of the personal movie cameras—and an extremely popular one—is the **GoPro camera**, or several varieties of these[29]. They are digital movie cameras, or camcorders, that are very small with many pixels, waterproof to a depth of 180 ft (60 m), and shockproof to a height of 3000 feet (about 1 km). The formats range from 2592 × 1944 to 3000 × 4000 pixels, each from 1.4 to 2.2 μm on a side. The data are stored on SD cards that have a capacity of up to 2 terabytes. One version has a field of view of 170° and an F/2.8 six element aspheric lens. The entire camera is about 1.6 × 2.4 × 1.2 inches and only 3.3 ounces—a little larger than the fictional Dick Tracy watch. The idea for GoPro came to Nick Woodman (1975–) when he was surfing in Australia in 2002, although he says it had been kicking around in the back of his mind for years[30]. He is now another one of those very young billionaires! The name arises from the fact that only the pros could take good pictures of surfers on their boards. So why not go like a pro: GoPro.

There have been great strides in camera development since the advent of the digital camera with its multimegabit detector arrays. It is hard to imagine larger formats or smaller pixel sizes, but more compact lenses, more versatile processing and better color rendition, redeye correction and the like are in store.

[29] GoPro Official Website and Wikipedia online.

[30] A variety of news and TV articles including 60 min.

IOP Publishing

Rays, Waves and Photons
A compendium of foundations and emerging technologies of pure and applied optics
William L Wolfe

Chapter 5

Optical communication—did you get the word?

The first examples of optical communication were undoubtedly smoke signals, or just people waving at each other. Semaphore may have come next. They are all optical in nature. You have to see them. Other forms of optical communication on a more proximate basis are a nod of the head and the wink of an eye, which happen mostly at Christmastime. Modern optical communication is used in two broad categories: free space communication and in fibers. Optical communication has a great advantage over radio VHF and UHF techniques in that it has much more bandwidth available due to its much higher frequency. That only occurred after the advent of the laser and its coherent radiation. The carrier frequency must be about ten times the information frequency. Since light has a frequency of about 6×10^{14} Hz and UHF (the frequency range of normal TV) has a frequency of about 3×10^{8} Hz, the advantage is obvious and great. Anytime you can get an advantage of more than a million, take it!

It is reported that the ancient Greeks used **fire**[1] and **reflections of the Sun** from their shields to send signals[2]; these certainly would not have been Morse code. Fire was also used as a signaling means on Hadrian's Wall in the second century[3]. **Heliographs** were next: they did use Morse code, and they used a shutter to interrupt sunlight reflected by a mirror. The first was invented in 1821 by Karl Friedrich Gauss (1777–1855), but it did not use Morse code and was used for geodetic purposes. It was called a heliostat. In 1866 two British engineers, Francis John Bolton (1831–1887) and Phillip Howard Coulomb (1831–1899), applied Morse code to the device[3]. Samuel Finley Breese Morse (1791–1872) invented the code of his name that translated the alphabet into dots and dashes. My initials in Morse code are

·– –, ·–· ·, ·– –.

[1] Homer, *The Iliad;* Aeschylus, *Agamemnon;* Herodotus, *The History of Herodotus.*
[2] Xenophon, *Helenica*, 450 BC.
[3] Sterling C 2008 *Military Communications: From Ancient Times to the 21st Century* (Santa Barbara, CA: ABC-CLIO).

During WWII the Germans used **Lichtsprechers** (light speakers), infrared communication devices[4]. Sprechen sie Licht?

Alexander Graham Bell (1847–1922) invented the **photophone** in 1880. It sent sound signals on a light beam[5]. He reflected sunlight onto a diaphragm that vibrated in response to an audio input and thereby modulated the light beam (figure 5.1).

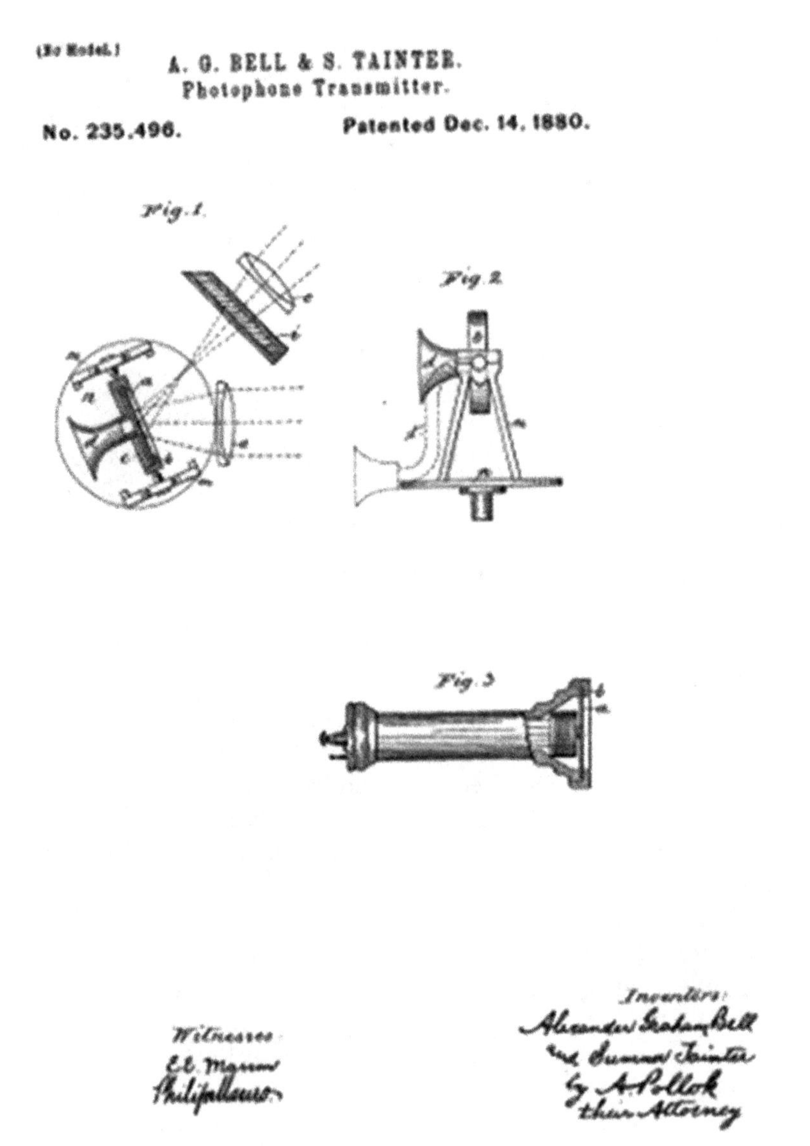

Figure 5.1. Photophone patent.

[4] Huxford W and Platt J 1948 *J. Opt. Soc. Am.* **38** 253.

[5] Bell A 1880 *Am. J. Sci.* **20** 305; Bell A 1880 *Nature*.

5.1 Photophone patent

The first suggestion of a coherent means of light communication was by O E DeLange in 1970, although his suggestion[6] never went very far, possibly because he proposed intensity modulation and direct detection rather than more sophisticated techniques like frequency or pulse modulation and heterodyne detection. It took some 10 years for these more sophisticated techniques to be applied[7]. These have been improved to the point where all the modulation and detection techniques used in radio science can be applied to light: pulse code modulation, digital encoding, phase sensing—all because lasers have become more stable, reproducible, and cheaper.

5.2 Fiber communication

Although communication in fibers had been done over short distances because the fibers did not transmit well enough, the real breakthrough seems to have been by Charles Kao (1933–), who recognized that one might be able to increase the transmission and decrease the attenuation by reducing the impurities in the fibers[8] (see Fiber optics, chapter 10). His first attempts provided an information rate of one gigacycle per second, or one gigahertz, with a (terrible) loss of 1000 db km^{-1}. But improvement was on the way.

The group at Corning soon showed, in 1970, that they could make fibers with 20 db km^{-1} (1% km^{-1} transmission) at 0.633 μm, the wavelength of a He–Ne laser. Then, with improved manufacturing processes and longer wavelengths, the distances between repeaters became greater. In 1977, the first field trials used a wavelength of 850 μm and got up to 45 Mbps (megabits per second) with repeaters separated by only a few kilometers since the loss was down to about 2 db km^{-1} (about 97% transmission). Not enough. Then at 1.3 μm, with 0.4 db km^{-1} (>99%), 90Mbps. Getting there. Bell Labs announced in 1980 their plans for 1988 for a transatlantic cable with 565 Mbps. Repeaters were about 50 km apart. By 2001, a cable could transmit about 100 wavelengths, each capable of 10 Gbps at once. That is 1 Tbps, a terabit per second[9]. Almost as big as our national debt! It was teraffic!

David N Payne (1944–) and Emmanuel Benoit Desurvire (1955–) invented an optical relay[10] to replace the optical–electrical–optical relay then in existence, and a more efficient system was realized. It consisted of erbium doping in the fiber and an independent light pump to generate laser action. It was an optical device that amplified the signal going down the fiber.

A continuation of design, material, and technique improvements has yielded impressive capability with optical fiber communication. It was announced in 2012 by NEC of America and Corning, Inc., that they had attained 1.05 peta baud—1.05×10^{15}

[6] DeLange O 1970 *Proc. IEEE* **58** 1683.

[7] Okoshi T and Kikuchi K 1980 *Electron. Lett.* **16** 179; Favre F and LeGuen D 1980 *Electron. Lett.* **16** 709.

[8] *Laser Focus*, April 1966; Kao C 1966 *Proc. Inst. Elec. Eng.*

[9] Hecht J 1994 *Laser Focus World*, November; online.

[10] Poole S *et al* 1989 *IEEE J. Lightwave Technol. Special Issue* **7** 1242; Mears R *et al* 1986 *Electron. Lett.* **22** 159; Desurvire E *et al* 1987 *Opt. Lett.* **12** 11.

bits per second—by combining spectral, polarization, and spatial mode multiplexing[11]. Others reported about the same capability over 52.4 km light paths, the distance between repeaters. It is generally accepted that the Library of Congress has about 10 terabytes of information in print. So, since a byte is 8 bits, and a peta is 1000 teras, this cable could transmit the contents of about 12.5 libraries of Congress every second! If we could scan it that fast. Truly, we would not have an information age like it is without optics.

The longest underwater fiber optic cable is from Norden, Germany, to Keoje, South Korea, some 39,000 km. The newest, finished in 2005, is 20,000 km from Marseilles, France to Yuhas, Singapore. It has a data rate of 1.28 Tbd[12].

5.3 Free space communication

The early devices of the Greeks, the American Indians, and Alexander Graham Bell were all free space optical communication methods. They, and their modern descendants, are limited by atmospheric effects and the curvature of the Earth, but they are very useful for short range terrestrial communication, about 300 km (180 mi)[13] and greater ranges between satellites. Laser-based systems have about the same information rate capability as fiber optics—without the relays. The German Lichtsprechers were probably the first to send a light beam that was voice modulated. They used filtered tungsten sources to send infrared light beams to thalofide detectors (see Infrared, chapter 12). Satellite-to-satellite communications do not have these limitations.

The Apollo spacecraft communicated with the Lunar Excursion Module by way of an optical link; it also used it for navigation[14].

It is widely accepted that optical free space communication is best used for short distances: the most widely used application is from the sofa to the TV set—the remote. The most recent development, primarily for military applications, accomplished 100 Mbps over 50 km[15].

[11] OSA Frontiers in Optics/Laser Science XXVIII, October 2012.

[12] *Optics and Photonics News*, March 2014.

[13] Turner C 2007 *Modulated Web site*; online.

[14] Personal involvement.

[15] *Photonics Spectra*, February 2014.

IOP Publishing

Rays, Waves and Photons
A compendium of foundations and emerging technologies of pure and applied optics
William L Wolfe

Chapter 6

Diffraction—diffraction didactations

This phenomenon is the redistribution of light when it encounters some sort of an obstacle, like a hole or a disc. By Huygens' principle, each point on a wave front becomes the source of another wave, he called a daughter wave. They then combine to form the ongoing wave. These can be seen at the edges of the obstructions. Diffraction limits the performance of imaging systems and gives rise to diffraction gratings that can perform as spectral dispersers—devices that separate light into its component colors.

The propagation of a wave may be thought of this way: each point on the wave front emits spherical waves that then combine to form a new wave front. This is shown in the figure for a plane wave. Of course, on the page the plane wave is be a line, and the spherical 'daughter waves' are circles. Then it can be seen or imagined that when there is any kind of obstacle, it will disturb the pattern—not in the middle, but at the edges. The dark lines are the edges of the aperture; the wave front, the locus of the fronts of the circles, is plane in the middle but diverges at the edges (figure 6.1).

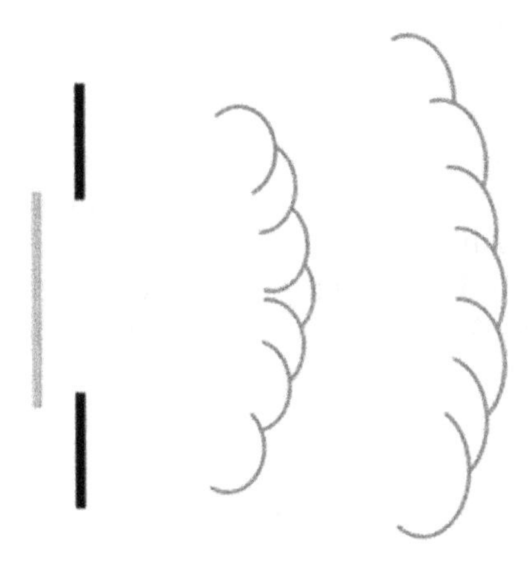

Figure 6.1. Diffraction representation.

doi:10.1088/978-0-7503-2612-4ch6

The **first observation** of diffraction may have been by Leonardo da Vinci (1452–1519), but observation and understanding are two different things. The first accurate description of it was by Francesco Maria Grimaldi (1618–1663) in 1665; published posthumously[1]. But this was still not understanding. It is true, there are many things we know, but fewer that we understand.

James Gregory (1638–1675) observed diffraction by a bird feather, and it might be considered the very first diffraction grating. Interestingly, he had much the same arrangement as Newton in his experiments on refraction: a hole in a window screen and a dark room. He observed colored rings produced by the feather that was illuminated with sunlight[2].

Long before Albert Einstein (1879–1955) showed that light consisted of photons, there was a raging argument during the period between the middle of the seventeenth century and the middle of the eighteenth about whether light was particles or waves. As chronicled in chapter 17 on light, Isaac Newton (1643–1727) believed in particles, but many others, especially in France, believed it was waves. One experiment that was convincing was the famous **double slit interference experiment** of Thomas Young (1773–1829), discussed in chapter 15 on interference; another was the diffraction of light.

One of the predictions of the wave theory of light and its diffraction was that directly behind an illuminated disc there would be a bright spot. Based on ray optics, the region behind such a disc would certainly be unilluminated, no bright spot. Based on Newton's theory of corpuscular optics, there would also be no bright spot. What would bend the corpuscles that were moving in a straight line? In fact, many scientists of the day considered this prediction to be ridiculous. Even today, the idea that there would be a bright spot behind an occulting disc is hard to grasp. But then Dominique François Jean Arago (1786–1853) showed that it was there[3]! A real coup for the wave advocates. This prediction of the spot was done by Simeon Denis Poisson (1781–1840), based on the diffraction theory of Augustin-Jean Fresnel (1788–1827); he thought it was absurd that such a spot should exist and was sure it would not since he was a believer in Newton's ideas. He calculated its existence and proclaimed at a meeting of the French Academy of Sciences that this was proof that Newton was right, since it was so absurd. The chairmen of the Committee, Arago performed the experiment very carefully and found the spot. It is ironic that it is now called by some the **Poisson spot**—but by others the **Arago spot**.

Fresnel's theory was based on the ideas of Christiaan Huygens (1629–1695) that a wave is propagated by the origination of many waves on its wave front, as described earlier. The integral formulation is described in more detail in appendix A. The final form is known as the **Fresnel–Kirchhoff diffraction** formula because Gustav Robert Kirchhoff (1824–1887) made some important additions in 1883[4].

---

[1] Grimaldi F 1665 *Physico de mathesis de lumine coloribus et iridae allsque annexis libri duo* (Bologna: Vittorio Bonati).

[2] Gregory J 1841 *Correspondence of Scientific Men of the Seventeenth Century* (Oxford: University Press) online.

[3] Fresnel A 1866 *Ouvres Completes* (Paris: Du Ministre Instruction Publique) online.

[4] Kirchhoff G 1883 *Ann. Phys.* **254** 663.

When the distances involved are large with respect to the wavelength, the Fresnel–Kirchhoff integral can be simplified greatly. Diffraction problems reduce essentially to evaluating Fourier transforms. This is called **Fraunhofer diffraction** in honor of Joseph von Fraunhofer (1787–1826), but it is not clear that he showed this. It is also not clear who did, since it is just a simplification of an integral. By the way, my mnemonic for this is that Fresnel diffraction is close and Fraunhofer is far because Fresnel is shorter than Fraunhofer—has fewer letters.

Jacques Babinet (1794–1872) derived his **principle of complementarity** in 1837[5]. It is useful in solving certain problems. It states simply that complementary screens, for instance an opaque circular disc and a circular hole of the same dimensions, produce identical patterns. The doughnut and the doughnut hole. (Babinet was a really smart guy; he gave up his study of law to become a physicist!) It has recently been shown that this principle needs some modification for incident waves that are not plane[6]. A more rigorous description of the principle involving electric and magnetic field vectors has been given by H G Booker in 1946[7]. Another principle of complementarity deals with light; it cannot be a particle and a wave at the same time. But it can be one or the other.

Reciprocity was described by Hermann von Helmholtz (1821–1894), perhaps by noting the symmetry in the diffraction equation of source and receiver. Thus, a point source at one point will produce the same pattern as a point source placed at the pattern; the pattern is then where the source was. This is often used in evaluating optical systems by assuming the light goes through them backwards. One example from my own background is evaluating cold shielding in infrared systems. Instead of considering where the light comes from, imagine your eye at the position of the detector. Look out to see what you see.

One of the very useful applications of diffraction theory and the phenomenon is **diffraction gratings**, used in modern spectrometers. In many ways they are superior to prisms for dispersing the light. They consist of either transparent or reflective plates on which are ruled many parallel lines. Or they may be a set of parallel wires. In either case, the diffraction caused by them is predictable and useful. The **first diffraction grating** of a very crude sort was made by David Rittenhouse (1732–1796) in 1786[8]—50 hairs laid across two screws at about 100 lines per inch that did show the colors and orders. This was followed by John Barton (1771–1834) in 1822 for the purpose of making fancy waistcoat buttons[9]. The first scientific accomplishment of gratings was made by Joseph von Fraunhofer in 1821 for his study of the spectrum of the Sun[10], but he still used wires—then later grooves on a (flat) mirror. Then came

[5] Babinet A 1837 *C. R.* **4** 638.
[6] Ganci S 2003 *Am. J. Phys.* **73** 84.
[7] Booker H 1946 *J. Inst. Electr. Eng.* **93** 620.
[8] Rittenhouse D 1786 *Trans. Am. Phys. Soc.* **2** 201.
[9] Grodzinski P 1947 *Trans. Newcomen Soc.* **26** 79–88.
[10] Fraunhofer J 1823 *Ann. Phys.* **74** 337.

the **reflection grating** by Anders Jonas Ångström (1814–1874) in 1869. Then somewhat later came the **concave grating** generated by Henry Augustus Rowland (1848–1901) at Johns Hopkins University in 1882, using his ruling engine[11]. JHU became the world's premier generator of spectrometer gratings with their very precise ruling engines. Robert Williams Wood (1868–1955) introduced **blazed gratings** there in 1944, and then John Donovan Strong (1905–1992) continued the tradition by developing a technique of vacuum deposition in 1935, which enabled gratings to be ruled on materials like aluminum-coated glass with a higher reflectivity than the speculum metal that had been used[12]. The figure shows a JHU grating given to me by Strong and a spectrum on my ceiling created by it from the Sun (figures 6.2 and 6.3). (A blazed grating is one with a shaped groove that increases the efficiency.) The work was later taken up by George Russell Harrison (1898–1979) at MIT during the years from 1948 to 1972[13]. One of the advances was the use of interferometric control of the ruling engine by Harrison and George Wilhelm Stroke (1924–),[14] although this was suggested earlier by Albert A Michelson (1852–1931) in 1927 but not realized[15].

Figure 6.2.

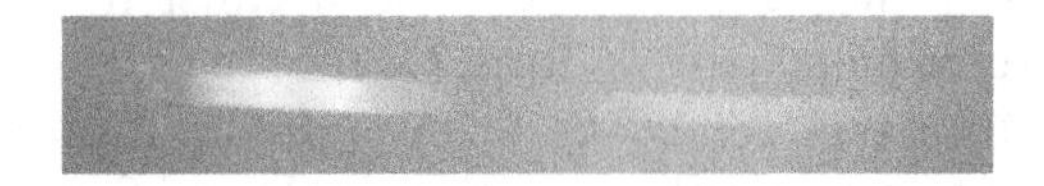

Figure 6.3. Diffraction pattern –– zero and +– orders.

A very important advance that made gratings much more available was replication, invented by J White and W Fraser in 1949[16]. This is analogous to artistic prints in which many **replicated gratings** can be made from one master. The latest practical technique for the production of gratings is an **interferometrically**

[11] Rowland H 1882 *Philos. Mag. Suppl.* **13** 469.
[12] Strong J 1935 *Phys. Rev.* **48** 480.
[13] Harrison G 1949 *J. Opt. Soc. Am.* **39** 522.
[14] Harrison G and Stroke G 1955 *J. Opt. Soc. Am.* **45** 112.
[15] Michelson A 1927 *Studies in Optics* (Chicago, IL: University of Chicago Press).
[16] White J and Fraser W 1949 *US Patent* 246748.

generated grating. The advent of the laser has made this a more attractive and realizable proposition, although it was suggested in 1915 by Michelson[17] and attempted by Burch in 1960[18]. With the laser, the technique became reasonably practical and was realized by several investigators[19]. An interference pattern is generated with a laser on a photosensitive material. Then it is etched away in a manner similar to the generation of integrated circuits. Interference produces sinusoidal gratings that are not blazed, but they are very well controlled and easy to generate. The figures below illustrate several types of gratings. The square grating, ruled by an engine, may have somewhat different than perfect square grooves. They could be more like vees or trapezoids. The blazed grating is ruled with a tapered diamond edge on the machine. The light is reflected as well as diffracted in the preferred direction. The sinusoidal grating, however, is very sinusoidal since it is generated by interference (figure 6.4).

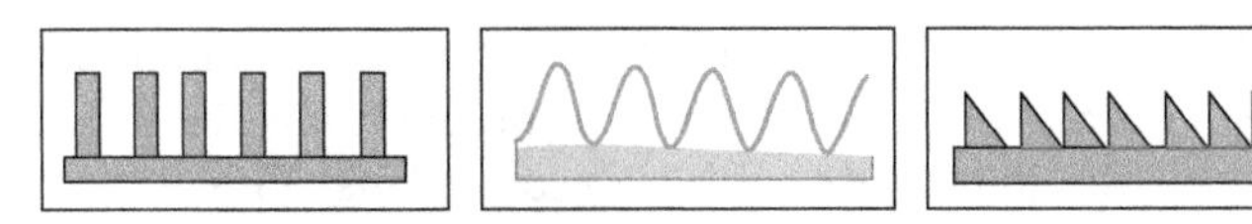

Figure 6.4. Square grating; sinusoidal grating; blazed grating.

Ruling engines are a marvel of precision mechanical engineering—and optics. The ruling they generate is a set of parallel lines at the spatial rate of from 20 lines per millimeter to almost 11,000 lines per millimeter less than one micrometer apart. The principal ruling engines have been those at MIT and JHU, as well as the Michelson and Mann engines.

The **resolution limits** of optical systems are ultimately limited by diffraction effects. A point source is not imaged as a pure point because the optical element is a form of disk and causes diffraction. The first description of an observation of this was by John Herschel (1792–1871) in 1828 in an article in *Encyclopedia Metropolitana.* He described the image of a star (a point source) as a central disc with several colored rings around it. Slightly later, in 1835 George Biddell Airy (1801–1892) provided the theory[20]. Due to diffraction by a circular disc, the light pattern is described by the so-called jinc pattern, or the Bessel function of the first kind divided by its argument. Each of the peaks would be a circle in three dimensions. For a rectangular aperture, like a spectrometer slit, the pattern would be the sinc function or the sine divided by its argument. They are both shown in figure 6.5. For the circular aperture, it turns out to be that the angle between those two, point sources that makes them just resolved is $1.22\lambda/D$, where λ is the wavelength and D is the diameter of the aperture. This shows why the rings are colored; their positions are determined in part by the wavelength of the light.

[17] Michelson A 1915 *Proc. Am. Philos. Soc.* **54** 88.

[18] Burch J 1960 *Research* **13** 2.

[19] Labeyrie A and Flamand J 1969 *Opt. Commun.* **1** 5; Rudolph D and Schmahl G 1967 *Umschau Wiss.* **67** 225.

[20] Airy G 1835 *Trans. Cambridge Philos. Soc.* **5** 283.

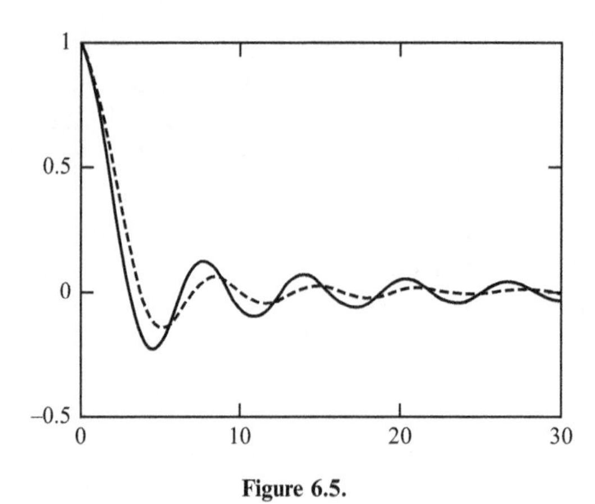

Figure 6.5.

Ernst Abbe (1840–1905) first determined that the diffraction limit of a microscope is $d = \lambda/(2n \sin \theta)$, where λ is the wavelength of the light, n is the refractive index of the medium, and θ is the half-angle of the objective lens. This is equivalent to the telescopic limit, as it should be. If the equation is inverted and $\sin \theta = \theta$, which it is for small angles, and $n = 1$ for air, then they are the same.

Another sort of diffraction is called Bragg diffraction. It consists of scattering or reflection of waves of light (and x-rays) from crystal planes. It is a form of three-dimensional diffraction. William Lawrence Bragg (1890–1971) reported his derivation of his law in 1912 that describes the condition for constructive interference. It is $2d \sin \theta = n\lambda$. When twice the separation times the sine of the angle of incidence is an integral number of wavelengths, the pattern has a maximum. He was a joint winner of the Nobel Prize in physics with his father. Bragg scattering—or reflection, or diffraction—is used in crystal analysis. If one can measure the angle of incidence of a beam of light or x-rays and the positions of the maxima, he can infer the lattice spacings. The scattering of x-rays and their interference was first observed by Max von Laue (1879–1960) in 1912, just a little before Bragg pronounced his rather simple formula[21]. These techniques have proved valuable in a wide variety of applications, including understanding the structure of DNA.

Diffraction gratings are now made by all sorts of simple, effective techniques that produce even miniature ones. They have become part of many smart phone accessories that are used for medical testing and nondestructive testing. This trend will continue and expand.

[21] von LaueM 1915 Concerning the detection of X-ray interference, *Nobel Lectures 1901–1921*, online; Friedric W, Kipping P and von Laue M 1912 *Sitzungsberichte der Mathematisch-Physikalischen Classe der Königlich-Bayerischen Akademie der Wissenschaftern zu Mönchen* 3.

IOP Publishing

Rays, Waves and Photons

A compendium of foundations and emerging technologies of pure and applied optics

William L Wolfe

Chapter 7

Displays—do you see what I see?

There are basically two kinds of displays: those that project an image from an object and those that create an image electronically. The former are the well-known slide projectors and their country cousins; the latter are things like our television sets that usually form some sort of raster pattern.

Heads up displays are those that do not require the operator to look down. Those are heads down displays! They are generally some form of display that is approximately horizontal and a reflector that sends an image to, for instance, a windshield. (Or a vertical display and two reflections.) The windshield then reflects the display and shows the scene superimposed upon it. Probably the first version was a gun sight of 1916 (see Eye tracking, chapter 9). In some cases, it is merely a display that is high enough on a dashboard that it obscures part of the windshield. HUD's, as heads up displays are abbreviated, originated with military aircraft for fairly obvious reasons[1]. Often they are superimposed on the windshield so that the driver can see the road through it and the information reflected from it. One example is the now defunct Cadillac night driving system[2].

Mechanical television systems predated our electronic ones. One version, the **Nipkow scanner** (figure 7.1), was invented by Paul Gottlieb Nipkow (1860–1940), who in 1884 used a rotating disc and a selenium detector to generate images[3]. The disc, as shown in the figure, had a spiral of holes in it, and as it rotated, each hole would allow light in. The holes generated an array of arcs as the disc spun. The first, leftmost lens focused the scene onto the disc. The second lens focused the disc onto a detector, but only the light that got through a hole got to the detector. There were far more holes than I have shown.

[1] Jarrett D 2005 *Cockpit Engineering* (Farnham: Ashgate Publications).

[2] Personal involvement.

[3] Reiman R *Who Invented Television*, online.

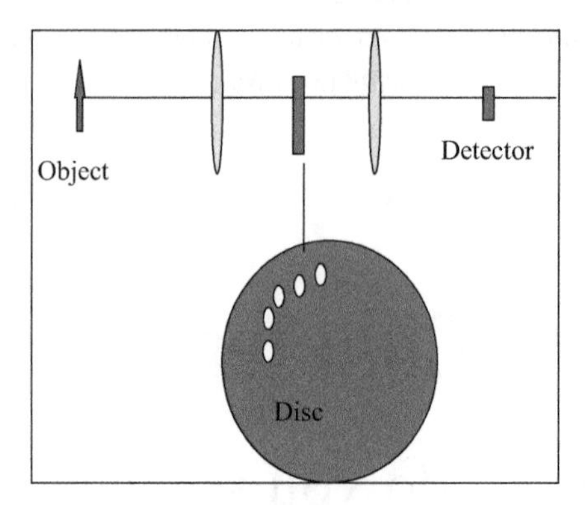

Figure 7.1.

Another was named the **Scophony system**[4]. It used lenses to form a beam that was then scanned in both the vertical and horizontal directions by mirror drums. The drum that generated the horizontal scan had 20 facets and scanned at about 30,000 rpm; the vertical drum had 12 facets and scanned at 240 rpm. Thus, there were 48 frames per second, each with approximately 480 lines. A Jeffree cell, an improved version of the Kerr cell (see Electro-optics, chapter 8), was used to modulate the light beam.

The Eidophor System[5], invented by Fritz Fischer (1898–1947) in 1939, is a display system meant to show bright images in large format, as in theaters. The key to the system is light modulation by irradiation of an electron beam on an oil. The oil film is placed where photographic film is usually placed in a normal projector. An electron gun shines on it and modulates the thickness of the film by depositing charges on the film. So, in essence, the Eidophor (image bearer) system is a normal film projector with the film replaced by an oil film that is illuminated with electrons. The rest of the process is to deflect the images past blocking strips by these changes.

CRT's, cathode ray tubes, have been ubiquitous display devices. But they have been almost completely supplanted by solid-state devices. They consist of an electron gun that fires electrons towards a phosphor screen that glows when so irradiated. The cathode is the negative electrode from which the electrons stream. The beam of electrons is the ray. This is a tube with a cathode and a ray, a cathode ray tube, or CRT. Electromagnets are used to scan the electron beam over the phosphor in a raster pattern and thereby generate an image based on the intensity and position of the beam. Three different phosphors and beams are used for the different colors for color TV. The first CRT was invented in 1897 by Karl Ferdinand Braun (1850–1918) as a laboratory oscilloscope. The principle of deflection of electron beams—and the electron itself—is credited to Joseph John Thomson (1856–1940) in that same year. Inventors are always eager to try something new. In 1907 Boris

[4] 1938 *Scophony Projection Set*, online, a reproduction of the operation manual http://www.tvhistory.tv/1938-Scophony-UK.htm.

[5] Labin E 1950 *SMPTE*; Sponable E 1953 *SMPTE*; Bauman E 1953 *SMPTE*; Yanczer P The Eidophor Television Project, Online; Hornbeck L 1991 *US Patent* 5061049.

Lvovich Rosing (1869–1933) was able to generate crude images with such a tube[6], but he still used mechanical scanning techniques. His student Vladimir Kosmich Zworykin (1888–1982) went on to introduce electronic scanning after seeing Farnsworth's developments in 1923[7]. Philo Taylor Farnsworth (1906–1971) was the first to demonstrate a complete electromagnetic scanning camera and receiver[8].

Figure 7.2. Magic lantern.

Slide projectors were used for… projecting slides! The earliest ones were the so-called **Magic Lanterns**, the first projectors of photographic images, generally attributed to Christiaan Huygens (1629–1695) in the late 1650s[9] (figure 7.2). The first projectors used oil lamps as sources. Then came limelight in 1870, and later arc lamps in 1890, and then electric bulbs. The slides, lantern slides, were pieces of glass coated with albumin and were replaced later with wet collodion plates. Starting in about 1950, the taking of 35 mm slides was very popular, and a means had to be made available for showing them. There were handheld versions, which usually had a small bulb and simple lens for magnification, but actual projectors could be used for groups of people to view at the same time. They have a light bulb, usually in front of a reflector to illuminate the slide. It is sufficiently intense that there is almost always a fan and a filter that rejects infrared radiation. The focusing optics project an image of the slide onto the screen. There are essentially two ways to illuminate the slide, called Köhler and Abbe illumination. The latter shines the source directly onto the slide; the former provides an extra lens that causes the slide to be illuminated uniformly. Abbe illumination is used if the source itself is uniform enough. Slide projectors can take many forms. Probably the best known is the **carousel projector** with the round tray of slides that drop into position sequentially. It was invented by David E Hansen (–2013) in 1965[10]. Carousels were discontinued in 2004, made obsolete by digital imagery. Bell and Howell made a **cube** projector in which the slides were all just stacked in a cube and dropped down sequentially to be projected. Others, like Hanimex, Argus, and Sawyer, made linear projectors, sort of like a straight-line carousel.

[6] German patent (1907); Scientific American.

[7] Zworykin V 1925 *US Patent* 1691324.

[8] *Popular Mechanics*, December 1934; Wikipedia.

[9] Pfragner J 1974 *Index, The Motion Picture: From Magic Lantern to Sound* (Folkstone: Bailey Brothers).

[10] Hansen D 1965 *US Patent* D201106 S filed in (1961).

Overhead projectors have been used in the classroom and for all sorts of presentations. They have a horizontal transparent plate upon which the transparency with imagery or text is placed. The light source below shines through the transparency, which is imaged onto a vertical screen by an appropriate set of optics. They were apparently invented by Roger H Appledorn in the early 1960s[11].

Opaque overhead projectors have been around since the beginning of the twentieth century; one of the first, if not the first, was patented by Philip Henry Wynne (1868–1919) in 1905[12]. It had essentially the same optics as modern overhead projectors, but the illumination had to be from above and then reflected, and therefore was more intense. These projected images of opaque objects.

The **telestrator** is a television-based device that allows the commentator to make marks on the television screen. It was invented in the 1950s by Leonard Reiffel (1927–), who used it to illustrate some of his science lectures on public TV. He has stated that it uses a sensitive transparent plate to determine the position of a stylus[13]. I assume that the signal is then transmitted to the TV image. The telestrator has been a boon to TV sports analysts—and viewers. Another is the yard marker.

The **first down yard marker** is a boon to viewers as well. It is created by a very complicated set of hardware and software. Each of the cameras has to have sensors that determine its angles and positions. The entire field has to be calculated. Then the line is constructed, usually in yellow, but on a pixel-by-pixel basis. It is replaced by the players' or referees' color so that the line does not go over them. On a pixel by pixel basis! Thirty times a second! For each camera! This was created by the staff at Sportvision and first used in 1998.

A compact modern **video projector** for computer-generated information was invented by Larry Hornbeck (1943–) in the early 1990s[14]. It consists of a large array —up to 8 million—of very small mirrors, micro mirrors, that deflect light from three diode lasers of different colors. The many million mirrors are driven piezoelectrically to scan the raster and to adjust the intensity (by the ratio of on time to off time 10,000 times a second). It is often called a DLP, or digital light projector.

LED's, **light emitting diodes**, are described in chapter 37 on sources. They are used in giant arrays to provide a display. The first flat panel TV display was accomplished by James P Mitchell in 1977[15]. It was monochromatic, but about thirty years later in 2009, with the advent of organic LED's, Sony was able to market a large screen, flat screen, color TV[16].

Liquid crystal displays are a form of spatial light modulators. Each pixel of an LCD has a polarizer in the front and one in the back of opposite orientation. In between is the special crystal that can rotate the plane of polarization an amount dictated by the voltage applied to it. Without any applied voltage, the cell is largely

[11] Minnesota Mining and Manufacturing Company 2002 *A Century of Innovation, the 3M Story* vol 2641 (Saint Paul, MS: 3M) p 2.

[12] Wynne P 1905 *US Patent* 803385.

[13] *Telestrator Invention Wins Emmy Just in Time for Super Bowl*, online.

[14] Hornbeck L 1995 *Proc. SPIE* **2641**; 2 US Patent 5061049 (1991) and related patents.

[15] Science Service, **97** (1978).

[16] Wikipedia, Sony XEL-1.

opaque since the two polarizers are crossed, but with increasing applied voltage the plane of polarization is rotated until there is full transparency. The amount of light is modulated, varied, according to the voltage applied to each pixel; it is a spatial light modulator. It is used in connection with an appropriate source of light as a display.

The history of the LCD may be considered to start with the discovery in 1888 by Friedrich Richard Kornelius Reinitzer (1857–1927) that cholesterol had two melting points and made colors—liquid crystals[17]. Otto Lehmann (1855–1922) explained how they worked and gave them their German name, *Flüssige Krystalle.* In 1911, Charles-Victor Mauguin (1878–1958) arranged them in thin layers between transparent plates. But it is not clear whether he applied voltages to change them. In 1927, Vsevolod Frederiks (1885–1944) devised a light valve that could be controlled electrically. The Marconi Company patented the first Liquid Crystal Light Valve in 1936. It was in 1964 that George H Heilmeier (1936–2014) created the first liquid crystal display[18].

The **virtual retinal display** projects a raster pattern, an image, directly onto the retina of the eye. It uses a source that is usually a laser diode—or three of them for color—that is intensity modulated and scanned across the retina. Usually a virtual image is used, and the exit pupil of the system is matched to the entrance pupil of the eye. This device was invented by Kazuo Yoshinaka in 1986[19].

Helmet mounted or **head mounted displays**, abbreviated **HMD's**, consist of eyeglasses that can project images to the eye but may also allow the real world to peek through. The displays are now usually made of LED's or LCD's for compactness. Some units allow the switching back and forth of the two views; some superimpose them. The first of these appeared in the 1969 on the VTAS, Visual Target Acquisition System[20], as an aid in aiming the Sidewinder AIM 9G (see Infrared, chapter 12).

Three-dimensional displays may be considered the holy grail of the display world. They come in essentially two varieties: those that require special glasses, and those that do not. One example of a 3D display is the rather popular movie *Avatar.* You needed a pair of polarized glasses (figure 7.3).

Figure 7.3. Polarized glasses.

[17] Reinitzer F 1888 *Monatsh. Chem.* **9** 421.
[18] Heilmeier G *et al* 1969 *Mol. Cryst. Liq. Cryst.* **8** 293.
[19] Display Device, Japanese publication JP1198892 (1986).
[20] Jagoe W and Radzelovage W 1974 *SAFE J.*.

Holography provides the second kind: no glasses, parallax, a true three-dimensional interference representation of the object. It is covered in the chapter of the same name.

One experiment has been in the Boulder Valley School District using the 3D DLP projector. It apparently uses sequential polarized images since children are shown with glasses and there is only one lens.

A simple, three-dimensional display technique has been used in three-dimensional advertising. The basic technique[21] is to project a real image using a mirror as shown in figure 7.4. The upside-down arrow below the axis is imaged as an erect arrow above the line—in 3D. Note that every image is three dimensional until it is put on film or a detector array. This simple apparatus has been used in a number of airport displays and was the subject of a patent before it was declared invalid[22]. The object, the lower upside-down arrow, is shielded from view, and the only part of the mirror that is necessary is above the axis.

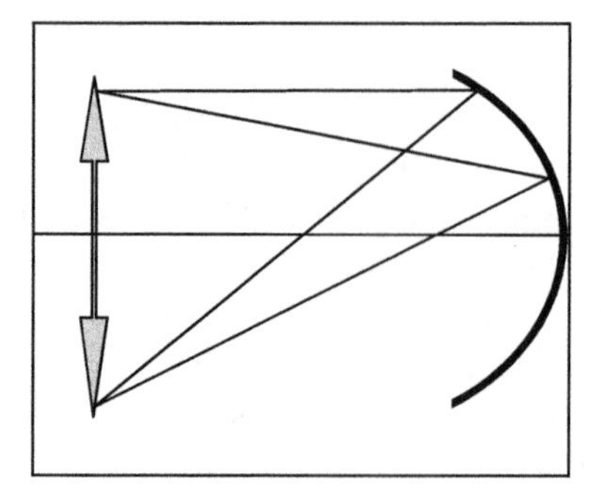

Figure 7.4.

We have graduated from CRT's to LED's, LCD's. and plasmas. The colors will get brighter with more dynamic range and greater rendition. Displays will become more intimate with AR and VR presented to us by way of eyeglasses. Displays will also become much larger with entire walls beckoning us to buy their products.

[21] Jenkins F and White H 1957 *Fundamentals of Optics* (New York: McGraw Hill).

[22] Personal participation in *Optical Products Development Corp v. Dimensional Media Associates.*

Chapter 8

Electro-optics and magneto-optics

This subdivision of optics deals with the interaction of electric and magnetic fields with light. Although one might include the detection and emission of light, like photodetectors and light emitting diodes, it does not. It deals primarily with how these fields modulate a light beam, change the transmission properties of materials, or even alter the spectral lines of atoms or molecules. Many of these effects are closely related and reversible. The effects separate generally into those that directly affect a light beam and those which interact with a medium that then affects the light beam.

The **Kerr effect** was discovered in 1875 by John Kerr (1824–1907). When an electric field is applied to a material, it becomes birefringent. That is, it has different refractive indices parallel and perpendicular to the applied electric field. The change in index is proportional to the wavelength and the square of the applied electric field. It has a different value for different materials, characterized by the Kerr constant, which is the constant of proportionality. The magneto-optical Kerr effect results when a light beam is reflected from a magnetic material and has a slightly rotated plane of polarization. The effect is often used to modulate a light beam by first polarizing it and then passing it through the Kerr cell, which is then turned on and off to modulate the beam at frequencies up to 10 GHz.

The **Faraday effect** is a magneto-optical one in which a magnetic field causes a rotation of the plane of polarization. It was discovered[1] in 1845 by Michael Faraday (1791–1867). It can be used to modulate a beam of light, much like the Kerr effect. The inverse Faraday effect induces a magnetic field with an intense, polarized light beam[2]. It was discovered in the 1960s. This is another example that almost everything is reversible in optics, although it is not part of the theory of reciprocity.

[1] Faraday M 1845 *Michael Faraday's Diary of Experimental Investigations*(London: George Bell).
[2] Popova D, Bringer A and Blügel S 2011 *Phys. Rev.* B **84** 214421.

doi:10.1088/978-0-7503-2612-4ch8

The **Pockels cell** is based on the effect of the same name and is named after Friedrich Carl Alwin Pockels (1865–1913), who discovered it in 1893. It is similar to the Kerr effect but is linearly proportional to the electric field and only occurs in crystals that do not have inversion symmetry. The applications are similar to those of the Kerr cell.

The **Voigt effect** was discovered[3] in 1902 by Woldemar Voigt (1850–1919). When a strong magnetic field is applied to a vapor, it generates double refraction. It also applies to films[4] and semiconductors[5].

The **Zeeman effect** was discovered by Pieter Zeeman (1865–1943) in 1896. He observed that the sodium D-lines were broadened in the presence of a magnetic field. Hendrik Antoon Lorentz (1853–1928) showed that they were actually split into two or three components depending on which way you looked at them. The electrons are affected by the magnetic field and produce circularly or plane-polarized light due to their rotation in the field. The inverse Zeeman effect occurs with absorption.

The **Cotton–Mouton effect**, discovered by Aimé Auguste Cotton (1869–1951) and Henri Mouton (1869–1935) in 1907, results in double refraction in a liquid in a transverse magnetic field. It is the magnetic analog of the electro-optic Kerr effect. The inverse C–M effect is the generation of a magnetization by a polarized optical beam in the presence of a magnetic field[6].

The **Stark effect** is the electric analog of the Zeeman effect. Named after Johannes Stark (1874–1957), it was discovered in 1913 and has been important in the development of atomic theory. Spectral lines split in the presence of an electric field and the light is polarized. There are linear and quadratic Stark effects, depending either directly upon the electric field or on its square. There are transverse and longitudinal effects depending upon which way you look at the lines—transverse, polarized, longitudinal, not. The inverse Stark effect occurs in absorption.

Many of the effects described above were important in the detailed understanding of atomic structure.

Electrochromic materials change their color with the application of a voltage. The color change may be just from transparent to dark gray or black. Such glasses can be used for *smart windows*[7]. The application of the electric field changes the molecular structure by reorienting electrons, causing various oxidation–reduction reactions[8]. Smart windows in one sense are those that change their transparency either by the introduction of a voltage or the incidence of more sunlight.

Smart windows can also be accomplished with suspended particles. Light absorbing particles are suspended between transparent layers. When electricity is applied, they line up and the sandwich becomes transparent. This requires that the electric field be maintained to maintain transparency.

---

[3] Voigt W 1908 *Magneto- and Electro-optik* (Leipzig: Teubner).

[4] Lissberger P and Parker M 1971 *J. Appl. Phys.* **42** 1708.

[5] Teitler S and Palik E 1960 *Phys. Rev. Lett.* **5** 546.

[6] Rizzo C *et al* 2010 *Eur. Phys. Lett.* **90** 64003.

[7] Passerini S *et al* 1989 *J. Electrochem. Soc.* **136** 3394.

[8] Oi T 1986*Annu. Rev. Mater. Sci.* **16** 185.

A third technique is the use of **liquid crystals**. These were examined by Friedrich Reinitzer (1857–1927) in 1888, who observed two different melting points and states of transparency[9]. He later discovered that these materials could also reflect circularly polarized light and rotate the plane of polarization. These were objects of curiosity until 1962 when RCA began investigating them for flat panel displays. In 1966, RCA got its first patent[10]. These panels consist of two crossed polarizers, between which is the liquid crystal, LC. When electricity is applied, the LC rotates the plane of polarization and prevents transmission, and vice versa. This technique can be used for TV displays or for entire windows. LC displays require a separate source of illumination; they are light modulators. In color TV, the various pixels are behind appropriate color filters. At least one corporation in San Diego has such windows[11].

---

[9] Reinitzer F 1888 *Monatsch. Chem.* **9** 421.
[10] Goldmacher J and Castellano J 1970 *US Patent* 3,540,796.
[11] Personal communication.

IOP Publishing

Rays, Waves and Photons
A compendium of foundations and emerging technologies of pure and applied optics
William L Wolfe

Chapter 9

Eye tracking—do you see where I see?

Over the years, various investigators have been tracking eyes. Surely the first eye trackers were other eyes. Our ancestors looked at their peers to see if they were looking down embarrassed or maybe shading the truth a bit. That would be determining where the eye seemed to be looking but not any real determination of its aim point—what it was looking at. Eye tracking has been used in the analysis of reading and other perceptual functions. More recently, advertisers have used it to analyze the effectiveness of their creations, and computer game designers and users have used it for directing their devices. Quadriplegics have used it to operate computers. The military has used it for guiding missiles and similar tasks.

Some of the earliest reports on eye movements were by Louis Émile Javal (1839–1907), who established in 1879 that we do not read in smooth scanning eye motions over the words, but by the saccadic motion of jumps and starts[1]. His experiments were accomplished by **direct observation** of eye movements. But it **was** eye tracking. Edmund Burke Huey (1870–1913) in 1908 built a crude tracker that consisted of a **contact lens with a hole** in it and a pointer attached to it[1]. Charles Hubbard Judd (1873–1946) and Guy Thomas Buswell (1891–1994) were the first to use a **non-contact** method, one that reflected a beam of light from the eye to a piece of film[2]. It should be obvious that accurate tracking—using a reflected beam like that—requires that the head be immobile, and many such trackers are used with chin rests or bite bars[3]. More sophisticated versions use a combination of the reflection from the front of the cornea and the rear of the lens of the eye[4]. When there is information from two points like this, the head can move. Another technique is the use of the corneal reflection and the center of the pupil of the eye.

[1] Huey E 1968 *The Psychology and Pedagogy of Reading* (Cambridge, MA: MIT Press) originally in 1908.
[2] Judd C 1915 *Genetic Psychology for Teachers, Psychology of Social Institutions and Psychology of High School Subjects* (Boston, MA: Ginn).
[3] Wikipedia, online.
[4] Crane H and Steele C 1985 Appl. Opt. 24 577.

doi:10.1088/978-0-7503-2612-4ch9

Modern systems all use a near infrared beam to reflect off the eye so as to not interfere with vision. Other techniques use other parts of the eye, like the retina in conjunction with the corneal reflection. Still others use more than one source or more than one camera[5]. One of the most heartening uses is the operation of a computer by the blink of an eye. An eye tracker follows the gaze of the user's eye and carries out the commands that are targeted and blinked upon[6]. Less heartening are the many commercial uses that determine our interests by the places we look and the duration of our stare[7]. Some car models now employ this technique to make sure we are attentive drivers.

The first helmet-mounted eye tracker is credited to Albert Bacon Pratt in 1916[8] (figure 9.1). It consisted of a rigid mount from the top of the helmet to a point out

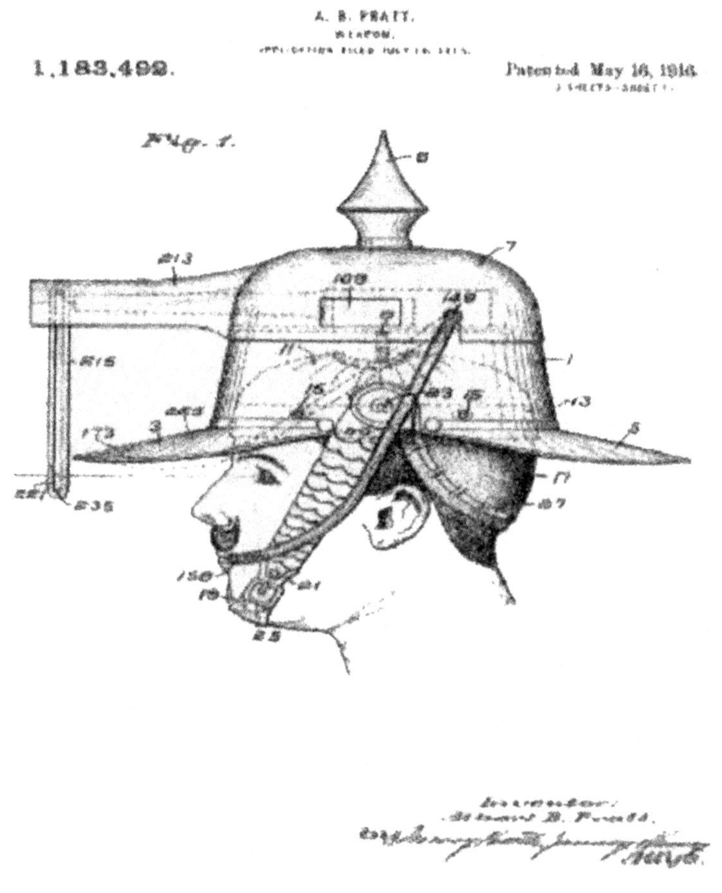

Figure 9.1. Pratt patent.

[5] Morimoto C *et al* 2000 *Image Vis. Comput.* 18 331.
[6] *University News*, Rochester Institute of Technology, December 2013.
[7] *The Economist*, December 2012; *SPIE Professional*, online October 2013.
[8] Pratt A 1916 *US Patent* 1183492.

front and an extension down to a point in front of the eye. The first HMD, helmet mounted display, appears to be in experimental aircraft of South Africa in a Mirage aircraft in the mid-1970s. This was followed by the Russian MIG-29 in 1985. The US used an HMD in a Cobra helicopter[9] and in a Phantom jet to aim AIM-9 Sidewinder missiles before they used their own infrared guidance system[10].

As the technology of tracking where the eye gazes improves by becoming smaller, cheaper and more mobile, it will be used in many more areas, such as the sequence of where an outstanding basketball guard looks, how and where an experienced surgeon gazes, the sequences of positions of pupils studying, analysis of depression and the design of buildings.

[9] Raybrook P 1998 *Armada Int.* **4** 44.
[10] Klass P 1972 *Aviation Week* **37**.

IOP Publishing

Rays, Waves and Photons
A compendium of foundations and emerging technologies of pure and applied optics
William L Wolfe

Chapter 10

Fiber optics—an inside job

Fiber optics have become very important in everyday life. Most of the TV programs we now receive come over fiber optic lines. They are optical signals that provide us with many more channels than VHF or UHF can provide. Some of our naval vessels use them on board for security and to reduce weight—that is, using light for lightness. Christmas trees now have strings of them. They are more energy efficient and reduce the chance of fire. Automobiles use them to route the many lights all around the cab and car. Physicians use them in endoscopes.

The transmission of light in a fiber depends upon the phenomenon of total internal reflection and the transparency of the fiber. When a ray of light exits a material of higher refractive index—at a large enough angle—into a medium of lower refractive index, it is totally reflected at 100%. It does not exit. The figure shows a fiber with two rays entering at different angles. The blue ray is at too steep an angle and exits the fiber; the red ray exhibits total internal reflection. Since a typical fiber has a core refractive index of 1.62 and a cladding refractive index of 1.52, that angle is almost 70°. Once in the fiber it repeats those reflections at the same angle millions of times until it is attenuated not by reflection losses, but by transmission losses. Modern fibers are very transparent, and before the signal is lost, amplifiers boost and relay it. See Communications for the history of the improvement in transmission and appendix B for details on refraction and internal reflection (figure 10.1).

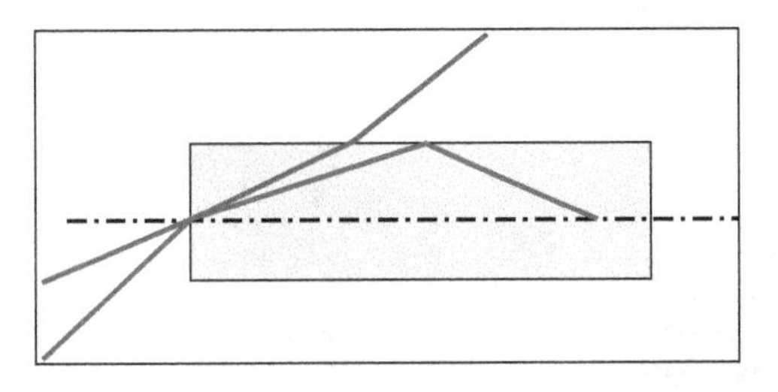

Figure 10.1. Fiber action.

Probably the earliest demonstration of guided light by internal reflection was in 1841 by Jean-Daniel Colladon (1802–93), who was trying to show the flow of water[1]. He was able to show light traveling in curved paths in the water—by total internal reflection. Total internal reflection had been described earlier[2] by Johannes Kepler (1571–630) in 1611. John Tyndall (1820–93) showed the same effect as Colladon in 1870[3], using containers of water and sunlight. It is interesting to note that the effect was used by the Paris Opera in Gounods *Faust* in 1853 to show a stream of fire from a wine barrel[4]. At about the same time, Jacques Babinet (1794–872) was able to guide light by internal reflection in glass tubes to illuminate the inside of the mouth[5].

William Wheeling in 1880 showed light could travel in hollow reflectorized (metallized) tubes for illumination. He illuminated a house with a light source in the basement and fibers all over to direct the light wherever he wanted it. His were not the solid fibers we use today but were hollow fibers that directed light. But they were fiber optics.

In 1888, Roth and Reuss illuminated body cavities with bent rods[6]. Henri Saint-Rene in 1895 used bent glass rods to relay an image in an attempt at TV.

In 1898, David D Smith applied for a patent on a curved dental illuminator.

In the 1920s, John Logie Baird (1888–946) and Clarence Hansell (1898–967) patented the idea of using arrays of fibers to transmit images for TV and facsimiles respectively[7]. Heinrich Lamm was the first to use it for internal medical examinations in 1930.

Probably the birth of modern fiber optics instrumentation was brought on by Brian O'Brien (1898–992), Harold Horace Hopkins (1918–94), Narinder Singh Kapany (1926–), and Abraham C S van Heel (1899–966). O'Brien was director of the Institute of Optics in America, van Heel in The Netherlands, and Hopkins in Great Britain. Kapany studied under Hopkins and then came to the United States. The critical time was 1954 when van Heel and Hopkins both published in *Nature*[8]. Although O'Brien did not publish, his discussions with van Heel were critical. They were the ones who first clad the outside of fibers with lower index material. Although the glass would have a higher index than air, thus limiting the incidence angle, it protected the fiber and reduced cross talk between adjacent fibers. Kapany coined the term 'fiber optics'. These groups in the US, UK, and The Netherlands pursued imaging and medical applications, whereas others saw opportunities in communication.

---

[1] J Colladon 1842 *C. R.* **15** 800.
[2] J Kepler, *Dioptrice* (1611), online.
[3] IHS Global Spec, online.
[4] J Hecht, *Daniel Colladon and the Origin of Light Guiding*, online.
[5] J Babinet 1842 *C. R.* **15**.
[6] About.com Inventors Fibers, online.
[7] *History of Fiber Optics*, Timbercon, online; Wikipedia, online.
[8] H Hopkins and N Kapany 1954 *Nature* **173** 39; A van Heel 1954 *Nature* **173** 39.

Charles K Kao (1933–) proposed long distance communication with fibers[9]. He probably would write it 'fibres.' He and George Hockham (1938–2013) did theoretical calculations in 1964 to show that low-loss fibers were possible if impurities in the glass were minimized[10]. This was suggested by Harold Rawson and proved true. Scattering was shown to be no problem, but impurities were. Note in the section on glass that there are all kinds, based on the 'impurities'—flint glass, borosilicate glass, lead, soda lime, etc. These varieties are wonderful for lens designers, but they introduced unwanted attenuation for those who wanted long range transmission in fibers.

In 1970 Corning Glass Works, now Corning, Inc., researchers Robert Maurer (1924–), Donald Keck (1941–) and Peter Schulz (1942–), developed single-mode fibers with sufficiently low loss for practical use (less than 20 db km^{-1}, or 10% km^{-1}) at 0.633 μm, the wavelength of a helium–neon laser. The rest of the story on optical communication by fibers is in the chapter on optical communications.

Although communications is probably the major application of fiber optics, there are other applications, both serious and somewhat frivolous. It is not clear who first used fiber optics on a Christmas tree, or for other decorations, but it is not very important either. Endoscopy is serious. Christmas tree lighting is somewhat frivolous. (See Scopes for the history of endoscopy.)

There are now thousands of miles of cable all over the world. There has been much done but we still need all-optical switching, more perfect cables for long haul transmission and better protocols for the handling of the data. They will come.

[9] Laser Focus, April (1966); C Kao, *Proc. Inst. Elec. Engrs.*, July (1966).

[10] J Hecht 1999 *The City of Light* (Oxford: Oxford University Press); J Hecht 1966 *Proc. Inst. Electr. Eng.*

IOP Publishing

Chapter 11

Glasses—some are half full

Not tumblers and not spectacles, but different types of glass materials. These and other optical materials that are used for lenses and windows are the subject of this chapter. Whereas glass in its various forms is the most used transparent material, and blanks for some mirrors, other materials are required in the ultraviolet and infrared parts of the spectrum. Most people understand that most glass is based on silica, also known as silicon dioxide or sand. But there are many other types, some without silica. Some are even plastic, but some people would not call them glasses. Some are crystals, and they may not be called glasses. But I have included glasses, glass-like materials, and other things used for glass in this chapter.

Safety glass, also called **laminated glass**, was invented by a mistake[1] by Éduoard Bénédictus (1879–1930) in 1903. But he should get credit for realizing what happened. He dropped a glass flask on the floor. It shattered but kept essentially its same shape. It had held cellulose nitrate, a liquid plastic, and had not been cleaned. Voila! This was incorporated in gas masks for World War I by Garrett Morgan (1877–1963). Soon this layered glass became standard for automobile windshields. The plastic commonly used today is polyvinyl butyral, or PVB[2], but others are used[3]. Note that those little, annoying rock dings can be repaired by drilling through the outside glass layer and injecting a resin that is then cured with ultraviolet light.

Bullet proof glass is essentially safety glass with several layers of laminate.

Tempered glass, also called **toughened glass**, is a form of safety glass that breaks into many very small pieces when dented, pierced, or scratched. Its most frequent use is probably for the side and rear doors on cars. It is not used for windshields because one of those rock dents would demolish it. Tempered glass was invented by Rudolph A Seiden (1900–1965) in 1938 and was first developed by Pittsburgh Plate

[1] Wikipedia and other online references.
[2] *The Autocar*, May 1939, 53.
[3] US patents: Iwamoto, 8628856 (2014); Yacovone, 8551600 (2013); Toyama, 8278379 (2012); etc.

doi:10.1088/978-0-7503-2612-4ch11

Glass Company, who called it Herculite. The tempering can be done by either heat or chemical treatment; it increases the compressive strength of the surfaces. In the heat treatment, the glass is heated and then rapidly cooled. The surfaces cool much faster than the interior, so their contraction causes a compressive strength of the glass. One can check the windows of a car for this glass with polarized sunglass by seeing if there are colors.

Float glass is not a different kind of glass; it is a different, and more efficient, way to make glass. The refined and molten glass material is poured on top of a bed of molten tin. It spreads out and is eventually cooled and removed. It is then sufficiently smooth that no grinding or polishing is necessary. The technique was developed by Pilkington glass company in 1959. One of my experiences years ago was with Ford. They could not figure out why their double pane windows had these imperfections. After some consideration, we determined that the imperfections were actually considerable perfection. The glass surfaces were smooth and parallel enough to cause interference patterns; they were float glass.

Plexiglass is one of those materials that some think is not a glass; it all depends upon what you consider a glass is—or what it isn't. Plexiglass, poly-methyl methacrylate, is a transparent plastic that is also called an acrylic glass, Acrylite, Lucite, and Perspex. It was patented by Otto Karl Julius Rohm (1876–1939) in 1901[4] but first brought to market in 1933 by Rohm and Haas, a company founded by two Ottos, Röhm and Haas (1916–1994).

Fused silica, also called **fused quartz** and **quartz glass**, is pure silica, silicon dioxide. Its kissing cousin, quartz, is crystalline; this is glass-like. It has a higher melting temperature, lower thermal expansion, and better ultraviolet and infrared transmission than normal glasses, but it is more expensive and therefore used in special situations; some are in lithography and mirror blanks. Crystalline quartz was discovered, not invented, many years ago in prehistoric times[5]. Jade and rose varieties were used for artistic carvings. Gaius Plinius Secundus, or Pliny the Elder (23–79), thought it was permanently frozen water, but it was Nicolas Steno (1638–1686) who first examined its structure[6]. Then in 1845 Karl Emil von Schafhäutl (1803–1890) first synthesized small quartz crystals[7], the forerunner to synthesizing quartz glass. General Electric started its manufacture in 1934[8]. Fused silica comes in a variety of trade names with slightly different formulations of impurities—or not. These include Suprasil, Homosil, and Ultrasil; these are low thermal expansion glasses with zero expansion at about 160 K and about 0.5 per million per degree over a broad range[9]. (These are very difficult measurements to make. Steve Jacobs used lasers and interferometers.)

[4] Röhm O, US patent.

[5] Driscoll K 2010 Understanding quartz technology in early prehistoric Ireland *Dissertation* University of Dublin.

[6] Steno N 1669 *De solido intra solidum naturaliter contento.*

[7] Schäufhatl K 1845 *Verlag der königlichen Akademie der Wissenschaften München* **20** 577.

[8] Rodney W and Spindler R 1954 *J. Opt. Soc. Am.* **44** 677.

[9] Jacobs S 1976 *Thermal Expansivity and Temporal Stability Measurements, College of Optical Sciences, University of Arizona* online; Jacobs S *et al* 2008 *J. Appl. Phys.* **47** 1683.

Zerodur is a glass ceramic (a polycrystalline material formed from a base glass) with a very low thermal expansion, thus making it ideal for large mirror blanks. It was developed by the Max Planck Institute in Heidelberg, Germany in 1968[10]. Although it is often touted as having a zero coefficient of thermal expansion, that is true at only two temperatures, 260 K and 300 K. However, the expansion is very low and less than one part in a million per degree over a reasonable range.

Cervit is another of the so-called ultra-low expansion materials, another glass ceramic—or in this case, a ceramic-vitreous material. Invented in 1967, it was discontinued in 1978 in favor of Zerodur and Sitall since it had some manufacturing flaws.

ULE is a registered trade name as well as shorthand for ultra-low expansion (glass), developed by Corning Inc. It is a titania silicate glass with expansion properties similar to Zerodur, Cervit, and Fused Quartz.

Pyrex is a brand name that originated with Corning Glass Works in 1915. Some say the name arose from the fact that the first product was a pie plate. Thus pie-rex, the king of pies. Others say it arose from the Greek *pyr*, meaning 'heat,' and rex, Latin for *king*—the king of heat. Nevertheless, it is a low thermal expansion borosilicate glass (about 3 parts per million per degree), patterned after the borosilicate glass invented by Otto Schott (1851–1935) in 1886. Eugene Sullivan (1872–1962) developed Nonex in 1908 based on Schott's Duran, and it was originally used for lanterns but eventually for cookware[11]. Corning removed the lead from Nonex and created Pyrex[12]. It was the blank for the Palomar telescope[13].

There are many, many types of **lens glasses** chemically developed for different refractive indices and dispersions. See chapter 25 on optical design for descriptions.

Infrared 'glasses' are, for the most part, not glasses, and certainly not silica glasses. Regular silica glasses become opaque at about 2 μm, so any system (most systems) will require the use of an alternate. There is a wide variety from which to choose, including crystals, special glasses, and thick and thin plastics. Early investigators found materials like salt—sodium chloride—were transparent in the infrared. In this case, out to about 20 μm. Other materials like potassium chloride, calcium fluoride, sapphire, and many others, also had the proper transparency, but they had unfortunate mechanical, chemical, and thermal properties[14]. Of course, salt is subject to attack by water and water vapor. One early attempt, in the 1930s, at developing infrared transparent materials was by Alexander Smakula (1900–1983), who combined two materials, potassium chloride and potassium iodide, and called it KRS-5. It was his fifth attempt at a mixed crystal, and the KRS stood for *Kristalle aus dem Schmelzfluss*, or crystals from the melt. They were first grown by R Koops in Smakula's lab in Jena in 1941[15]. Arsenic trisulfide glass was an early attempt at

[10] Döhring T *et al* 2009 *Proc. SPIE* **7281** 728103.

[11] *A Brief History of Pyrex*, online.

[12] Gantz C *Corning Pyrex Bakeware* Industrial Design History.com, Auburn University, online.

[13] Caltech Astronomy History 1908–1949; online.

[14] Wolfe W and Zissis G 1978 *The Infrared Handbook* (Washington, DC: US Government Printing Office).

[15] Koops R 1948 *Optik* **3** 298; Smakula A *et al* 1953 Inhomogeneity of thallium halide mixed crystals and its elimination *Technical Report Laboratory for Insulation Research* MIT. Available online.

having a true glass for the infrared. It is a so-called chalcogenide glass, meaning it consists of elements from the fifth and sixth columns of the periodic table. This led to the investigation of a family of chalcogenide glasses with different combinations of elements from these two columns. Although Boris Timofeevich Kolomiets (1908–1989) may not have discovered them, he was instrumental in bringing them to the attention of the world, largely for their semiconducting properties[16]. Kodak introduced a line of Irtan materials in about 1960. They were hot pressed versions of some good materials, like ZnS. The limitations were scattering since they were hot-pressed compacts and size due to the limitations of the presses. I think the name stands for infrared transmitting. Cleartran materials ZnS and ZnSe were introduced by Dow a little later. You guess at the name! They are created by chemical vapor deposition. Therefore, they have much less scatter and can be formed into a variety of shapes; they have a broad transmission range, covering the infrared to almost 15 μm. Chemical vapor deposition is much like evaporation: material is deposited as a chemical reaction in an appropriate chamber.

Ultraviolet 'glasses' have some of the same restrictions as infrared glasses, but at the other end of the spectrum. Silica-based glass does not transmit at wavelengths below about 0.35 μm. So, again natural crystals of various materials were explored that would be transparent in this region of the spectrum. Crystal quartz was one of them.

Optical glasses can be considered those that are intended for use as lenses, although I guess just about any glass is an optical glass. These were first just whatever was available. Crown glass derived its name from the fact that it was used in the 'crowns' above doors. The next step probably was making a glass with different constituents and then determining what its characteristics were—a trial and error approach. It was Otto Schott and Adolf Winkelmann (1848–1910) who first calculated, or at least estimated, what properties a given composition would have. It was their law of mixtures or the additive model that changed things in 1893. Some 30 years earlier, Hermann Franz Moritz Kopp (1817–1892) had described a similar law to obtain the specific heat of minerals and other solids[17]. Winkelmann described a method of adding oxides to a glass mixture to obtain a certain specific heat capacity[18]. Messrs Schott, Zeiss (Carl Zeiss, 1816–1888), and Abbe (Ernst Abbe, 1840–1905) used the technique to build better microscopes—and a first-rate company[19]. Since then, many improvements have been made to the original formulation, including weighting factors and non-linearities[20]. The technique has been used to design glass that does not change focus with temperature, (almost) zero expansion glass, and many forms of varying refractive index and dispersion.

It is hard to imagine improvements in low expansion glasses; it appears they are good enough, although a reduction in cost would be very helpful. It also seems that

[16] Kolomiets B 1964 *Phys. Status Solidi* **7** 359.
[17] Kopp H 1865 *Philos. R. Soc. Lond.* 155 71.
[18] Winkelmann A 1893 *Ann. Phys. Chem.* **49** 401.
[19] Dragic P and Ballato J 2014 *Opt. Photon. News*.
[20] Volf M 1988 *Mathematical Approach to Glass* (Amsterdam: Elsevier).

just about every formulation of glasses for lenses has been done, and there are advances in lens design that do not require additional types of glass. Glasses will combine with electronics, thereby generating interactive displays and functions. Techniques for strengthening, like Gorilla glass, will continue to improve.

IOP Publishing

Rays, Waves and Photons
A compendium of foundations and emerging technologies of pure and applied optics
William L Wolfe

Chapter 12

Holography—is it real or is it a …?

Holo—entire and graph—drawn or pictured, together make up holograph or hologram and the practice of holography. It is the process of generating a true three-dimensional image of an object by the use of interference and not the use of lenses. The process can be understood (I hope) by imagining that a light source illuminates an object and a mirror at the same time. The reflected light from both are brought together and caused to interfere, and the pattern is recorded. By the reciprocity relations in optics, when the recording is then illuminated by another source, it will generate the original image in three dimensions with parallax. That is, you can look around the sides of the image. The first holograms were made with monochromatic, coherent light. Later, reflection holograms and rainbow holograms allowed the reconstruction in white light. They work on different principles, but each separates the different colors for appropriate interference.

Dennis Gabor (1900–1979) was the **inventor** of the technique in 1947[1] but could not really implement it until the advent of the laser, a coherent source of light. The laser was first demonstrated in 1960 by Theodore Maiman (1927–2007) (see Lasers, chapter 16). In 1962, Emmett Norman Leith (1927–2005) and Juris Upatnieks (1936–) at the Willow Run Laboratories of the University of Michigan made the **first hologram**[2]. They were working on the optical processing of side-looking radar—synthetic aperture radar, or SAR—signals for Lou Cutrona, the head of the radar lab there. Since they had most of the equipment available, they decided to try it out. Yuri N Denisyuk (1927–2006) also tried it in 1962[3], but his was a **reflection hologram** that was generated with white light. In 1968, Stephen A Benton (1941–2003) at Polaroid made the first white-light transmission hologram, or **rainbow hologram**[4], by a

[1] Gabor D 1948 *Nature* **161** 777.
[2] Leith E and Upatnieks J 1962 *J. Opt. Soc. Am.* **52** 1123 and the Ann Arbor newspapers.
[3] Denisyuk Y 1962 *Dokl. Akad. Nauk* **144** 1275.
[4] Benton S 1977 White light transmission holography *Applications of Holography and Optical Data Processing* ed E Marom *et al* (Oxford: Pergamon).

doi:10.1088/978-0-7503-2612-4ch12

somewhat complicated process used to separate the colors. Lloyd Cross came up with the first **holographic cinematography** by using successive white light holographic frames in 1972[5]. Then in 1978, Leith, Hsuan Chen, and James Roth produced a hologram that could be generated in either mono-chromatic or white light[6].

It is one thing to have holographic motion pictures, but quite another to have **holographic TV**. For the movies, one can prepare all the frames at leisure (well, at least ahead of time) and then project them at the required rate. For TV, the frames must be taken and projected at TV frame rates. Currently, that is 30 frames per second. So a medium that can be recorded and erased at this rate to generate successive holographic frames must be available. Two organizations are currently working on this: MIT[7] and the University of Arizona[8]. Both are approaching the required frame rate.

The first exhibition of holographic art was at the Cranbrook Academy of Art in Bloomfield Hills, MI, in 1968 by Lloyd Cross[5], who also worked at Willow Run.

Although much interest in holography is based on the display of three-dimensional images, as noted above, other applications may be more important. Two of these are the prevention of illicit use of credit cards and forged currency, as well as massive data storage. Most of our credit cards now have some sort of holographic impression on them that produces an appropriate image and is hard to forge. Some countries are doing the same with their currencies, like Brazil[9].

It was announced by Inphase Technologies in 2005 that they would soon sell holographic storage devices that could store 300 gigabytes of data with access times that are ten times as fast as DVD's. They also report that such a process could store 1.6 terabytes and read it at 120 megabits per second. This is 340 times the capacity of a DVD and 20 times the data rate. It was first suggested in 1963 as a means of data storage, probably by IBM's Almaden Labs[10].

It is simple to say. Holographic displays will gradually replace their two-dimensional cousins in all sorts of applications. These include, for instance, medical training wherein holos will replace cadavers, automobile development of concept cars, even boardroom visual displays. The military will unroll holo maps that show the terrain in length width and height. Our TV sets will one day be holographic, three dimensional with parallax. There is no telling what may come of digital holograms, 3D representations of digital designs, no prior image. One thing we can tell is that this will probably come from computer aided designs. From CAD to a three-dimensional figure.

[5] Cross L *Rough Draft of Multiplex* on the net via Wikipedia reference 2 (with its warts).
[6] Leith E, Chen H and Roth J 1978 *Appl. Opt.* **17** 3187.
[7] Holographic TV on the net.
[8] Peghambarian N private communication.
[9] Wikipedia, online.
[10] Knight W 2005 *New Scientist*, online.

IOP Publishing

Rays, Waves and Photons
A compendium of foundations and emerging technologies of pure and applied optics
William L Wolfe

Chapter 13

Infrared—do you see what I can't?

The discovery of the infrared part of the spectrum was by William Herschel, nee Friedrich Wilhelm Herschel (1738–1822) in 1800[1]. Of course, people before this felt the heat radiated by fires and the Sun, but they did not associate it with optical radiation. Herschel was making measurements of the visible spectrum with a thermometer and noticed that it read even higher when it was moved just out of the red region. I believe his phrase was *beyond the range of refrangibility*. He speculated that infrared radiation was the same stuff as visible radiation and that there was little or no ultraviolet since it did nothing to his thermometers. Infrared measurements and instruments have since been important in both military and civilian applications, although many of the more modern advancements have been facilitated by military advancements. A noted infrared spectroscopist, Gordon Brims Black McIvor Sutherland (1907–1980), said in 1957[2], *There is little doubt that the problem of detecting a military target by the passive method of detecting the heat radiating from it was primarily responsible for the remarkable advances in the photodetection of infrared radiation. Yet the applications of these new photo detections have been far greater in the pure and industrial science than in the military field.* It is equally true today, if not more so.

Most of the ensuing developments—following the German-born Englishman Herschel—were done in Germany and were closely related to the advancement of detection techniques, of detectors. Detectors are of two main types: thermal and photon. The thermal detectors predated the photonic ones, but they are less sensitive and slower—although they cover the whole spectrum. Early applications, if they may be called that, were mostly spectroscopic. Significant steps include the invention of the slide rule by Marianus Czerny (1896–1985) for the calculation of blackbody radiation in a finite spectral region (since the integration of the blackbody spectrum was not possible—the integral was non-integrable) (figure 13.1), the development by

[1] Herschel W 1800 *Philos. Trans. R. Soc.* **31** 284.
[2] Sutherland G 1957 *Proc. IRE* **105** 306.

doi:10.1088/978-0-7503-2612-4ch13

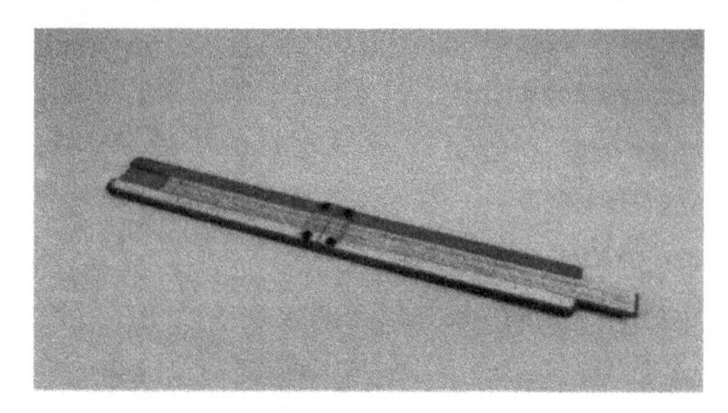

Figure 13.1. Czerny slide rule.

the Germans of Lichtsprechers, infrared communication devices, and ship detectors, called Kiels. At the end of World War II, some of these German scientists emigrated to the US and so did their technology. Then came the development of increasingly competent detectors, although one could only operate in the long-wave infrared (8–12 μm) with very cold detectors. This changed with the advent of the mixed crystal mercury cadmium telluride. Until the invention of the charge-coupled device by Willard Sterling Boyle (1924–2011) and George Elwood Smith (1930–), only single detectors and one-dimensional arrays could be used. This enabled the use of fairly large arrays of both photon and thermal detectors. The latter, which can be operated at room temperature, have made infrared equipment much more affordable (although an infrared array still costs much more than a visible one). These detector advances were accompanied by similar strides in optical materials. Today, there is now a fair variety of glasses, compacts, and manufactured crystals that transmit throughout the region. In addition to the military applications of missile detection and guidance, infrared devices perform medical, industrial, forensic, and other functions.

Thermal detectors had their origin based on the discovery of the thermoelectric effect by Thomas Johann Seebeck (1770–1831) in 1821. In 1826, Leopoldo Nobili (1784–1835) and Macedonio Melloni (1798–1854) applied this to what is now known as a thermocouple—the junction of two dissimilar metals that generates a signal with the incidence of heat[3]. The invention of the bolometer (throw meter) is generally attributed to Samuel Pierpont Langley (1834–1906) in 1878. It was reported that it could detect a cow at a distance of over 1000 feet[4], giving rise to a limerick:

> Langley invented the bolometer
> A funny kind of thermometer.
> It could detect the heat
> From a polar bear's seat
> From a distance of half a kilometer.

The metal (platinum) bolometer of Langley was greatly improved with the so-called thermistor bolometer invented in 1946 by J A Becker and Walter Houser Brattain (1902–

[3] Nobili L 1830 *Bibliotheque Universelles des Sciences Belles Litres et Arts* **44** 225.
[4] Langley S 1888 *The New Astronomy* (Houghton: Mifflin).

1987) using semiconductor material[5]. The temperature coefficient of resistance was much higher. For many years, just about every thermal detector system used thermistor bolometers[6]. However, the first thermistor bolometer is properly attributed to Michael Faraday (1791–1867) in 1833. He observed the effect in a sample of silver sulfide.

Some were immersed to the back of a hemisphere[7], and even hyperhemispheres were considered[8]—based on the aplanatic Weierstrass sphere. Both were generally made of germanium since it has a high refractive index of about 4, providing the best gain of any infrared material. The hemispherical lens provides a gain, image reduction, proportional to the refractive index, whereas the hyperhemispherical lens provides the square of the index (figures 13.2 and 13.3). Considerable work was undertaken by many investigators in order to improve and develop the use of thermocouples[9]. A thermopile is a series combination of thermocouples[10]. One form of modern bolometer that is very useful in astronomy was developed by Frank James Low (1933–2009)[11]. It is a doped germanium detector (one with intentional impurities) operated at a few degrees above absolute zero. Other investigators have examined thermal detectors based on the transition from normal conductivity bolometer to superconductivity[12]. They promise high responsivity because the transition is so abrupt, but they are tricky to use for the same reason. Marcel J E Golay (1902–1989) invented a thermal detector based on the expansion of a gas in a tube— the **Golay cell**[13].

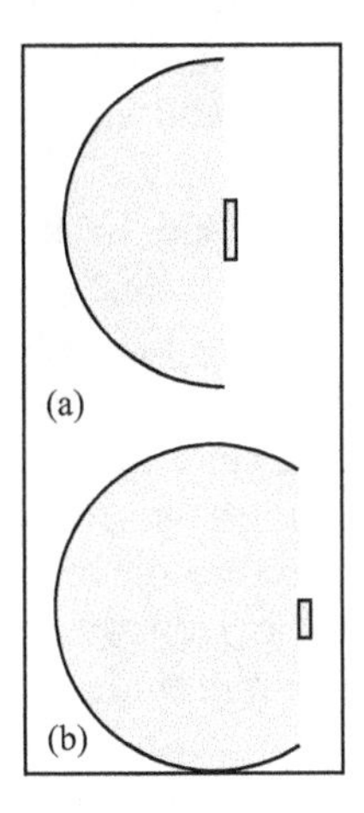

Figure 13.2. (a) Hemisphere. (b) Hyperhemisphere.

[5] Becker J and Brattain W 1946 *J. Opt. Soc. Am.* **36** 354.
[6] Becker J et al 1945 *Final Report on the Development and Operating Characteristics of Thermistor Bolometers* OSRD 5991.
[7] Dewaard R and Weiner S 1967 *Appl. Opt.* **6** 1327.
[8] Wolfe W 1962 *Proceedings of the Conference on Optical Instruments and Techniques* (London: Chapman Hall).
[9] 1964 *Platin. Met. Rev.* **8** 1.
[10] Hornig D and Okeefe J 1947 *Rev. Sci. Instrum.* **18** 474.
[11] Low F 1961 *J. Opt. Soc. Am.* **51** 1300.
[12] Andrews D *et al* 1942 *Rev. Sci. Instrum.* **13** 281.
[13] Golay M 1947 *Rev. Sci. Instrum.* **18** 357.

Figure 13.3. Bolometer.

While these thermal detectors were all available shortly after the discovery of the infrared part of the spectrum, **photon detectors** did not come into being until sometime later. The photoconductive and photovoltaic effects were discovered by Alexandre-Edmond Becquerel (1820–1891) and Willoughby Smith (1828–1891) in 1839 and 1873, respectively[14]. Apparently little attention was paid to these advances until the advent of World War I, or at least its impending occurrence. The **thalofide cell**[15]—consisting of thallium, oxygen, and sulfur—seems to be the turning point. And then we saw developments of a variety of **lead sulfide**, PbS, cells, and those of **lead telluride**, PbTe, and **indium antimonide**, InSb. The lead sulfide response reaches to almost 3 μm and needs moderate cooling (e.g. dry ice), and lead telluride to almost 6 μm and needs to be cooled to 77 K (−320 °F), as does indium antimonide. Lead sulfide was developed by Robert 'Bob' Cashman (1928–1988) at Northwestern University as an evaporated detector. He sometimes bemoaned the variability of the materials, citing the circumstance that the cows from which the materials were extracted changed pastures and therefore diets[16]! Although Bob also worked with lead selenide and lead telluride, it was Henry Levinstein (1919–1986) who had the most impact on the latter. Indium antimonide eventually replaced these detectors. It goes to almost 6 μm like PbTe and must be cooled to 77 K (−320 °F), but it is more reliable since it is a crystal rather than a chemically or evaporatively deposited film. It was developed in the UK by Avery and colleagues[17] and in the US by the group at the Chicago Midway Laboratories[18]. These detectors still did not cover the all-important long wave infrared region of 8–12 μm, where the radiation from the Earth's background peaks. For this, doping silicon and germanium were attempted[19]. Such detectors were sensitive to more than 12 μm but had to be cooled

[14] Becquerel E 1840 *C. R.* **9** 561; Smith W 1873 *Nature* **7** 303.

[15] Case T 1917 *Phys. Rev.* **9** 305; Case T 1920 *Phys. Rev.* **15** 289; Lovell D J 1971 *Appl. Opt.* **10** 1003.

[16] Personal communication.

[17] Avery D, Goodwin D and Rennie A 1957 *J. Sci. Instrum.* **34** 394.

[18] Mitchell G, Goldberg A and Kurnick S 1955 *Phys. Rev.* **97** 239.

[19] Burstein E 1949 *Nottingham Photoconductivty Conf.*.

to about 20 K (−423 °F). Then came **mercury cadmium telluride**, a mixed crystal of, yes, mercury telluride and cadmium telluride. It is variously called mercad, HCT, MCT, and cadmium mercury telluride. It was conceived by Bill Lawson and colleagues in the UK[20] and brought to the US by Paul Kruse[21]. The mixture can be adjusted so that it is a useful detector anywhere from the visible to about 14 μn, and it only requires cooling to about 77 K. As of this writing, the vast majority of infrared detectors are indium antimonide, mercury cadmium telluride, and bolometers.

The next major move in infrared detectors was the advent of two-dimensional **arrays**, which was made possible by the invention of the charge-coupled device, or CCD[22]. Then arrays of HCT, InSb, and bolometers were constructed generally by joining the detector slab to a silicon processing slab. This is easy to write, but the technology was difficult (e.g. lining up all the 'bump bonds' to within about 1 μm). The so-called microbolometer technology was fostered by the Night Vision Labs of the US Army in the late 1980s under a project called HIDAD, High Density Array Development. They funded Honeywell and Texas Instruments for the development, and today it is a major industry. Both have patents on the technology[23]. Although the Hornbeck patent of TI appears to precede that of Higashi of Honeywell by several years, both were under security wraps for some time. I believe that the real originator of this technology was Paul Kruse, a colleague of Higashi (and me) and the same Paul Kruse who brought mercury cadmium telluride to the US. Subsequently Honeywell licensed its procedure to a variety of other manufacturers.

Detector specifications[24] were a sensitive topic (excuse the pun) for many years in the infrared field. A detector could be labeled more sensitive, but what did that really mean? The responsivity of a detector was defined as the ratio of the output to the input, often in volts per watt. The detectivity, D, was the ratio of the signal-to-noise ratio, S/N, to the input, usually in reciprocal watts. But these could be functions of the size of the detector or the bandwidth of use. So a specific detectivity was defined as the *specific detectivity*, D^*, which normalized the S/N to the bandwidth and area[25]. Another such normalization took into account the background[26], the D^{**} or D^*_{BLIP}. These became the norms for describing the performance of detectors and comparing one to another.

Infrared applications include, among other things, imaging, communication, temperature measurement, missile defense, and missile homing. Probably the first infrared imaging device was the **evaporograph**, invented[27] by Czerny after the ideas of John Frederick William Herschel (1792–1871)[28]. This Herschel was the son of the

---

[20] Lawson W, Nielsen S, Putley E and Young A 1959 *J. Phys. Chem. Solids* **9** 325.

[21] Personal experience.

[22] Boyle W and Smith G 1970 *Bell Syst. Tech. J.* **49** 587.

[23] Hornbeck L 1991 *US Patent* 5021663; Higashi R 1994 *US Patent* 5300915.

[24] Jones R 1952 *Nature* **170** 937; Jones R 1952 *J. Opt. Soc. Am.* **42** 286; Jones R 1953 *Adv. Electron.* **5** 27.

[25] Jones R 1957 *Proc. IRIS* **2** 9.

[26] Jones R 1960 *J. Opt. Soc. Am.* **50** 1058.

[27] Czerny M 1929 *Z. Phys.* **53** 1.

[28] Herschel J 1840 *Trans. R. Soc. Lond.* **131** 52.

discoverer of infrared. The evaporograph was later developed by workers at Baird Atomic Company[29], notably Bruce Hadley Billings (1915–1992). Other early viewers included the sniperscope, which was an active infrared viewing device (see Scopes, chapter 35). A rather rudimentary device called the Kiel IV was developed by Werner Weihe (1903–1989) and his group at Jena in Germany during World War II. It consisted of a bolometer detector and an optical mechanical scanner that scanned the detector image over object space. It was used to detect shipping in the English Channel. A pioneer in the use of **infrared imaging** was R Bowling Barnes (1906–1981), with a company named after him, Barnes Engineering. He led the way in early diagnostic **thermography**[30] based on thermistor bolometers. The Barnes thermograph had a simple optical system and a single flat mirror that scanned a raster; it took fifteen minutes to get a decent picture[31]. It was too slow for practical application, but technology improved that.

Prior to the invention of the CCD, all infrared imagers were based on either single detectors or linear arrays and optical mechanical scanning devices. Many of these, about 75, have been summarized[32]. One class of imagers is air-to-ground **strip mappers**. Many of these were developed in the 1960s. They usually consisted of a linear array that was swept mechanically from side to side as an aircraft or satellite moved forward. I had the pleasure of supervising the design of the first one of these that used mercury cadmium telluride detectors. It later became the AN/AAD-5. These were replaced by so-called FLIR's as two-dimensional arrays became available. FLIR stands for Forward Looking InfraRed. There is even a company by that name, FLIR. Originally, it was because they could look forward—that is, get an image towards the front of the aircraft—but with the company, it may now mean an eye to the future.

The Germans developed **Lichtsprechers** (light speakers) communication devices that modulated infrared beams, but their lives were short[33].

A major development during World War II was the **Sidewinder missile** with its infrared guidance. Its predecessor was the **Dove missile**, designed around 1944 at Polaroid by Edwin Herbert Land (1909–1991) and Dave Grey[34], and transferred to Eastman Kodak in 1946. The **Firestreak missile** was developed in the late 1940s and early 1950s in the UK[35]. The Sidewinder used a folded reflective optical system, a reticle, and a cooled lead sulfide detector (figure 13.4). It went through many changes over the years and is now named the AM-RAAM, or Advanced Medium Range Air-to-Air Missile. It started with William 'Bill' McLean (1914–76) at what was then the Naval Ordnance Test Station in 1947 at China Lake, California, in the Mojave Desert, and is now the Naval Weapons Center in the same place. The first

[29] Personal experience and *Proc. IRE* **47** 1423 (1959).

[30] Barnes R 1964 *Ann. N. Y. Acad. Sci.* **121** 34; Wolfe W 1964 *Ann. N. Y. Acad. Sci.* **121** 57; Lawson R N 1956 *Can. Med. Assoc.* **75** 309.

[31] Personal observation, 1958.

[32] Wolfe W (ed) 1958 *Optical Mechanical Scanning Devices* (Ann Arbor, MI: The University of Michigan).

[33] Huxford W and Platt J 1948 *J. Opt. Soc. Am.* **38** 253.

[34] McKenny V 1998 *Insisting on the Impossible: the Life of Edwin Land* (Reading, MA: Perseus); Friedman N 1983 *US Naval Weapons* (London: Conway Maritime Press).

[35] Boyne W (ed) 2002 *Air Warfare* (Santa Barbara, CA: ABC-CLIO).

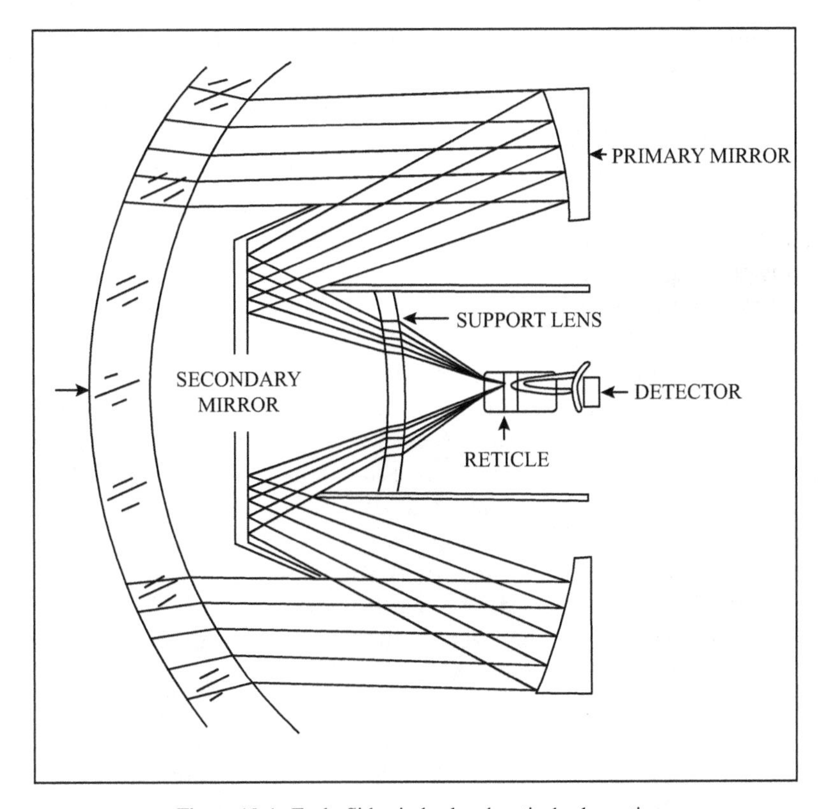

Figure 13.4. Early Sidewinder head optical schematic.

launch was in 1953. The reticle was a so-called rising Sun with one half fifty percent transparent and the other half consisting of alternating fully transparent and opaque radial spokes (figure 13.5). The focal spot from the telescope was rotated around the reticle or the reticle was spun around the fixed spot, and the phase and amplitudes of

Figure 13.5. Rising Sun reticle.

the signal signified the two angles of direction to the target. It worked so well that the target airplane drones eventually just had flares on their wingtips to preserve them[36]. It is hard to tell how many 'improvements' there have been. The version just before the AMRAAM was the AIM-9X. So there were at least nine, but there is the whole alphabet, only parts of which were used. The AMRAAM now has a coudé system (see Telescopes, chapter 42) to get the detector array and coolant off the gimbal[37]. The reticle has been replaced by an array of detectors and computer algorithms for determining the angles.

Infrared systems were used for detecting the launch of intercontinental missiles and for their interception should that be necessary. The launch detectors sensed the radiation around 3 μm from the rocket plumes from geosynchronous orbit. Then long wave detectors, mercury cadmium telluride, are designed to sense the payload in midcourse since it would be a warm (about room temperature) object with a very cold background—space, at a temperature equivalent to about 60 K (−350 °F). The earliest detection systems were studied during the presidency of Dwight David Eisenhower (1890–1969). The Russians had launched Sputnik in 1957, and its appearance signaled that the USSR could soon launch ICBM's; we had to do something. Although we had a BMEWS (Ballistic Missile Early Warning System) of radars in the northern regions, they were limited by line of sight and the curvature of the Earth; we needed a way to sense the launch that allowed enough time to launch our bombers and missiles. The concept was developed by ARPA, the Advanced Research Projects Agency, and dubbed **MIDAS**, the **Mi**litary **D**efense **A**larm **S**ystem[38]. The sensor was okay, once a solar scatter problem had been resolved, but satellite and communication technology constraints limited the effectiveness of MIDAS. However, it paved the way for the **DSP**, the **D**efense **S**upport **P**rogram, which started in 1966. The MIDAS program was developed by what is now Lockheed Martin Company with the sensor developed by the Aerojet Corporation.

DSP uses satellites in geosynchronous orbit to sense launches—and nuclear explosions. The first launch was in 1970 and the last in 2007, when it gave way to **SBIRS**, the **S**pace **B**ased **I**nf**R**ared **S**ystem. A linear array of detectors at the focus of a Schmidt telescope scans over the Earth six times a second. Whereas reconnaissance satellites need high resolution on the ground for identification purposes, these are merely warning devices. They need just enough of a footprint so that the radiation from the launching rocket is sufficiently higher than the background. It apparently has this resolution, since launches of Scud missiles were detected during Desert Storm[39]—and these are not ICBM's. The satellites are made by Northrop Grumman Aerospace systems, then TRW, with the sensors made by Northrop Grumman Electronic Systems, formerly Aerojet Electronics Systems.

SBIRS, which replaced the DSP starting in about 2007, is a complicated system combining the DSP with an elliptical orbiting geosynchronous sensor and a low

[36] Personal observation.
[37] Personal involvement.
[38] Personal experience and Wikipedia.
[39] 1991 *Aviation Week and Space Technology* 19 January 28.

altitude one[40]. It has received various improvements and adaptations over the years.

These studies and designs occurred during the Strategic Defense Initiative started in 1983 by President Ronald Reagan[41] (1911–2004). It was affectionately called 'Star Wars' by some and cynically by others. Eight hit-to-kill intercepts were demonstrated over the Pacific Ocean[42]. Hit to kill or kinetic energy vehicles, KEV's, are required as a result of the treaty that prohibits any nuclear arms in space.

Civilian applications include, among others, night driving, forensics, intrusion detection, spectroscopy, lie detection, breast cancer screening, and other medical applications. The remarks above by Sutherland, a renowned spectroscopist, have been realized in spectroscopy, and that story is in that chapter. However, there are many areas that Sutherland could never have envisioned. Several of these are described below.

Cadillac introduced an infrared **night driving system** in 2000. The first design used mercury cadmium telluride detectors at the relatively high temperatures that could be attained with Peltier coolers[43]. They were soon replaced by a microbolometer array, as a result of the HIDADS program discussed above. Note that this is another example of military developments leading to civilian applications. It was discontinued in 2004 for cost, but others, like Audi, Mercedes, Lexus, and BMW, are entering the fray since the arrays are becoming less costly[44]. A little insight into the process is in order. GM carried out customer surveys with expensive, existing units. People loved the night vision. GM carried out, via Raytheon, studies on really cheap optics. (GM owned Raytheon at the time.) This included various replication techniques that did not generate good enough results for the optics. They considered where to put the sensor since it could not go behind the windshield, for the glass is not transparent to the infrared radiation. Putting it near the engine is dicey because of the heat. If it is in the grill, it is subject to possible flying stones. They had to consider the display. A heads-up display would be good for driving but would eliminate use during heavy mists and fogs during the daytime.

Simple **intrusion detectors** are now available in many stores and in many varieties[45]. One consists of an infrared-transparent plastic Fresnel lens (see Lenses, chapter 16) and a pair of thermal detectors[46]. They usually can come with floodlights so that when anything warm and moving is detected, the lights go on. The military has considered much more complex versions that can scan the perimeters of airfields and the like.

Lie detection, without the many connectors to the body, can be done in the infrared in at least two ways. One is to shine near infrared radiation onto the scalp. It penetrates a reasonable distance into the brain, reflects back, and thereby indicates levels of activity in various regions[47]. Infrared imaging of the face can detect minor

[40] Several online sites.
[41] All the news media of the time.
[42] Barrie A 2013 *Missile Defense Agency Successfully Tests 'kill vehicle' to block ICBM'S* Foxnews.com.
[43] My design. A Peltier cooler is a solid state device with no moving parts.
[44] Wikipedia: *Auto Night Vision* with many references.
[45] About 20 patents from 1976 on.
[46] The one I have.
[47] Scott B 2006 *International Patent* PCT/US2006?040847.

changes in temperature due to minor changes in blood flow due to tension generated by lying—sort of blushing in the infrared[48].

Screening for **breast cancer** is based on the fact that any tumor will have an increased temperature of a few degrees, but carcinogenic ones will be even higher, about a degree higher. Thus, a good infrared camera with enough thermal sensitivity should be able to not only detect tumors of the breast but differentiate the cancerous from the benign since the sensible temperature difference of a good camera is a small fraction of a degree. This technique has been with us since the mid-1950s[49]. M Grout has written a nice review of thermography versus mammography, including decisions by medical and insurance organizations[50]. Debbi Walker and Tina Kaczor have written a more modern assessment[51] that relegates the modality to an adjunct procedure.

Another medical application that is less complicated but more pervasive is the infrared **ear thermometer**. It consists of what is essentially a light funnel and a thermal detector[52]. The device is put in your ear and aimed at the ear drum, the tympanum, from which it receives radiation. It is calibrated and thereby records a good internal temperature—with less invasion than either an oral or anal device! There are several different ones on the market; the reference cited here seems to be the earliest, at least the earliest patented. There is a host of different types of covers.

Theodore Hannes Benzinger (1905–1999) invented the ear thermometer in 1964, according to *The New York Times*, and he filed for a patent a year later[53]. It was a simple device by today's standards, consisting mainly of a thermocouple and funnel. Others have improved upon it; Jacob Fraden invented the world's best-selling ear thermometer, the Thermoscan® Human Ear Thermometer.

Infrared cameras are useful in firefighting, both in buildings[54] and in forests[55]. In buildings they can 'see' through fairly dense smoke. They can assess where the heart of the conflagration lies. They can monitor walls to see if there is roaring fire on the other side. The first documented civilian life saved was in 1988[56]. Most forest fires occur in the western United States in the afternoon as a result of thunderstorms. The smoke rises and obscures the true extent of the fires. It can be more than embarrassing to drop the smoke eaters into the fires rather than just outside them. Overhead infrared reconnaissance can prevent this. It is also useful in cleanup operations for finding smoldering remains. San Diego started using this system in 2009 and has improved upon it since then. The fact that these images can be relayed

---

[48] Pavlidis I 2004 *Proc. SPIE* **5405** 270.

[49] Lawson R 1956 *Can. Med. Assoc. J.* **75** 309.

[50] Grout M 2007 *Thermography for Breast Cancer Screening* (Scottsdale, AZ: Arizona Center for Advanced Medicine).

[51] Walker D and Kaczor T 2012 *Nat. Med. J.* **4** 7 and online.

[52] O'Hara G 1986 *US Patent* 4602642.

[53] Benzinger T 1965 *US Patent* 3416973.

[54] *Thermal Firefighting Applications*, online; personal involvement.

[55] *Infrared Eye in the Sky Aids Firefighters, San Diego Center*, August 2012.

[56] Scott R *et al* 1999 *An Overview of Technology and Training Simulations for Mine Rescue Teams* (Washington DC: National Institute for Occupational Safety and Health).

to a ground crew has greatly increased their capability to direct crews properly. A development that is not historical but current is the use of holographic infrared cameras to fight fires[57].

Other applications include assessing power lines, industrial equipment, home inspection, criminal apprehension, and even marijuana detection.

Several firms[58] perform aerial **inspection of power lines** using infrared and visual cameras. When power line transfer stations, or even just connections on poles, are not operating as they should, it is because they have higher resistance and therefore more heat or because they are disconnected, open, and have less heat and continuity. These cameras are usually of the microbolometer variety with 0.3 megapixels and 80 mK temperature sensitivity.

Infrared cameras have been used by various police forces for the **apprehension of criminals**. They can be seen by virtue of their heat radiation in woodsy areas, behind obstacles, by virtue of the radiation reflected around them and in similar circumstances. One example, as shown in figure 13.6, is the sensing of Dzhokhar Tsarnaev hiding in a boat after the Boston Marathon bombing[59]. Police and firefighters have also used thermal imaging to find victims of tragedies in the rubble, notably in the Oklahoma tornado tragedy[60]. Infrared cameras have also been used to find those lost at sea[61].

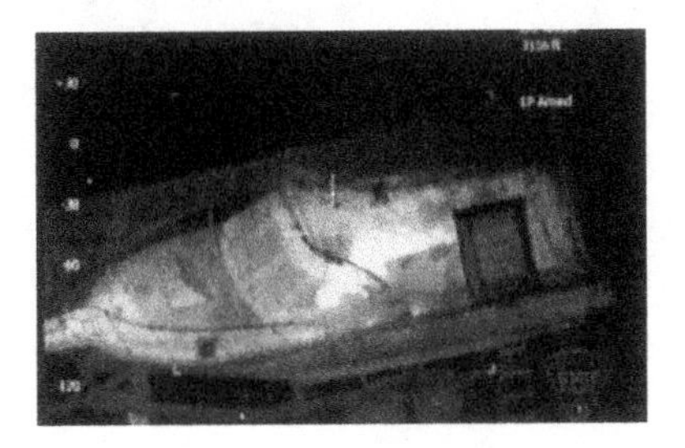

Figure 13.6. Infrared image of the Boston Bomber.

Weather mapping from satellites has gone from a very basic beginning to sophisticated imaging and probing. The first attempt was Vanguard[62], meant to measure cloud cover (visibly), but it did not work very well[63]. Neither did the Explorers. TIROS-1 was the first weather satellite considered to be a success. The first of the really useful infrared satellites were the GOES series—Geostationary Orbiting Environmental Satellite. They orbit at about 22 300 miles in a stationary position over the equator. Each satellite thus views the same spot on the Earth all the time. Several are in different positions around the Earth to monitor different places.

[57] Locatelli M *et al* 2013 *Opt. Express* **21** 5379.
[58] For instance: Oregon Infrared, Helivison and Certified Infrared, on the net.
[59] On the net: Infrared police chopper images show Boston marathon suspect.
[60] American Society for Nondestructive Testing, online.
[61] Thermal imaging cameras used by search and rescue teams, online.
[62] Hill J 1991 *Weather From Above: America's Meteorological Satellites* (Washington, DC: Smithsonian Institution).
[63] Vanguard, NASA.

The original design used a pair of MCT detectors (one for backup), a telescope, and a scanning mirror. It took about 15 min to scan the segment of the Earth it was assigned. The detectors were cooled by radiation cooling. This system mapped the temperature distribution of its field of view. From that, cloud altitudes and other phenomena were obtained. GOES 12, the earliest still in operation, was launched in 2001. Polar weather satellites travel over the poles in orbits that are Sun synchronous—they are over the same spot each orbit at about the same time of day. They are much lower, 517 miles high, but they do not observe the same spot continuously. They have been designated as NOAA POES—that is, National Oceanic and Atmospheric Administration Polar Orbiting Environmental Satellites. Europe, Japan, and China have their equivalents of both polar and geosynchronous weather satellites. Many images can be obtained from the net.

The future of infrared applications is hard to see. There will be more civilian applications as microbolometer technology advances, the arrays become cheaper with larger formats and smaller pixels. Multiwavelength applications will abound as filtering techniques improve. They may not be fully spectral but will use more than one wavelength band for temperature inference. Moisture detectors will appear on agricultural drones. Chlorophyll detectors too. Breast cancer screening may finally come into its own. Necrotic tissue and circulation imaging as well.

IOP Publishing

Rays, Waves and Photons
A compendium of foundations and emerging technologies of pure and applied optics
William L Wolfe

Chapter 14

Optics institutions—are you institutionalized?

I don't think a history of optics would be complete without a discussion of some of the more important optics facilities throughout the years and throughout the world. Some of the earliest work was done by individuals or by academies that had optics as a minor part of their activities, which may have been physics, astronomy, or natural philosophy. The optical institute is a rather modern phenomenon, but I consider it important since I was part of one; I was institutionalized. I describe here only the most well known of the optical institutes, whether they be universities, government establishments, or other labs. They are not listed alphabetically, chronologically, in order of importance, or any other logical way; they are in random sequence, generated by the random institute generator inside my head. Numbers of students and faculty are approximate and change rapidly—at least annually. I find it fascinating that several of these institutions arose for military reasons. The Germans, largely due to Schott and Abbe, had a thriving optical industry. So the French started one in their defense. And then the British. And the US started it with the Institute of Optics and added the College of Optical Sciences.

Ècole supérieure d'optique was opened in 1920 as part of *Institut d'optique théorique et appliquée* to train engineers. It grew from a need for optical engineers, those competent in optical instruments, outside of Germany, a long time and present rival of France. It currently has about 240 students, of which 40 are doctoral candidates. It was founded by Armand de Gramont (1879–1962), an industrialist, and Henri Chrétien (1879–1956), an astronomer. Interestingly, it is located on Avenue Augustin Fresnel in Paris.

The Optical Engineering Department of Imperial College London was formed in 1917 with F J Cheshire as Director, A E Conrady (1866–1944) as Professor of Optical Design, and L C Martin as lecturer. It was established in part to fulfill the growing need for optics competency in England since most of it was in Germany, at that time a foe in World War I. This was the same motive the French had. It was later subsumed into the physics department as the Technical Optics section, with Martin at the head.

In the United States the first and foremost (and only) optical institution for many years was **The Institute of Optics** at the University of Rochester in Rochester, NY. Excellent histories of it have been written by Hilda Kinslake[1] (1902–2003) and C R Stroud[2]. It was founded in 1929 with a grant from Eastman Kodak and Bausch and Lomb in order to reduce US reliance on German optics after World War I (again, like the French and British) and due to the increased optical activity in the US. It might be said that it actually started at Columbia University by William Southall, who was on the faculty there and had experienced the problems with optics in England and France. George Eastman (1854–1932) sent a letter to the president of the University of Rochester suggesting that such a facility should be established, but Eastman thought it should be in Rochester. Edward Bausch (1854–1944) and Adolph Lomb (1866–1932) got into the act, and together they all established the Institute. It is not bad to have Eastman Kodak and Bausch and Lomb behind the establishment of something in optics! It currently has 30 professors, eight with joint appointments, five adjuncts, and twelve research scientists. There are about 100 each of graduate and undergraduate students. It offers the BS, MS, and PhD degrees.

Directors of the Institute over the years have been: Brian O'Brien, 1920–1954; Bob Hopkins, 1954–1965; Lem Hyde, 1965–1968; Brian Thompson, 1968–1975; M Parker Givens, 1975–1977; Nicholas George, 1977–1981; Ken Teegarden, 1981–1987; Duncan Moore, 1987–1993; Dennis Hall, 1993–2000; and Wayne Knox, 2000–2011.

[1] Kingslake H G 1979 *The Institute of Optics: the First Fifty Years, 1929–1979* (Rochester, NY: College of Engineering and Applied Science, University of Rochester); Kingslake H G 1987 *The Institute of Optics 1929–1987* (Rochester, NY: College of Engineering and Applied Science, University of Rochester).

[2] Stroud C R (ed) 2004 *A Jewel in the Crown, 75th Anniversary Essays, The Institute of Optics, The University of Rochester* (Rochester, NY: Melliora Press).

CREOL, the College of Optics and Photonics in Orlando, FL, has 30 faculty, 40 researchers, and 160 students. It was founded by the vision of several citizens of Florida, and M J Soileau became its founding director in 1987. He arrived from Texas with colleagues Eric Van Stryland and David Hagen, who both became deans. They started work in a double wide trailer. It presently offers only graduate degrees, but you do have to work for them! It was founded as a center in 1986 and became a college in 2004. There is a good online description of CREOL and its history at *CREOL—The College of Optics and Photonics at the University of Central Florida.*

The College of Optical Sciences at the University of Arizona was established in 1964 in Tucson, AZ, with contracts from the Department of Defense, which realized a need for more optical engineers. It was not Germany this time; it was the growing use of optics in the military. Its first director, when it was the Optical Sciences Center, was Aden B Meinel (1922–2011), who moved from being chairman of the Astronomy Department. It became a college in 2005. It currently has a faculty of about 75, many of whom have joint appointments, several adjunct professors, and about 250 graduate and 250 undergraduate students. The organization has had directors until the middle of Jim Wyant's tenure in 2005, and then deans. Directors and deans were as follows: Aden Meinel, 1964–1968; Peter Franken, 1968–1978; Bob Shannon, 1978–1988; Dick Powell, 1988–1998; Jim Wyant, 1988–2010; and Tom Koch, 2010–present.

The National Institute of Astrophysics, Optics and Electronics (*El Instituto Nacional de Astrofisica,* Óptica *y Electrónica,* or INAOE) was founded in 1971 by governmental decree. It opened its doors a year later. Guillermo Haro (1913–1988) was its first director, and Daniel Malacara (1937–) its first technical director. Thirty-two of its 131 faculty members specialize in optics. It offers only graduate degrees.

The Vavilov Optical Institute (or State Optical Institute), named after Sergei Ivanovich Vavilov (1891–1951), a co-discoverer of Cerenkov radiation, was established in St. Petersburg, Russia, in 1918. It is a combination teaching, research, and fabrication facility, with a large staff of about 10 000 faculty, researchers, and technicians.

Centre d'optique, photonique et laser at the Université Laval was founded over 50 years ago. It resides in Quebec City, Ontario. It has 20 professors and the same number of researchers, as well as 100 students.

The Shanghai Institute of Optics and Fine Mechanics was established in Shanghai (of course) in 1964. It has almost 500 graduate students but does not offer an undergraduate degree. The faculty consists of 90 professors, of whom 52 supervise PhD candidates.

The Xi'an Institute of Optics and Fine Mechanics in Shaanxi, China, was established by the Chinese Academy of Sciences in 1962. The founding director was Gong Zutong (1904–1986). There are 190 researchers, associate researchers, and deputy senior technical staff, and 393 graduate students.

The Willow Run Laboratories of the University of Michigan should get a mention, partly because I worked there, partly because it was the origin of practical holography (see Holography, chapter 12), and partly because it had a first-rate infrared laboratory. The Labs were established in 1946 by William Gould Dow (1895–1999), then chairman of the EE department at Michigan, and scientists at Wright Patterson Air Force Base to investigate an anti-ballistic missile system, called WIZARD. It was dubbed the Michigan Aeronautical Research Center and was a collaborator with Boeing in the design of the BO-MARC missile. In 1950, it renamed itself the Willow Run Laboratories, after its location at the Willow Run Airport serving Detroit, since it had expanded its activities. It then renamed itself once more in 1972 to be the Environmental Research Institute of Michigan and eventually separated from the university. The two main laboratories there were the radar lab, headed by Lou Cutrona and the infrared lab headed by Lloyd Mundie, who was succeeded by Gwynn Suits and Mike Holter.

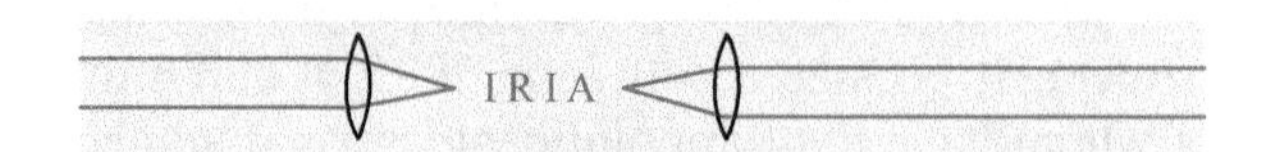

IRIA, the **Infrared Information and Analysis center** at the University of Michigan Willow Run Labs, was established by the Office of Naval Research to aid in the transfer of useful infrared information among government agencies and contractors. It was established in 1954. Its first director was Gilbert Kelton, who left in 1956. I was the second director and lasted until 1966, when I left. I was followed by George Zissis, Joe Acetta, and Dave Shumaker. It is still in existence, although it has metamorphosed into Sensiac at Georgia Tech, and then as part of the Defense Information Center. It has broadened its scope to cover information about other sensors.

The Institute of Optics at Rochester is stable with about the same number of students, but can now claim a Nobel laureate alumna, Donna Strickland. The

College of Optical Sciences is now The *James C. Wyant College of Optical Sciences* and can boast many new scholarships and endowed chairs. CREOL, the College of Optics and Photonics, now boasts some 28 core faculty members plus those who are joint with other departments and visiting and courtesy faculty. The INAOE, Institute of Astrophysics, Optics and Electronics has increased its faculty by five; it seems to be relatively stable in its growth. How could the Vavilov grow? IRIA has become part of the Defense Systems Information Analysis Center but still expedites infrared military information. We expect to see growth in most of these institutions as optics becomes a more essential, ubiquitous part of everyday life.

IOP Publishing

Chapter 15

Interference and interferometers—some constructive thoughts

The origins of interference and interferometry are veiled in antiquity. The ancients must have seen and wondered about the colors of thin water or oil films and soap bubbles. Interference is also intimately involved with diffraction. Often objects diffract or break up a light wave into many 'daughter waves' that then interfere.

Isaac Newton (1643–1727) observed what are now called Newton's rings, even though he believed in the corpuscular nature of light and could not explain them[1]. About 100 years later, both Christiaan Huygens (1629–1695) and Augustin-Jean Fresnel (1788–1827) were dealing with the wave theory of light, but it was the double-slit experiment of Thomas Young (1773–1829), reported in 1804[2], that showed that light was indeed a **wave motion** (Einstein showed it was also a particle many years later—in 1905)[3]. See chapter 18 on light for modern interpretations of this experiment.

Interference occurs when two or more waves interfere with each other, but they have to be more or less coherent. Waves that occur at random times and/or have many different wave lengths or frequencies do not interfere. So the idea of coherence enters when one considers interference. See appendix C for technical details.

Coherence and partial coherence have been the subject of many investigations. One of the pioneers of this area of investigation was Pieter Hedrik van Cittert (1889–1959), who, with Frits Zernike (1888–1966), established the theorem bearing their names: the Fourier transform of the mutual coherence function of a distant incoherent source is equal to its complex visibility[4]. This complicated idea is explained more thoroughly in appendix C, but it means, for instance, that we can obtain the brightness distribution of some distant astronomical objects. This was

[1] Newton I 1704 *Optiks*.
[2] Young T 1803 Bakerian Lecture *Phil. Mag.* Nov. 24.
[3] Einstein A 1905 *Ann. Phys.* **17** 132.
[4] Cittert P V 1934 *Physica* **1** 201; Zernicke F 1938 Physica **5** 785.

doi:10.1088/978-0-7503-2612-4ch15

used by Michelson before it was formally proven. He measured the visibility of fringes with his stellar interferometer, described below.

Anti-reflection coatings were patented[5] in 1938 by Katharine Burr Blodgett (1898–1979). Her original coatings were soap films, but since then many other substances have been used and techniques have been improved drastically. However, Harold Dennis Taylor (1862–1943) had the idea in 1904 to use acids to alter the surfaces of lenses[6], and Alexander Smakula (1900–1983) had a 1936 German patent on the process[7]. The idea led to what has been called *invisible glass.* It is used on eyeglasses and picture glass most notably. The idea is to interfere the wave reflected from the front surface with that reflected from the back surface with just enough phase delay that the interference is destructive. Then there is no reflected light. This, of course, applies to only one angle and one color, but it is pretty good for others. Others have developed computer programs for much more complicated layers of interference filters and films and for multiple angles and wavelengths. See interference filters later in this chapter.

Experimenters developed theories of both multiple wave and two-wave interference and have based various instruments on these phenomena. Interferometers have been very useful in all sorts of precise measurements, from determining the figure on the Hubble telescope to replacing the gyroscope. There are too many variations of these instruments to be described; only the classical and main ones are described.

The **Michelson interferometer** was invented in 1881 by Albert Abraham Michelson (1852–1931), the first American Nobel laureate in physics, to test for the existence of the luminiferous ether[8]. He and Edwin Williams Morley (1838–1923) fortunately found that there was no such thing. At least their results convinced others that there was no such thing—Einstein notably, who was convinced that the speed of light was constant in all inertial frames of reference. They did find that there was no difference in the speed of light in direction or time of year, position of the Earth in orbit, or anything else they could measure. The Michelson interferometer has perhaps seen its best days as a spectrometer (figure 15.1) (see Spectroscopy, chapter 40).

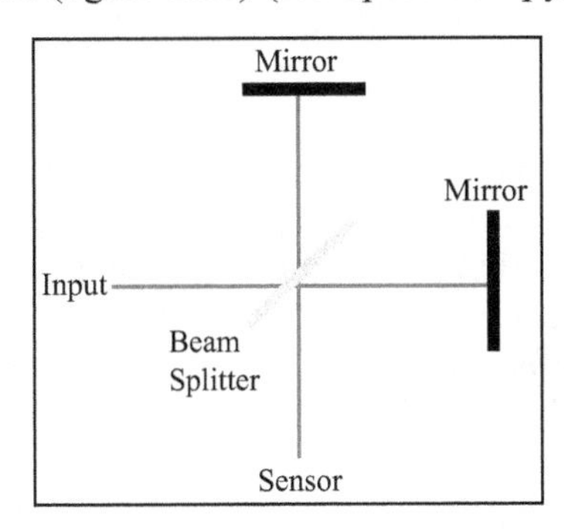

Figure 15.1. Michelson interferometer.

[5] US Patent 2220660 (1938).

[6] Taylor H 1904 *British Patent* 29564/04.

[7] Smakula A 1936 *German Patent* 685767.

[8] Michelson A 1881 *Am. J. Sci.* **22** 120.

The **Twyman–Green interferometer** is a version of the Michelson that uses collimated light. It was conceived and patented by Frank Twyman (1876–1959) and Arthur Green in 1916. This device is used mostly for testing optical parts. The Michelson and Twyman–Green interferometers are two-wave devices, since only the two waves interfere. The Fabry–Pérot interferometer and those like it are multiple wave interferometers. Two-wave devices have a sinusoidal line shape, whereas the multiple-wave instruments have a sharper shape—due to additional interferences.

The **Fabry–Pérot interferometer** was invented by Maurice Paul Auguste Charles Fabry (1867–1945) and Jean-Baptiste Alfred Pérot (1863–1925) in 1897[9]. It is a multiple-wave interferometer in which the waves repeatedly interfere with each other, as shown in figure 15.2, but with the light incident normally. (It can only be illustrated with the oblique incidence.) The transmission has the unusual property that it is highest, when it has a value of one, when the reflectivity is total, also a value of one. The FP interferometer was originally used to examine the fine structure of spectral lines (because of its sharper line shape), but it has also been used as a part of interference filters, laser cavities, fine distance measuring, and even in gravity-wave detection.

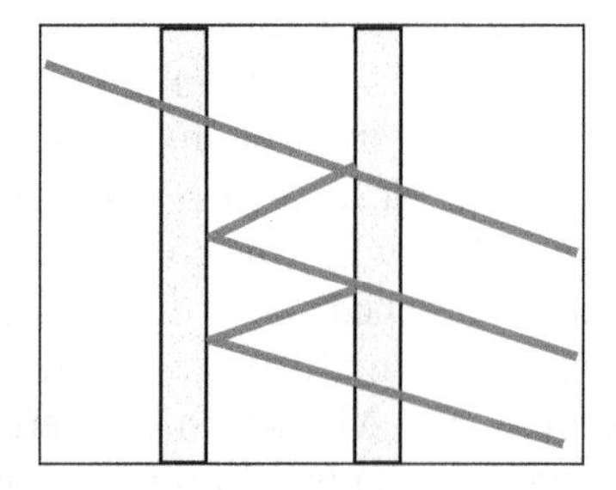

Figure 15.2. Fabry Pérot interferometer.

The **Michelson stellar interferometer** is not the same as the Michelson interferometer, although both were invented by Michelson. The stellar interferometer was used by Michelson and Francis Gladheim Pease (1881–1938) to measure the diameter of the star Betelgeuse in 1920[10]. They found it was about 300 times that of the sun. The stellar interferometer uses a pair of outrigger mirrors that obtain two independent beams from the star. These are combined on a single focal plane where they interfere. The quality of the interference, the fringe contrast, is a measure of the degree of coherence, which, in turn, is a measure of the size of the source (figure 15.3). (A larger source has more individual emitters than a smaller one.)

[9] Fabry C and Perot A 1897 *Ann. Chim. Phys.* **12** 459.
[10] Michelson A and Pease F 1921 *Astrophys. J.* **53** 249.

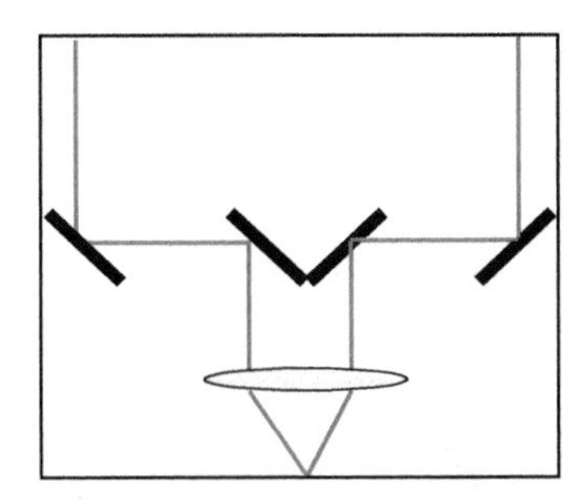

Figure 15.3. Michelson stellar interferometer.

The **Lummer–Gehrke** interferometer[11] or plate, invented by Otto Lummer (1860–1925) and Ernst Gehrcke (1878–1960), is a simple rectangular plate through which light is propagated. Light is incident at a fairly sharp angle and reflects repeatedly at an angle a little less than that for total reflection—so the rays get out. They are then gathered by a lens and superimposed where they interfere (figure 15.4).

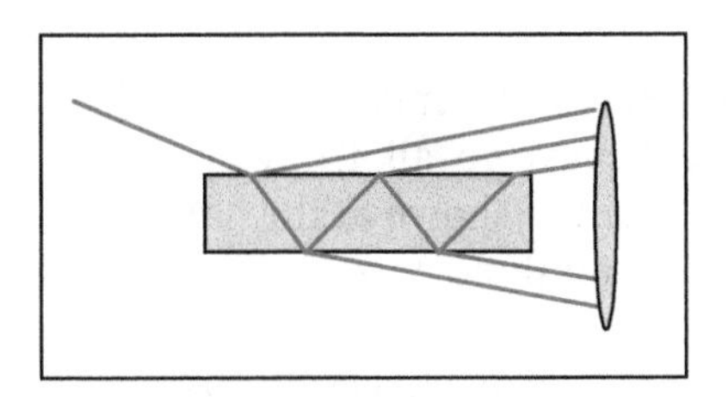

Figure 15.4. Lummer–Gehrke plate.

The **Rayleigh interferometer** was invented by the great Lord Rayleigh (1842–1919) and is similar to others that use two beams and recombine them after they traverse different paths (figure 15.5) (see Speed of Light, chapter 41).

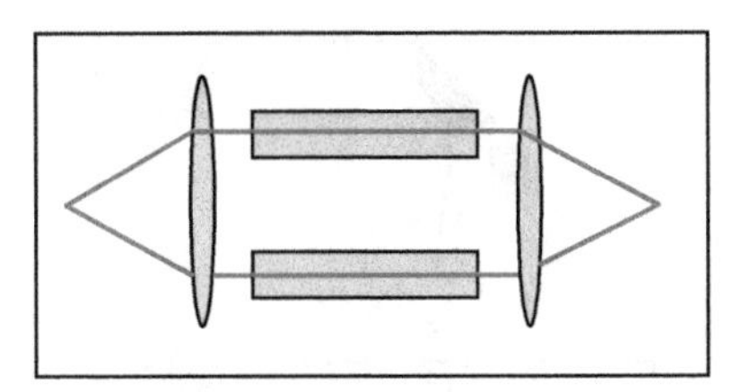

Figure 15.5. Rayleigh interferometer.

The **Mach–Zehnder interferometer** was first described by Ludwig Mach (1868–1951) and Ludwig Louis Albert Zehnder (1854–1949) in 1891 and 1892[12]. This Mach is the son of the famous Ernst Mach of mechanics (and optics) fame. The Mach–Zehnder

[11] Lummer O 1901 *Verh. Deutsch. Phys. Ges.* **3** 85; Lummer O and Gehrke E 1903 *Ann. Phys.* **10** 457.
[12] Mach L 1891 Z. *Instrum.* **11** 275; Zehnder L 1892 Z. *Instrum.* **12** 89.

interferometer uses an arrangement similar to the Michelson, but it does not use the beam splitter twice—as a beam splitter and beam combiner. The light is collimated and comes from the lower left; it is split by the beam splitter (in color) and goes to the mirrors (in black) on the right and upper left; it is combined by the beam combiner on the upper right and interferes (figure 15.6).

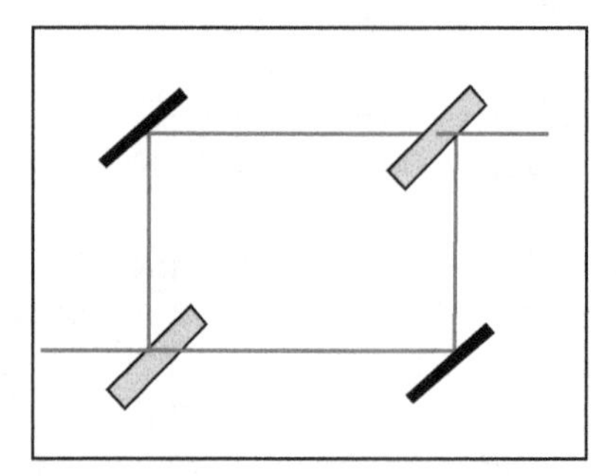

Figure 15.6. Mach–Zehnder interferometer.

The **Sagnac interferometer** looks much like the Mach–Zehnder, but it has a significant difference: the beams go all the way around in both directions and interfere back at where they enter (shown at the bottom of figure 15.7). There is only one beam splitter/combiner and three mirrors. This round trip makes it a candidate for being a gyroscope, and such was the case. It was the forerunner of the FOG, the Fiber Optic Gyroscope. George Sagnac (1869–1928) first used it to investigate the luminiferous ether, as did Michelson and Morley[13]. The interference arises because, if the interferometer is rotating, the light going in one direction must go a little further than in the other direction. Thus, it can be used to measure the rotation rate of the interferometer. In FOG's, the fixed mirrors are replaced by coils of optical fibers (see Optics Olio, chapter 26).

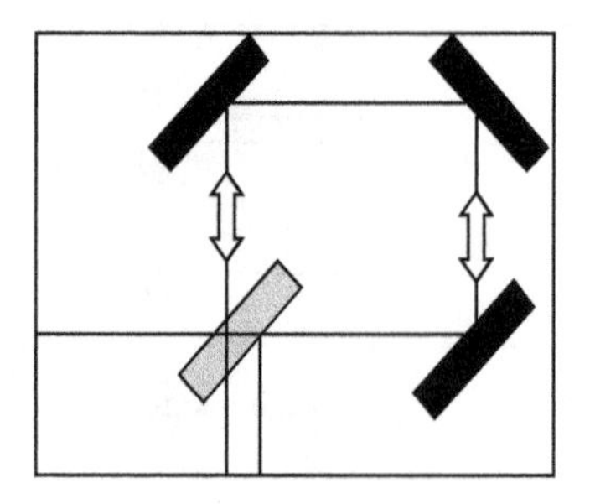

Figure 15.7. Sagnac interferometer.

Newton's rings were studied by him and are a result of interference—in spite of the fact that he did not believe in waves. If a spherical shape is placed on a flat base, the

[13] Sagnac G 1913 *C. R.* **157** 1410.

resulting interference pattern will appear as a bull's-eye with a series of concentric rings of different widths, radii, and spacings. They are used today for testing optical components. The spacings of the rings reveal information about the curvature, and any deviations from circularity show variations from sphericity (figure 15.8).

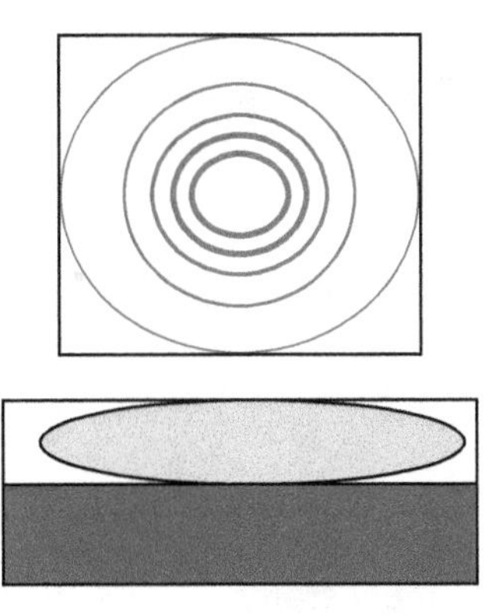

Figure 15.8. Newton's rings, experimental arrangement.

Phase-shifting interferometry was probably introduced in 1966 by Carre[14]. Lots of folks worked on the concept and the implementation, but only a few publications arose[15]. It really bloomed with the advent of CCD's and small, powerful computers. It is used mainly for the measurement of surface quality and surface roughness.

Interference filters may be considered to have had their origin with the anti-reflection coating discussed above, but they later became much more sophisticated than simply the use of a single layer. Some meant for narrow passbands use the thin film version of a Fabry–Pérot interferometer.

Can there be still another form of interferometer? I doubt it. They will be miniaturized and appear in smart phones.

[14] Carré P 1966 *Metrologia* **2** 13.

[15] Wyant J *et al* 1975 *Appl. Opt.* **14** 2622; Crane R 1985 *Appl. Opt.* **24** 3049; Bruning J *et al* 1974 *Appl. Opt.* **13** 2693.

IOP Publishing

Rays, Waves and Photons

A compendium of foundations and emerging technologies of pure and applied optics

William L Wolfe

Chapter 16

Lasers–a coherent discussion

Lasers were once called by the acronym LASER for **L**ight **A**mplification by **S**timulated **E**mission of **R**adiation, but laser is a household word and probably the most exciting source of light since the Sun. Some think lasers had their nascence with Albert Einstein (1879–1955), who produced equations for spontaneous and stimulated emission, thereby introducing the notion of stimulated emission[1]. I doubt he had in mind the laser and its coherent emission, but he did understand about excited electrons being stimulated into producing photons. He understood a lot of other stuff too! Years later, after their work on masers (Microwave Amplification by Stimulated Emission of Radiation) Charles Hard Townes (1915–2015) and Arthur Leonard Schawlow (1921–1999) propounded the idea of a laser[2]—that is, a light maser. At about the same time, Gordon Gould (1920–2005), a graduate student at Columbia where Townes and Schawlow were on the faculty, proposed a similar idea[3]. After considerable work on masers, Townes asked a student to initial his notebook to claim priority on a light box with thallium inside and mirrors on four of the walls on September 14, 1957. Then thallium was replaced by potassium and the four walls became a Fabry–Pérot cavity (see Interference, chapter 15). A patent request was filed in March of 1960[4] with a priority date of 1958, and a paper was presented to the Physical Review (see[2]). During this same period Gould got interested in the laser. He had his lab notes notarized on November 16, 1957. As best as I can tell, his patent was filed in 1967 but with a priority date of 1959[5]. Note that these priority dates are always subject to litigation. It would appear that the three men all had essentially the same idea at the same time. It is enough in this history to note that it is not clear who should get the credit for the first idea of the

[1] Einstein A 1916 *Verh. Dtsch. Phys. Ges.* **18** 318; Einstein A 1917 *Phys. Z.* **18** 121.
[2] Townes C and Schawlow A 1958 *Phys. Rev.* **112** 1940.
[3] American Physical Society This Month in Physics History online.
[4] Townes C and Shawlow A 1960 *US Patent* 2929922.
[5] Gould G 1968 *US Patent* 3609570.

doi:10.1088/978-0-7503-2612-4ch16

laser—maybe all of them. They approached it from different backgrounds and had ample opportunity to share ideas at Columbia, intentionally or casually. Gould gets the credit for the first use of the acronym in his notes of 1957 and at the conference of pumping in Ann Arbor, MI, in June 1959[6].

It is clear who first made the laser work: Theodore Harold Maiman (1927–2007) of Hughes Research Labs in Malibu, CA, in 1960, using a ruby rod. Once the ideas of stimulated emission, excited states, and cavity resonators were proven feasible for the fabrication of a laser by Maiman, others joined the fray, if I may call it that. The idea is that other materials could be used to do the same thing and get different wavelengths, powers, and configurations. Ali Javan (1926–2015) produced the **first gas laser**[7], consisting of helium and neon, in 1960. Javan had worked with Townes at Columbia but moved on to Bell Labs after a four-year post grad stint at Columbia. He first conceived the **He–Ne laser** in 1958, about one year after the Townes–Shawlow–Gould announcements and before Maiman demonstrated the first ruby laser. Javan operated this first continuous wave laser in 1960. It still is one of the most used lasers in the lab.

Chandra Kumar Naranbhai Patel (1938–) produced the first **carbon dioxide gas laser**[8] in 1964. It was notable for working in the infrared region around 10 μm—actually on any of a number of the carbon dioxide bands in this general region. It was also an extremely powerful laser. Whereas the He–Ne's produced milliwatts, the CO_2's produced kilowatts! Patel got only 1 mW from his first demonstration, although he was able to produce both continuous and pulsed operation. The carbon dioxide laser is used for welding because of its power and in surgery since it immediately cauterizes its incisions.

Another major laser was neodymium doping in yttrium aluminum garnet—or **Nd:YAG**—first demonstrated in 1964[9] by J Geusic and his collaborators at Bell Labs. Since then many different materials have been used to get different wavelengths and powers. A wide variety of dopants has been tried. The main output is at about 1 μm, but other wavelengths are available.

The first **diode or semiconductor laser** was demonstrated by Robert Hall (1919–2016) and his group at General Electric in 1962[10]. They were closely followed by a group at IBM, who apparently did it independently[11]. They used gallium arsenide and generated light in the infrared. The first visible diode laser was created by Nick Holonyak (1928–) in the same year. Then everybody got into the act and created a host of them with at least 30 different wavelengths from 375 nm to 3330 nm. There are surely more diode lasers in existence than any other kind.

Ultraviolet lasers, generally called excimer lasers, were invented by Nicolai Gennadiyevich Basov (1922–2001), V A Danilychev, and Yu M Popov (1929–)

[6] Hecht J 1994 *Winning the laser patent war* Laser Focus World.
[7] Javan A and Bennett W 1964 *US Patent* 3149290.
[8] Patel K 1964 *Phys. Rev.* **136** 1187.
[9] Geusic J *et al* 1964 *Appl. Phys. Lett.* **4** 182.
[10] Hall R *et al* 1962 *Phys. Rev. Lett.* **9** 366.
[11] Nathan M *et al* 1962 *Appl. Phys. Lett.* **1** 62.

in 1970[12]. Their version used a xenon dimer, Xe_2, excited by an electron beam that gave an output at 172 nm. Others' improvements used noble gases[13]. These lasers have outputs in the milliwatt range and wavelengths from about 125 nm to 350 nm, and they are used for the lithography of microcircuits and for eye surgery called LASIK, Laser Assisted *In Situ* Keratomileusis. These short wavelengths interact with human tissue and disintegrate it without heating, and in electronics they provide the ability to make smaller circuits because of their shorter wavelengths.

X-ray lasers have been postulated and demonstrated. The initial work was at Livermore Labs by Lowell Lincoln Wood (1941–) and George Chapline (1942–) initially for use as a weapon. It is reported that in 1983 in a nuclear test in Nevada that an x-ray laser was produced with excitation by an atomic bomb, although this result is not documented, perhaps for security reasons[14]. Free electron lasers that generate x-rays have been demonstrated in Germany at the Deutsches Elektronen-Synchrotron in 2005 using a synchrotron. Others are working on longer pulses, greater output and the like, but it is clear that such lasers are far from useful industrial tools.

Lasers have had application in many, many phases of our lives. They handle our data in Google searches, weld the cars we ride, position the car parts in assembly, do surgery, catch us speeding, act as pointers in lectures—and even make jobs for optics professors! I won't try to list the references here; just go to the net and search a bit. Yes, they are truly ubiquitous. So is optics.

They will be used in all sorts of different applications, but there is only so much we can do to improve them. One is increased intensity. Another is getting more things to lase so that we can use them in specially directed ways.

[12] Basov N *et al* 1970 *Zh. Exp. Phys. Tech. Pisma Red* **12** 473.
[13] Ewing J and Brau C 1975 *Appl. Phys. Lett.* **27** 350; Tisone G *et al* 1975 *Opt. Commun.* **15** 188; Ault E *et al* 1975 *Appl. Phys. Lett.* **27** 413; Searles S and Hart G 1975 *Appl. Phys. Lett.* **27** 243.
[14] Hecht J 2008 History of the x-ray laser *Opt. Photon. News* May.

IOP Publishing

Rays, Waves and Photons

A compendium of foundations and emerging technologies of pure and applied optics

William L Wolfe

Chapter 17

Lenses—focusing in

Lenses are used for a variety of tasks. The early uses were for telescopes and microscopes, but perhaps they were predated by Nero's use of a gem in the arena. They are still used for telescopes and microscopes, but also for eye glasses, binoculars, lithography, cameras, peep holes, smart phones, and more. When many people think of optics, they think of lenses.

The **earliest known lens** is called the Nimrud lens because it was found in the Nimrud ruins in Assyria, now Iraq, by Austen Henry Layard (1817–1994)[1]. It is a piece of quartz about 1.25 cm (0.5 inch) in diameter that apparently had been ground and polished. It is about 3000 years old. Egyptian hieroglyphs depict the use of meniscus lenses[2]. Aristophanes (446–386 BC) mentions a burning glass in *The Clouds*[3]. Pliny The Elder (23–79) makes it clear that lenses were used in old Rome[4]. He also wrote that Nero used an emerald as a lens to better watch the gladiators, as mentioned above. That would be a spectacle for a spectacle! Although that may have helped Nero see better, spectacles were not really available until the 1280s[5]. They were developed by lens makers in Venice and Florence, and somewhat later in Holland. It is somewhat ironic that Galileo had to go to Holland to get a telescope made by Hans Lippershey (1570–1619)—sometimes written Lipperhey—that was then transported back to Pisa, instead of from Florence or Venice to Pisa. During the ensuing years, there were many experiments to get good performance from single lenses, all doomed to failure. His belief that one could not correct for chromatic (color) aberration led Isaac Newton (1643–1727) to design the first all reflective

[1] Layard A H 1853 *Discoveries in the Ruins of Nineveh and Babylon* (London: John Murray); Gasson G 1972 *The Opthalmic Optician* 1267.

[2] Kriss T C and Kriss V M 1988 *Neurosurgery* **42** 899.

[3] Aristophanes, *The Clouds*, (424 BC).

[4] Pliny The Elder, *Natural History*; on the net.

[5] Glick T F, Livesey S J and Wallis F 2005 *Medieval Science, Technology and Medicine: An Encyclopedia* (New York: Routledge).

telescope. These simple lenses all had spherical or plane surfaces, with the spherical surfaces being either concave or convex. A spherical surface is automatically generated when two pieces of glass are ground together. Any other figure requires special grinding and polishing.

Following these earliest of days, creative engineers and scientists devised many different lenses for many different tasks. Only the major forms will be dealt with here, since there are literally tens of thousands of different lens forms.

Simple, single element lenses consist of biconvex (or double-convex), plano-convex, (or convex-piano), meniscus, plano-concave, and biconcave (figure 17.1). Note that these can all be reversed, but they will all be called by the same name. Optical people do not differentiate between plano-convex and convex-piano unless they are being picky!

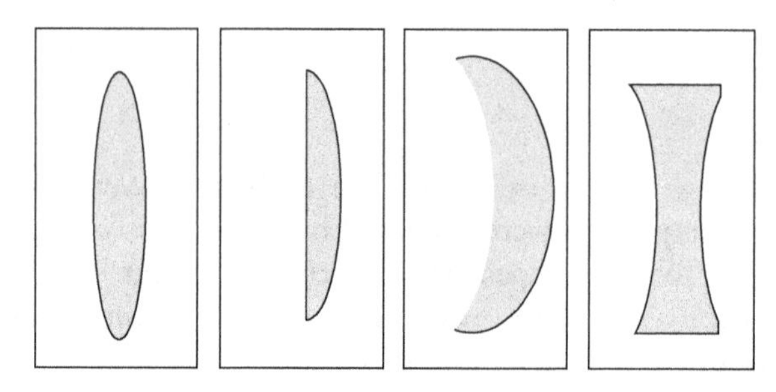

Figure 17.1. Single element lenses: biconvex; plano convex; meniscus; biconcave.

The first **achromatic lens**, one that is designed to focus two different colors together, was invented by Chester Moore Hall (1703–1771) in 1733, and independently invented and patented by John Dollond (1706–1761) in 1758[6]. It is made by two pieces of glass that have different dispersions—variations of refractive index with wavelength. They now come in both contact and separated varieties. An achromatic lens corrects for spherical aberration and for two colors. It was generally understood at that time, due to Newton, that all glasses had the same dispersion. Leonhard Euler (1707–1783) suggested in 1747 that you could make an achromatic lens of glass and water. Dollond doubted this, but in the tradition of a good scientist he set out to disprove it—and thereby proved it; he made an achromatic lens of glass and water. Recall that although Newton had the law of sines (refraction) correctly, he had the wrong reasoning, based on his corpuscular theory of light.

Achromatic lenses were superseded by **apochromatic lenses** that corrected for three different colors, typically red, green, and blue—short, medium, and long wavelengths of the visible. The first was probably that of Ernst Abbe (1840–1905) and Paul Rudolph (1858–1935) of Zeiss, their lens being a thick contact triplet between two menisci, which was patented in 1890[7]. Unfortunately it suffered from severe astigmatism and did not last long.

---

[6] Dollond J 1758 *Philos. Trans.*; Dollond J 1911 *Encyclopedia Britannica* **8** online.

[7] German patent 55313; US patent 435271; British patent 6029/90 all (1890).

There are probably more different types of **photographic lenses** than any other category, perhaps because there is a large variety of camera types and they are the largest consumer product. They include, in addition to the standard ones, telephotos, portraits, macros, landscape, zooms, and more. The classic discussion of these is by Rudolf Kingslake (born Rudolf Klickmann, 1903–2003)[8], listed in the bibliography. Thousands of different lenses are described in patents and in lens design libraries. These are often used as starting points for new designs. So many lenses, so little space! By definition, it seems that the **first photographic lens** had to come after the invention of photography, although similar ones were used with camera obscuras (see Cameras, chapter 4).

Louis Jacques-Mandé Daguerre (1787–1851) is generally credited with the invention of the first practical camera in 1826[9]. Shortly thereafter Charles Chevalier (1804–1859) invented a lens for his camera that was an achromatic landscape lens of relative aperture F/15 with a flat field[10]. His lenses for the camera were at first just doublets of double convex and double concave elements (figure 17.2(a)), but then he combined the two cemented doublets to make a separated doublet of cemented doublets! Cemented doublets are also called contact doublets (figure 17.2(b)). He was just in time for the *Société d'e encouragement pour l'industrie nationale* competition in 1840. He won. But Josef Petzval (1807–1991) designed a better, faster version at about the same time—an F/3.6 with a cemented doublet and a separated doublet—the famous Petzval portrait lens (figure 17.2(c)). He came in second. It has been reported that he had the calculational help of 13 artillerymen to do those ray traces[11] (see Optical Design, chapter 25). It might not seem like much, but the separation that Petzval used gave him an additional surface to bend and a separation to adjust. The photographic film of the day was not fast, so the lens had to be. F/3.6 is about 18 times faster than F/15. You can judge for yourself whether the award went to the right person, but note that this was a French Society; Chevalier was a member, and Petzval was Viennese[12].

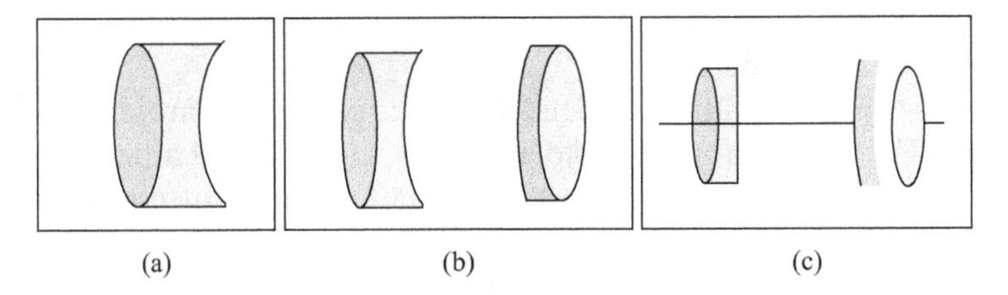

Figure 17.2. (a) Chevalier lens, (b) cemented doublet, and (c) Petzeval lens.

[8] Kingslake R 1989 *A History of the Photographic Lens* (New York: Academic).

[9] Daguerre L 1839 *Historique et description des procedes du Daguerreotype et du diorama* (Paris: Alphonse Giroux et Cie).

[10] Chevalier C 1841 *Nouvelles Instructions sur l'Usage du Daguerreotype* (Paris: Chez l'Auteur).

[11] Kingslake R 1841 *Nouvelles Instructions sur l'Usage du Daguerreotype* (Paris: Chez l'Auteur).

[12] Colucci D 2014 *The Petzval Lens* Photographic Historical Society of New England and online.

It is unfortunate in a way that an almost universal way to cite the speed of a lens is by its F/number, the ratio of the focal length to the aperture diameter (figure 17.2). A ‘faster’ lens is one that focuses more light on the film plane. This happens when the cone of light is larger, steeper; the cone angle is fatter. That happens when the focal ratio is smaller. So a ‘faster’ lens has a smaller F/number. Lesser is more. One glaring example of this unfortunate use is in K Kaprelian’s ‘Objective Lenses of f/1 Aperture and Greater’ in *SMPTE.* He means f/1 and smaller, like f/0.7. Another is the alternate use of f/ and F/. Lens designers often call this the relative aperture—the aperture relative to the focal length—and even just the aperture, as in *faster aperture.* And a faster lens allows for a slower shutter speed!

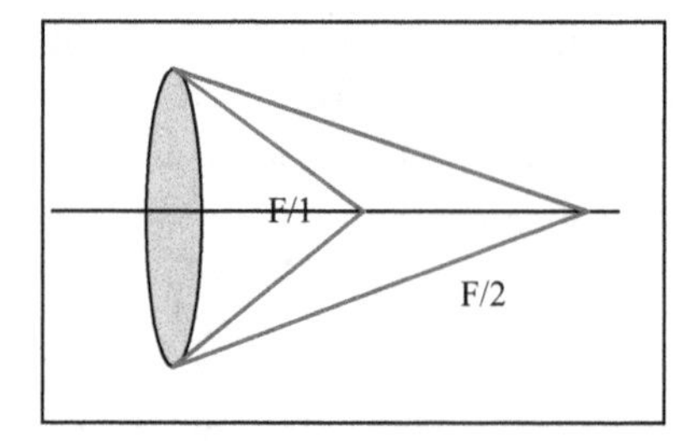

Figure 17.3. Optical speed F/numbers.

The **Chevalier** and **Petzval lenses** both suffered from barrel distortion at the edges of the field, so symmetrical objectives were designed to overcome this: the **Panoramic lens** of Thomas Sutton (1819–1875) in 1859; the **Periskop** of Carl August von Steinheil (1801–1870) in 1865; and the **Orthoscope** lens of C C Harrison in 1860[13]. The Globe and the Periskop lenses consisted of two two-element concave facing menisci with their outer spherical surfaces on the same sphere and a stop at the center of the sphere. Sutton’s panoramic lens was filled with water.

Next came additional correction with the **Steinheil Aplanat** by Carl August von Steinheil and the **Dallmeyer Rectilinear** by Thomas Rudolphus Dallmeyer (1859–1906), both in 1888.

The **Cooke triplet**, or rather triplets, were designed by Harold Dennis Taylor (1862–1943), who worked for T Cooke and Sons. He obtained a patent for the first design in 1893[14]. It is a double concave high index element surrounded by convex--piano and plano-convex elements.

The **double Gauss lens** is the predecessor of most fast lenses of today; at least 70 are listed by Cox[15]. Of course, a double Gauss lens consists of two single Gauss lenses, which are two menisci (a better plural than meniscuses). Just as obvious is the fact that a Gauss lens was invented by Carl Friedrich Gauss (1777–1855) in 1817. In 1888 Alvan Graham Clark (1832–1897) took out a patent on using a pair of them

[13] Harrison C 1862 *US Patent* 35605 available on the net.

[14] Taylor D 1893 *British Patent* 22607.

[15] Cox A 1974 *Photographic Optics* (New York: Amphoto).

back to back[16]—or at least in opposing orientation because I do not know which is the back. Modern examples are the **Topgon** and the **Metrogon**. About twenty more variations on this theme have been made and sold.

Anastigmats were similar to double Gauss lenses in that they were symmetrical about a stop, but the two lenses were not menisci; they were contact doublets consisting of high-index crown and low-index flint glasses. Some of the variations no longer appear very symmetric. It appears that the first of these was patented by H H Schroeder in 1888[17]. Many design variations were generated, including the stigmatic lenses of Aldis, the Unar, and the famous Tessar, of which millions have been made. A nice history and discussion of the double Gauss is online at Wikipedia with that title.

Eyepiece lenses, or just **eyepieces**, are lenses that are used on devices like telescopes or microscopes to get the image to the eye (figure 17.4). They are always close to the eye and usually reasonably close to the focus of the objective lenses of the instruments. There are several classic ones: Ramsden, Kellner, Huygens, Erfle, Plossl, and Abbe. The earliest of these was a simple convex lens used by Zacharias Janssen (1580–1638) in 1590 shortly after the invention of the microscope by his father and Hans Lippershey.

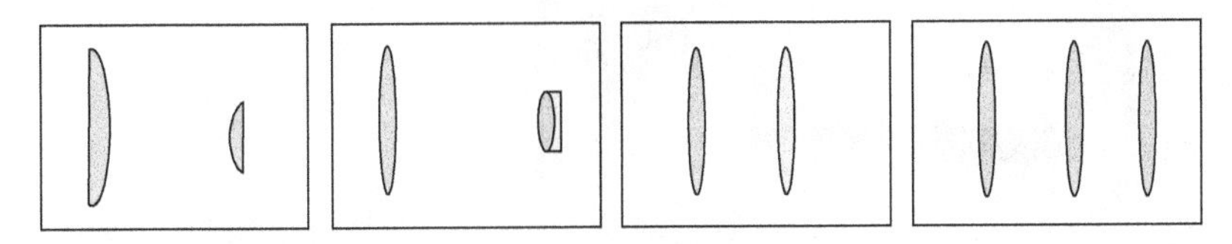

Figure 17.4. Eyepiece lenses: Ramsden; Kellner; Plossl; Erfle.

Sky or Fish-eye cameras cover the entire overlying hemisphere. The term 'fisheye' was coined by Robert W Wood (1868–1955) because he thought a fish on the surface would see the entire sky[18]—also why it is called a sky lens. The first was by R Hill[19]. Others followed, some with a bit less than the entire hemisphere. Perhaps the most famous is the **Biogon** designed by Ludwig Jakob Bertele (1900–1985) in 1951[20]. There is now a family of lenses with that moniker.

Zoom lenses come in two varieties: parfocal and varifocal. The parfocal variety is a true zoom lens that maintains focus while its focal length changes. Of course the varifocal does not, or it would not be different. Note that when the focal length changes, so does the magnification and the field of view. The first parfocal, true zoom lens was designed by Clile C Allen in 1902[21] (figure 17.5). The movie industry adopted it and produced it in 1927 with Clara Bow starring. Bell and Howell produced the Varo in 1932.

[16] Clark A G 1888 *US Patent* 399499.
[17] Schroeder H 1888 *British Patent* 5194/88 and *US Patent* 404506.
[18] Wood R W 1905 *Physical Optics* (London: Macmillan).
[19] Hill R 1923 *British Patent* 225398.
[20] Bertele L 1951 *US Patent* 2721499.
[21] Allen C 1902 *US Patent* 696788.

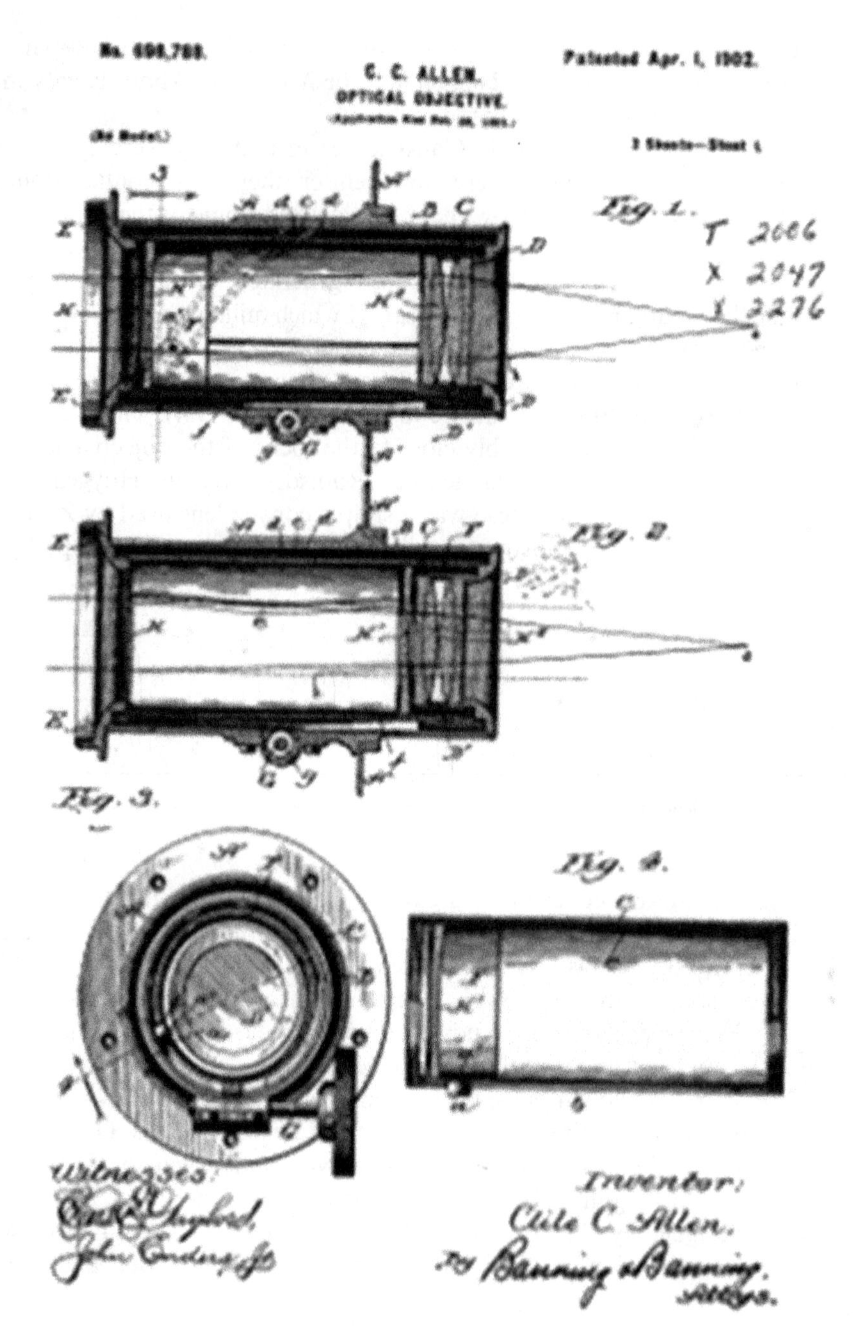

Figure 17.5. From Allen's patent.

The **contact lens** was first imagined by Leonardo da Vinci (1452–1519) in 1508, and John Herschel (1792–1871) figured out how to make one in 1823, but it was not until 1887 that one was made and fitted to an eye. They were first made of plastic in 1939, and soft ones came upon the scene in 1971.

Gradient index lenses were first conceived by man in 1854 by James Clerk Maxwell (1831–1879)[22], the same Maxwell of the famous equations, but gradient index optics preceded that in Nature. The mirages that we see on the road ahead are caused by the gradient index of air caused in turn by the gradient in temperature. The atmosphere itself allows us to see the Sun shortly after it has set and before it rises because of the gradient in the index of the atmosphere because of its density gradient. The first gradient index, or GRIN, lens was created by God: the human eye. The index varies from about 1.406 at the center to about 1.386 nearer the edge[23]. He also provided trout, octopi, squid, and others with gradient index eye lenses[24].

The **Maxwell fisheye lens** is a sphere with a varying refractive index, the equation for which $n = n_0/(1 + r^2)$, where n_0 is the maximum index and r is the fractional radius. It is probably the oldest of the gradient index lenses, dating to 1854[25]. Some years later, in 1905, Robert W Wood created a GRIN lens using gelatin[18]. In 1944 Rudolf Lüneburg (1903–1949) published his findings of the now-called **Luneburg lens** that focuses incident parallel light beams from any direction onto the anterior surface[26]. Figure 17.6 represents both lenses. The Luneburg lens has a refractive index profile given by $n = n_0\sqrt{(2 - r^2)}$. The two refractive index profiles are shown in figure 17.7 in which n_0 is 1.4.

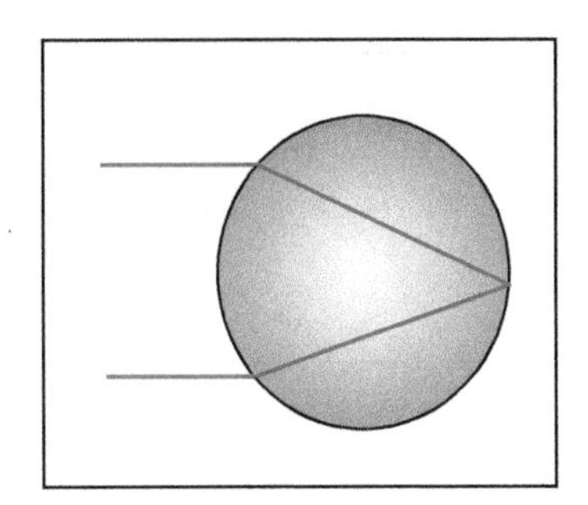

Figure 17.6. Fish eye and Luneburg lens.

[22] Maxwell J 1854 *Camb. Dubl. Math. J.* **8** 188.
[23] Hecht E 1987 *Optics* (Reading, MA: Addison Wesley).
[24] Jagger W and Sands P 1996 *Vis. Res.* **36** 2623; Jagger W and Sands P 1999 *Vis. Res.* **39** 2841; Kroger R and Gislen A 1996 *Vis. Res.* **44** 2129.
[25] Freres N 1854 *Camb. Dubl. Math. J.* **9** 9; Niven W (ed) 1965 *The Scientific Papers of James Clerk Maxwell* (New York: Dover).
[26] Luneburg R 1944 *Mathematical Theory of Optics* (Providence, RI: Brown University).

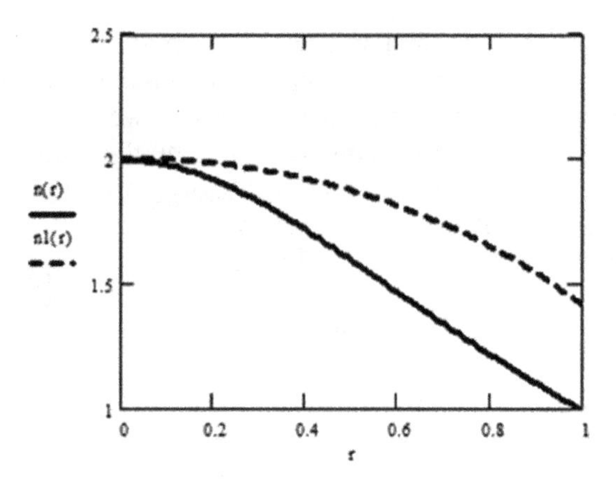

Figure 17.7. Relative refractive indices.

A Kissin' cousin of these, although not a gradient index lens, is the Weierstrass sphere (figure 17.8). It is a simple sphere of refractive index n that provides magnification better than that of a hemisphere[27]. A ray that would have come to the axis at a distance nR outside the sphere does so at a distance R/n inside the sphere, where n is the refractive index of the sphere and R is its radius. The figure shows a solid line coming to a point on axis without the sphere, and the dashed line inside the sphere.

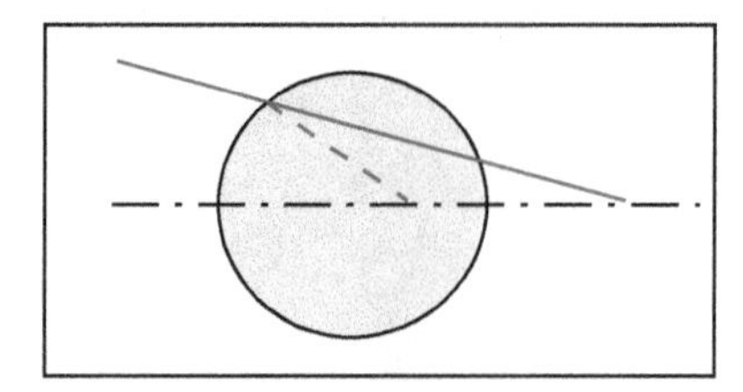

Figure 17.8. Weierstrass sphere.

The earliest patents on modern, practical **GRIN lenses** are those of R S Moore, J R Hensler, and Donald B Keck (1941–); and Robert Olshansky—1973, 1975, and 1975[28]. But it seems Robert W Wood created the first one, as impractical as it was[18]—a gelatin cylinder.

The **Fresnel lens**, (with a silent s—in Fresnel not lens!) was originally designed for lighthouses, but it has other modern uses. I have one on the rear window of my RV that covers a wide field. It is also used in infrared intrusion detectors. I have one in a

[27] Wolfe W 1962 *Proceedings of the Conference on Optical Instruments and Techniques* (London: Chapman Hall); Fletcher D et al 2000 *Appl. Phys. Lett.* **77** 2109.
[28] Moore R 1973 *US Patent* 3,718,383; Hensler J 1975 *US Patent* 3,873,408; Keck D and Olshansky R 1975 *US Patent* 3,904,268.

device at the top of my garage. The little gismo on top, not the flood lights (figure 17.9). It consists of a plastic that is transparent in the infrared—in front of a pair of IR detectors. And the Fresnel lens is still used in lighthouses. It was first constructed by Augustin-Jean Fresnel (1788–1827) in 1821,[29] but he was preceded conceptually by Nicolas de Condorcet, marquis de Condorcet (1743–1794) in 1748,[4] Georges-Louis Leclerc, Comte de Buffon (1707–1788) in 1780,[30] and independently by David Brewster (1781–1868) in 1811[31]. But it is still called a Fresnel lens and consists of segments of a full lens. Originally this was accomplished by manufacturing the separate parts and assembling them; today they are made by forming a plastic. This arrangement drastically reduces the size and weight of large ones. The diffraction limit still exists (for each segment) so it does not have good resolution, but it has been invaluable for use in lighthouses—and pretty good for RV's.

Figure 17.9. Infrared, Fresnel lens intrusion detector.

Axicons are special conical lenses that are used for just a few applications (figure 17.10). They are circularly symmetric with a plane circular surface and a conical one, and can be either refractive or reflective. They do not make images; they create images of on-axis point sources that are lines and some laser beams into circles. They were first proposed by John Mcleod in 1954[32].

Figure 17.10. Axicon.

---

[29] Encyclopedia Britannica, online.

[30] Fresnel Lens, Wikipedia, online.

[31] *Appletons Dictionary of Machines Mechanics Engine-work and Engineering*, Appleton (1874); available online.

[32] Mcleod J 1954 *J. Opt. Soc. Am.* **44** 952.

Lenticular lenses, also called **fly's eye lenses** and **integral lenses**, were first proposed by Gabriel Lippmann (1845–1921) in 1908. His early demonstration was with a small array of spherical rod ends. Later he used an array of some twelve lenses and reported, 'One no longer sees individual microscopic images; they are replaced by a single (integral) image…' Thus, integral lenses. A fly has hundreds of such lenses that allow it a limited resolution, but wide field of view. Thus, fly's eye lenses. Lenticular means shaped like a lens, so I do not know how it came to mean an array of them; I guess we need to say lenticular array. These arrays of lenses were used to generate three-dimensional imagery[33] and for other optical uses, notably the Shack–Hartmann test for the quality of optics[34].

Most **anamorphic lenses** are cylindrical. The first such appears to be that of Henri Chretien (1879–1956) in 1926[35]. He wanted a wider field of view for tanks. Another case of military applications leading to civilian ones. In cinematic use, an anamorphic system can be more than just a cylindrical lens. But those are often used to make a laser diode beam circular.

Anti-reflection coatings were first imagined by Harold Dennis Taylor, of the Cooke triplet[36]. He noticed that some old, tarnished lenses actually transmitted light better than their pristine counterparts. He figured it was that the coating was of a lower refractive index (see Interference, chapter 15). He tried to accomplish this with new lenses by using acids, but the results were too erratic. Alexander Smakula (1900–1883) invented the evaporative coating process while he was still at Zeiss[37].

We have seen the numerical control advances that enable all sorts of aspherics at reasonable cost. Now general shapes and nano lenses. All the techniques of microwave lenses can be applied to nano lenses.

[33] Roberts D and Smith T *The History of Integral Print Methods*, online.

[34] A lenticular array is positioned at the aperture and the individual images portray the wavefront.

[35] Schneider *Optics*, *History of Anamorphic Lenses;* Wikipedia, both online.

[36] Taylor H 1904 *British Patent* 29564/04.

[37] Smakula A 1936 *German Patent* 685767.

IOP Publishing

Rays, Waves and Photons
A compendium of foundations and emerging technologies of pure and applied optics
William L Wolfe

Chapter 18

Light—the light of our lives

What is light? The short answer is that we do not really know[1]. It is polarized, acts like a particle, behaves like a wave, is all over at once, and cannot be localized. The longer answer is that many great physicists have, over the years, pondered this question. The answers have varied, been contradictory and interesting. The trail is interesting to follow.

18.1 The early days

Some ancients believed that we saw by shining light from our eyes[2]. This required some dancing around to explain why we could not see at night, but there were arguments. Empedocles (490–430 BC) believed that Aphrodite put the fire in our eyes. Euclid (mid-300 to mid-200 BC) asserted that light travels in straight lines[3]. Certainly, a reasonable geometric concept. He also questioned the eye-emission concept (emission theory or extramission theory). One of his arguments was to ask how we see the stars immediately. This also came up in the discussion of the speed of light—whether it was infinite or not. Titus Lucretius Carus, commonly known as Lucretius (95–55 BC), was an atomist and thought that light was minute atoms with infinite speed[4].

18.2 Corpuscles

Isaac Newton (1643–1727) thought that light consisted of small, solid bodies, and he developed some powerful reasons for believing this[5]. He also believed these corpuscles of light had sides. This was his way of explaining double refraction in

[1] Roychouduri C and Roy R 2003 The Nature of Light: What is a Photon, *OPN Trends*.
[2] Wikipedia, online ed Lingberg D 1976 *Theories of Vision from Al-Kindi to Kepler* (Chicago, IL: University of Chicago Press).
[3] Euclid, *Optica*, Oπτικα ~ 300 BC; available online.
[4] Lucretius T 1656 De rerum natura translation at Project Gutenberg.
[5] Newton I 1704 *Opticks*.

doi:10.1088/978-0-7503-2612-4ch18

the Iceland spar (calcite, $CaCO_3$) crystal. He did get refraction right, the law of sines, but his ideas required that light travel faster in denser materials than in less dense ones (wrong). His corpuscular theory of light was not the same as our current photon ideas. His particles, corpuscles, were small, hard, and had mass.

18.3 Waves

Christiaan Huygens (1629–1695) believed that light was a wave motion[6], and he communicated this to the French Royal Academy of Sciences in 1678. For many years, the two theories of light existed side by side, on opposite sides of the Channel, but most believed Newton's—partly because he had reported on so many experiments, and partly because of his most impressive reputation. It was not until 1801 that Thomas Young (1773–1829) reported his double slit experiment and finally put the corpuscular idea to rest[7]. It is ironic that one of the phenomena that Newton could not explain with his corpuscular theory is now known as Newton's rings, an interference phenomenon.

18.4 Electromagnetic radiation

James Clerk Maxwell (1831–1879) unified the electrical laws of Carl Friedrich Gauss (1777–1855), André-Marie Ampére (1775–1836), and Michael Faraday (1791–1867) in 1865[8] into four vector differential equations that many claim to be the most elegant in physics. When they are manipulated correctly, one gets the wave equation, describing light and other electromagnetic radiation. (See appendix D.)

In 1866, Heinrich Rudolph Hertz (1857–1894) unequivocally proved the existence of electromagnetic waves[9]. He used a spark gap that produced a wave with a frequency determined by a capacitor and inductor and a receiver that was a small gap in a conductor. The spark that was generated in the transmitter induced a resultant spark in the receiver. The waves were radio waves, and he later showed that they traveled with the speed of light and that they reflected and refracted like light.

18.5 Quantum mechanics

In 1900, Max Karl Ernst Ludwig Planck (1858–1947) showed that the vibrations that produced light were quantized[10]. It is ironic that, although Planck revolutionized physics by introducing quantization, his former physics professor advised him against going into physics, saying, 'In this field, almost everything is already discovered, and all that remains is to fill a few holes.' Après moi, le déluge!

The story of how this happened is fascinating. Toward the end of the seventeenth century, the late 1800s, there was great interest in so-called blackbody radiation because it did not depend upon any material properties. In a sense, it was pure

[6] Huygens C 1690 *Traité de la lumiere*.

[7] Young T 1804 *Philos. Trans. R. Soc. Lond.* **94** 1.

[8] Maxwell J 1865 VIII. A dynamical theory of the electromagnetic field *R. Soc.* **155** 459.

[9] Hertz H 1887 *Ann. Phys.* **267** 421.

[10] Planck M 1901 *Ann. Phys.* **4** 553.

radiation; it was a function only of temperature and wavelength. A blackbody is an object that absorbs everything; it is black, a perfect absorber. Gustav Robert Kirchhoff (1824–1887) found that something that is a perfect absorber is also a perfect emitter. The technical jargon is $\alpha = \varepsilon$; absorptivity equals emissivity. That is Kirchhoff's law[11]. He also established that the distribution of such radiation depends only upon temperature and wavelength[12]. The realization of such a so-called blackbody is a fairly large chamber with a small hole in it. Think of a shoe box with a hole punched in it by a pencil. If you look into it, it will appear black. This blackbody and its radiation were considered both theoretically and experimentally. Everyone calculated the so-called density of modes—how many waves of different frequency there could be in the blackbody cavity per unit volume. The question then was how much energy did they carry.

Two things were already known about blackbody radiation: the total amount, encompassing all frequencies, and the wavelength of the maximum. These are, respectively, the Stefan–Boltzmann law and the Wien displacement law. $M = \sigma T^4$ and $\lambda_m T = 2898$ μmK. The total radiation is a constant times the fourth power of the temperature and the wavelength of the maximum of the spectral distribution times the temperature is another constant. Joseph Stefan (1835–1893) announced the law in 1879[13], and Ludwig Boltzmann (1844–1906) did so independently in 1884[14]. If you care, $\sigma = 5.67 \times 10^{-8}$ W m^{-2} K^{-4}. Wilhelm Carl Werner Otto Fritz Franz 'Willy' Wien (1864–1928) developed the displacement law, probably by observation[15].

Lord Rayleigh, John William Strutt (1842–1919), and James Hopwood Jeans (1877–1946) derived an expression for the energy that was simply kT, the Boltzmann constant times temperature. It was a classical way to describe amounts of energy, kT per degree of freedom. It led to an expression for the spectral distribution of blackbody radiation that was $M(\lambda) = c_1 kT/\lambda^5$. I think it is clear that this leads to infinity as the wavelength gets short enough. It was called the ultraviolet catastrophe and doomed this version. The constant c_1 is called the first radiation constant and is $2\pi c^2 h$.

The other version of the times, attributed to Wien and Planck, assumed energy went as the exponential $e^{u/kT}$, where u is the energy and it is normalized by that factor kT. Thus, the equation would be $c_1/\lambda^5 e^{u/kT}$. While these theorists were busy thinking good things, the experimentalists, especially Wien, were improving their techniques. They certainly found that the spectrum did not go to infinity at short wavelengths, but they also found that the second formulation was just not quite right. It was Wien who did the measurements that showed that his and Planck's

[11] Kirchhoff G 1860 *Ann. Phys. Chem.* **109** 275; Kirchhoff G 1860 *Philos. Mag. J. Sci.* **20**; Kirchhoff G 1860 *Pogg. Ann.* **109** 275.

[12] Kirchhoff G 1882 *Gesammmelte Abhandlungen* (Leipzig: J. A. Barth).

[13] Stefan J 1879 *Über die Beziehimg zwischen der Wärmestrhlung und Temperatur, Bulletin of the Vienna Academy of Sciences*; Stefan J 1879 *Sitzungsberichte der mathematische-naturwissenschaften Classe der kaiserlichten Akademie der Wissenschaftern* **79** 391.

[14] Boltzmann L 1884 *Ann. Phys. Chem.* **22** 291.

[15] Wien W 1893 *Sitzungsberichte d. Akad. d. Wissensch. Berlin* **9** 55; Wien W 1894 *Wiedemann's Annal.* **52** 132.

theory, using the exponential in the denominator, was not quite right at long wavelengths. It was these measurements at longer wavelengths that helped. Wien's Nobel address discusses many of these issues[16]. The Wien expression and the measurement results are represented in figure 18.1. The theory predicted values just a little too low in the longer wavelengths, the tail of the curve).

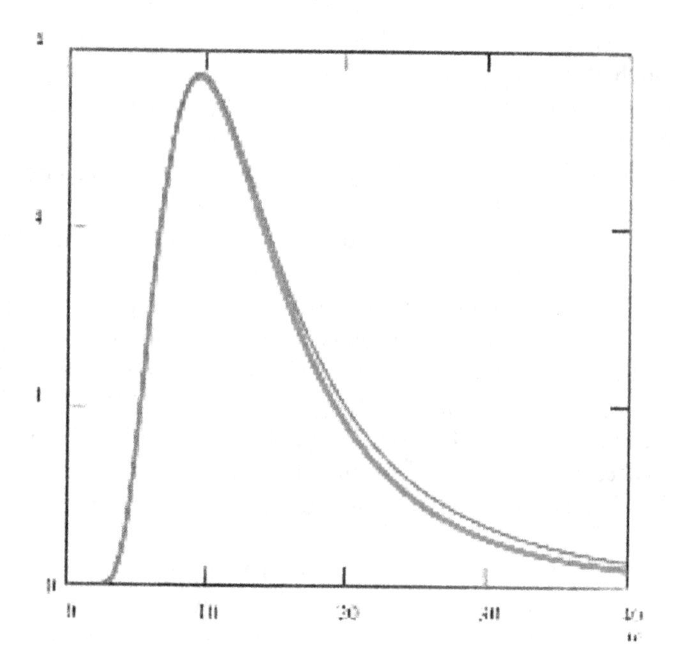

Figure 18.1. Blackbody curve discrepancies.

It has been reported that when Planck saw these results by Wien, presented at a meeting, he immediately started to think about how to fit the curve. He had the reputation of being a good physicist, good mathematician, and good curve fitter. So how could you get the tail to increase without changing the other part of the curve? If you multiply the numerator by a positive number, it increases the entire curve. Same thing if you add a positive number to the numerator. If you multiply the denominator by a fraction, it is the same as multiplying the numerator by a positive number. If you subtract a number from the denominator, it will also increase the entire curve. But if you subtract a positive number from the exponential term (by putting it in parenthesis), it increases the curve where the exponential has lower values, where the wavelength is longer. So do it, subtract a positive number from the exponential term in parenthesis (figure 18.2). Start with one. That did it, but what is the physics behind it? Planck was a good physicist and would not be satisfied with a result that was obtained by simple curve fitting. It is said that this was one of the

[16] Wien W 1967 On the laws of thermal radiation *Nobel Lectures Physics 1901–1921* (Amsterdam: Elsevier).

most trying periods in his life, getting the physics right to agree with the new curve. The form $1/(r-1)$ can be recognized by mathematicians as the sum of an infinite geometric series, where r is the ratio of terms. So if $e^{u/kT}$ is the ratio, it all works fine. The energy of the modes is the sum of $e^{u/kT}$, $e^{2u/kT}$, $e^{3u/kT}$, ... But this means the energies are discrete, they are digitized, they are quantized. The quantum age has begun!

At least one investigator, Douglas Hofstadter (1945–) of Stanford University, thinks Planck made the analogy between the energy distribution of an ideal gas and the photon distribution in the blackbody curve. These two curves, as shown, are similar, but surely not alike. The red line is the Planck curve, the blue line, the normal curve of an ideal gas. (I have diddled the constants to make the curves line up.) Perhaps this was also part of his effort to explain the spectral distribution of blackbody radiation.

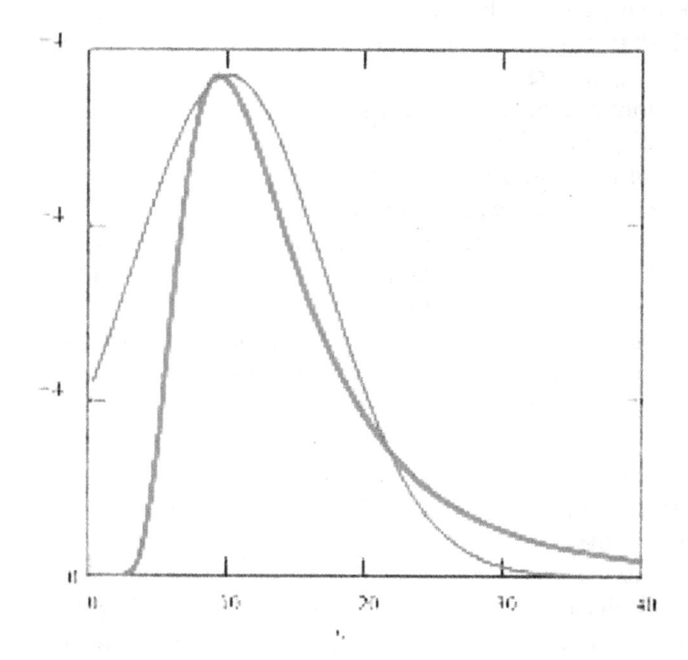

Figure 18.2. Blackbody and ideal gas distributions.

Planck quantized the material vibrations that give rise to the radiation, not the radiation itself.

I have to make one last note. The classical equation for the distribution of blackbody radiation is $c_1/\lambda^5 e^{c2/\lambda T}$. The Planck expression that introduced all of quantum mechanics is $c_1/\lambda^5(e^{c2/\lambda T} - 1)$. That little stinkin' one caused all this havoc, and it made the expression non-integrable so Czerny had to invent a slide rule to calculate it. But my quantum optics friends love it.

18.6 The photon

It is reasonable to ask, and people probably did, 'If the material vibrations are quantized, is the light itself quantized?' Albert Einstein (1879–1955) interpreted the results of the photoelectric effect in 1905 and showed that light itself was indeed quantized[17]. But they were not the corpuscles that Newton envisioned; they had no mass and even had wavelike properties. As a result, both the photoelectric effect and interference could be observed to show that light was quantized, that it consists of what we now call photons—that also had wave properties. Einstein was a theoretician; he did not do the experiment but considered the results of the photoelectric effect: when light is shone onto a photoemissive surface, electrons are expelled. It was found that the more intense the light, the greater the electron current, and light of shorter wavelengths produced more energetic electrons (because they went faster). This is consistent with the more photons, the more electrons, and the more energetic the photons $E = h\nu = hc/\lambda$, the more energetic the electrons. Einstein is credited with quantizing light, and it is said that Gilbert N Lewis (1875–1946) coined the word *photon* in 1926[18]. I would be remiss if I did not interject that this was Einstein's *annus mirabilis*, his miracle year, his *wunderjahr*. He quantized light; he introduced special relativity; he described Brownian motion; and he produced what is probably the best-known equation in all of physics, $E = mc^2$. The article on Brownian motion helped to establish the kinetic theory of fluids and the existence of the atom.

But Young's double slit experiment that generates an interference pattern is still valid. Light is also a wave. This leads to the ideas of **duality** and **complementarity**. Everything is a wave and a particle, but one cannot invoke both properties at the same time. *But what is light really? Is it a wave or a shower of photons? There seems no likelihood of forming a consistent description of the phenomena of light...we must use sometimes the one theory and sometimes the other*[19].

Young's double slit experiment, carried to extremes, has also provided a very unusual insight into how light travels. When a beam of light shines on those two slits, an interference pattern is generated. When one slit is closed, it disappears. This can be interpreted either by the waves interacting or by the probability that photons will go in the same places as the waves. But, when the beam is reduced to a single photon that appears every once in a while, independent of the others, the interference pattern is still generated if both slits are open but not if one is closed. How does that first photon, or the independent ones that follow it after a considerable time, know that there are two open slits? It was proposed by Richard Phillips Feynman (1918–1988)[20] that the photon explores all possible paths before it takes the most probable one, the one of least action. It must go faster than the speed of light to do this searching. How do you prove it? If this is true, is our universe the most probable of all possible universes? Another possible explanation is non-localization of the photon. It is wide enough to sense both slits, or just one. I surely do not know.

It is still the same. I doubt if we will ever know what it really is. A photon? A wave? A photon with a pilot wave? A corrugated hot dog?

[17] Einstein A 1905 *Ann. Phys.* **17** 132.
[18] Lewis G 1926 *Nature* **118** 874.
[19] Einstein A and Infeld L 1938 *The Evolution of Physics* (New York: Simon and Shuster).
[20] Feynman R 1948 *Rev. Mod. Phys.* **20** 367.

IOP Publishing

Rays, Waves and Photons

A compendium of foundations and emerging technologies of pure and applied optics

William L Wolfe

Chapter 19

Optical lithography—I walk the line

Optical lithography[1], also called photolithography, is primarily a method for putting down a pattern of leads and other devices on integrated circuits, although it also has a broader range of applications. The process starts with the application of a thin (about 1–2 μm thick) layer of photosensitive material. A photographic image is imposed on the material and the exposed regions can then be washed away (the opposite effect is also possible wherein either positive or negative photoresist material is used). Part of the challenge of this technique is to produce smaller and smaller patterns of lines and the like. This enables the creation of smaller electronic devices. Current practice gets down to about 50 nm, using ultraviolet light from excimer lasers (see Lasers, chapter 16).

Photolithography for integrated circuits took a leaf from regular, artistic lithography. Lithography for copying text and pictures was invented by Johann Alois Senefelder (1771–1834) in 1796[2]. He found that his crayon marks on limestone slabs lasted after many inkings. Originally, an object was drawn in wax or oil on a smooth limestone or metal plate. An acid was applied that etched away the other areas, and ink was applied and then transferred by contact. Modern systems use hydrophobic and hydrophilic areas to create an image on plastics from a negative, and still use ink and physical transfer.

Optical lithography for modern integrated circuits grew from a need for making them economically and in large numbers. The integrated circuit grew from the transistor of John Bardeen (1908–1991), Walter Houser Brattain (1902–1987), and William Bradford Shockley (1910–1989) in 1947. It was the brainchild of Jack St. Clair Kilby (1923–2005) of Texas Instruments in 1958 (who received the Nobel Prize in 2000). The next year, Robert Norton Noyce (1927–1990) introduced the idea of

[1] Flagello D 2007 *Proc. SPIE* **6520** 652004; Flagello D 1997 *Proc. SPIE* **3050**.

[2] Metropolitan Museum of Art, online.

doi:10.1088/978-0-7503-2612-4ch19

adding a metal layer at the end and etching away the parts that were not needed to form proper interconnections, the forerunner of photolithography[3].

In the early days of circuit lithography, the very early 1960s, masks were made manually by surgically cutting a master made of rubylith, an acetate with a red layer atop it. The patterns were cut by hand, later with a controlled tool, and imaged one-to-one onto the silicon disc. The challenges in projection printing of the mask onto the wafer were obtaining good resolution over large object and image fields and separating the mask plane from the wafer surface. Definition down to about 5 μm was about the best that could be obtained. This was aggravated by the production of larger and larger silicon wafers. One of the first of these imagers was designed by J Dyson; it was a system consisting of both lenses and mirrors, as shown in figure 19.1[4]. (The lens shapes are not shown true to form, but there were three of them.) The light shines from the left through the mask to the beam splitter (which is really a cube), then up through the optics to the concave mirror and back down through the beam splitter to the wafer. Each lens is therefore used twice. The next major step, I think, was the introduction of the Perkin Elmer Micralign optical system based on a design by Abe Ofner (figure 19.2)[5]. The system shown here uses the concave primary twice and the convex secondary once. Since the convex

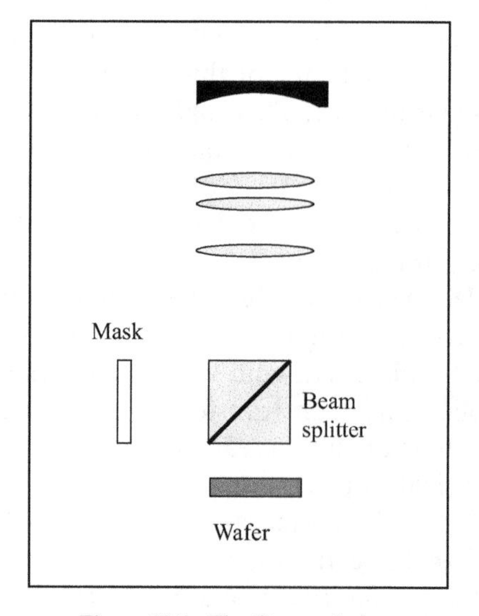

Figure 19.1. The Dyson design.

[3] Noyce R N 1959 *US Patent* 2981877.
[4] Dyson J 1959 *J. Opt. Soc. Am.* **49** 713.
[5] Ofner A 1975 *Opt. Eng.* **14** 130.

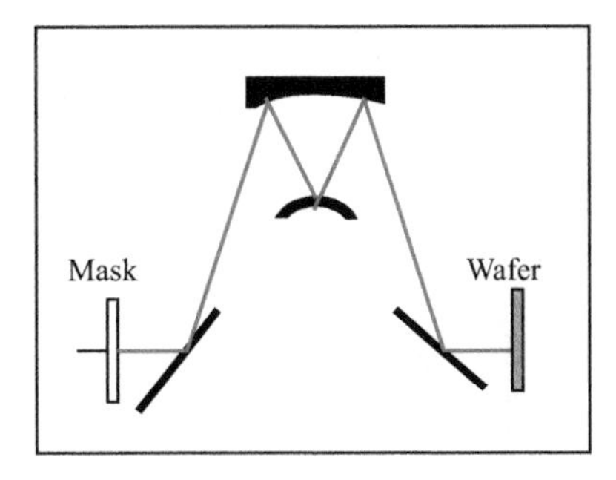

Figure 19.2. The Ofner design.

secondary that is used once has twice the curvature of the concave primary that is used twice, the system has zero Petzval sum (total curvature of field) and therefore a flat field. It images one to one and was used for about a decade in this industry. It was a much simpler design than the Dyson. The first reduction lithography in which the image on the wafer was smaller than the object mask was created in 1978. It incorporated a stepper so that part of the object was projected onto the wafer at a 10:1 reduction ratio, and then it stepped to the next part and repeated the process—stepper repeaters.

From that time on, until the mid-nineties, the progress was in better optics, larger wafers, better sources, more pixels, and better throughput. The sources went from mercury arcs at about 400 nm to excimer lasers at 248 nm. The wafers increased from diameters of 5 cm to those of 20 cm, the optical systems from about 10 elements to twice that. That was the reason for better production speed and better coatings. There has been a tremendous market for solid state electronics of all sorts, so there were considerable expenses incurred in the instruments to make them—in production. The use of the wavelength of 248 nm precluded the use of simple glass.

This era also saw the combination of stepping and scanning. A step-scan system could scan an image using a basic optical element of relatively small field of view, and then step it and scan again.

The next period of the history of photolithography saw the conversion to immersed optics and a wavelength as short as 193 nm and a numerical aperture as low as 1.2. From 1994 to 2012, the feature size reduced from about 350 nm to 50 nm[6]. Figure 19.3 shows the incredible increase in complexity of the optical systems to accomplish these feats[7]. And these are not even the immersion optics versions!

[6] Ku P 2006 *EECS 598-002 Nanophotonics and Nanoscale Fabrication.*
[7] Bruning J 2012 *SPIE Trans.* 6329.

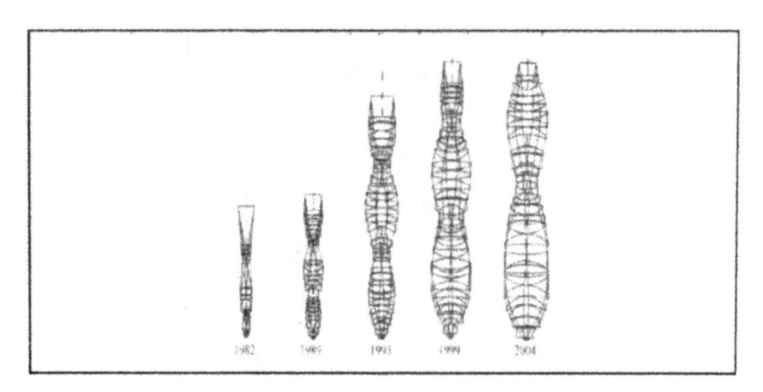

Figure 19.3. Photolithography optical systems.

When you hold that smartphone or camera with 25 megapixels in your hand, give thanks to the guys who design these incredible optical systems for photolithography.

The lenses get more complicated. The wavelengths get shorter and the lines get smaller. Extreme ultraviolet techniques seem to be the last step in optical lithography. Electron and x-ray techniques will follow.

IOP Publishing

Rays, Waves and Photons
A compendium of foundations and emerging technologies of pure and applied optics
William L Wolfe

Chapter 20

Medical optics—aaaah

There is a wide range of applications of light in medicine. These include devices for diagnosis, for treatment, and remote applications. Of course, the practice of medicine is about as old as civilization, and most applications have long roots. Optics is about the same age. I have divided this chapter into three sections: diagnosis, treatment, and telemedicine.

Two of the simplest tools for medical diagnosis are the eye and the hand. The eye can see pimples, rashes, eye dilations, and a hang-dog appearance. The hand can feel a pulse and a temperature. But these are not optical instruments, although it can be argued quite successfully that the eye is the best optical instrument of all. This chapter deals with man-made, optical instruments.

20.1 Diagnostics

One of the simplest optical diagnostic tools is the **otoscope**. Its history is described in chapter 36 on scopes. The **endoscope** is described there too. The **stethoscope** is not an optical device. **Bronchoscopes**, **sigmoidoscopes**, **colonoscopes** and the like are all variations on the endoscope. The **colposcope**, used for examining the vagina, is also described there.

The **ear thermometer**, sometimes called a **thermoscope**, is an infrared device that is inserted in the ear and reads the temperature of the tympanum, the ear drum. It is described in chapter 13 on infrared. Thermography, infrared or heat imaging, has been used for many diagnostic applications: breast cancer screening, delineation of burns, fever, circulation disorders and the like. The techniques are also described in chapter 13.

Colorimetry is what goes on in the medical laboratory with the blood or urine sample you have given to your doctor or nurse, although another meaning is the measurement of the colors in an object. In the present case, sometimes called chemical colorimetry, it is the use of a specific color or wavelength to assess the amount of material, such as the concentration of a certain chemical in a solution. In

doi:10.1088/978-0-7503-2612-4ch20

the chapter on spectroscopy, it is noted that every substance has a unique spectrum; it absorbs at certain wavelengths and transmits at others. Once that spectrum is known for a given chemical, a specific wavelength (or color) can be used to test how much of it is in a solution by measuring the transmission (or absorption) at that wavelength. Another way to look at it is that a colorimeter is a spectrometer that works at only one wavelength and makes absolute transmission (absorption) determinations. Over seventy percent of the tests in the clinical lab are done colorimetrically[1]. The basis for the measurement goes back to the Lambert–Bouguer law of Johann Heinrich Lambert (1728–1777) and Pierre Bouguer (1698–1758) in 1760 and 1729 [2] respectively, which states that the transmission is exponentially related to the concentration ($\tau = e^{-\alpha c x}$), where τ is the transmission, α is the absorption—proportional to the concentration—and x is the thickness of the sample. The first mention of using this technique for measuring concentration was by Müller in 1853. The first instrument was that of Louis Jules Duboscq (1817–1886), invented in 1870[3]. The first photoelectric colorimeter was described by Arnoldus Goudsmit Jr (1909–2005) and William Henry Summerson (1887–1976) in the 1930s[4]. Modern instruments measure many different quantities at a time using mechanical peristaltic flow techniques in multiple paths[5].

Fluorescent cancer detection was pioneered by R Alfano (1941–)[6] in the 1960s, and variations have been patented[7] but apparently are not yet in practice. The technique uses a violet or ultraviolet source and then analyzes the resultant fluorescent output of the skin or other body part. Since cancerous cells are different from benign ones, there is a slight difference in their spectra. This has been demonstrated with promising results and can be used for examining surgeries to see if all malignant tissue has been removed at the end of the surgery.

Infrared imaging has been used to determine the extent of a burn. There is a temperature difference between necrotic tissue and healthy tissue that has more intimate contact with the internal heat of the body. This has been done in the very near infrared (about 1 μm)[8] and in the middle IR at 3–5 μm[9]. Another technique is the evaluation of blood flow using a doppler lidar technique[10]. A laser with a wavelength in the near infrared where the skin and tissue are fairly transparent is used to illuminate the area. The flow of blood will cause a doppler shift in this wavelength, giving an indication of its speed based on the amount of the shift.

Photoacoustic spectroscopy may be the first step toward the analysis that Bones did on *Star Trek*. An infrared laser is shone on the skin. Its light is absorbed by

[1] Nahm K private communication.
[2] Lambert *Photometrica Augustae Vindilorum* (1760).
[3] Duboscq J and Mene C 1886 *C. R.* **67** 1330.
[4] Goudsmit A Jr and Summerson W 1935 *J. Biol. Chem.* **111** 421.
[5] Personal observation.
[6] Alfano A 1984 IEEE *J. Quantum Electron.* **20** 1507.
[7] Wolfe W 2001 *US Patent* 6256530.
[8] Jackson D 1953 *Br. J. Surg.* **40** 588.
[9] Patel M et al 1991 *Eng. Med. Biol. Soc. Proc.* **13** 316.
[10] Stern M 1975 *Nature* **254** 56.

glucose molecules 10 μm under the skin. They make a sound; its level is a measure of their concentration. It is not quite like the *Star Trek* version in that it needs to be in a protective enclosure to prevent false signals from variations in atmospheric pressure, temperature, and humidity[11]. But it is a way to monitor diabetics without sticking it to them!

There are other modalities that also purport to measure **glucose concentration without needles:** near-infrared spectroscopy, Raman spectroscopy, and scatter measurements. The first uses wavelengths from 700 nm to 1000 nm where the skin and flesh are relatively transparent. The illumination goes to a depth of as much as 50 μm, but is subject to a variety of troubles. Surface measurements are subject to various contaminations. Transmission methods were attempted on the ear lobe, finger webs, and cuticles, and reflection methods on the finger and cuticle by R Marbach and T H Kochinsky in 1989[12]. Some investigators used attenuated total reflection infrared spectroscopy[13]. Light scattering methods were employed by others[14].

20.2 Treatment

Photodynamic therapy is the use of light to heal. Typically it uses certain compounds that are light sensitive to attack cancer cells. It has also been used as a treatment for acne. Early accounts cite the use of certain leaves that are applied to the skin to encourage such things as repigmentation and rashes. In the first few years of the 20th century, the laboratory of H von Tappeiner (1847–1927)[15] showed that oxygen was essential in the use of this photodynamic action, now called photodynamic therapy[16]. It is the action of singlet oxygen that makes it work.

SAD, or seasonal affective disorder, is the blues that my wife used to get when we lived in New England! It is a true disease that has been found to be caused by the lack of sufficient sunlight. Research on it began with an engineer, Herb Kern, who noticed his depression in winter. Scientists at the National Institute of Mental Health devised a light box to treat him; he felt better in just a few days[17]. Norman Rosenthal (1950–) was the first to examine it systematically[18]. Bright light for 30–60 min is not the only treatment, and experiments are ongoing to see if there is a preferred spectrum. Currently the solar spectrum is preferred for obvious reasons.

Carbon dioxide lasers have been used to sterilize and cauterize giant cell tumors[19]. These lasers have considerable power, in the range of 10 to 100 watts, and can easily burn skin and other tissue. And they have been used to ablate basal and squamous

[11] Photonics Showcase, March 2014; RSI 10/1063/1.4816723.

[12] Marbach R and Kochinski T Z. *Fur angew. spektr.*

[13] Heise H *et al* 1989 *Anal. Chem.* **61** 2009.

[14] Bruusema J *et al* 1996 *OSA Conf. on Biomedical Optical Spectroscopy and Diagnostics.*

[15] Von Trapeiner H and Jodbauer A 1904 *Deut. Arch. Klin. Med.* **8** 427.

[16] J Toth coined the term.

[17] Fiona M and Cheevers P 2003 *Positive Options for Seasonal Affective Disorder* (Alameda, CA: Hunter House).

[18] Rosenthal N *et al* 1984 *Arch. Gen. Psychiatry* **41** 72.

[19] Kenan S *et al* 1988 *Bull. Hosp. Jt. Dis. Orthop. Inst.* **48** 93; Kirby K 1990 *Clin. Orthop. Relat. Res.* **253** 231.

cell carcinomas[20]. They have also been used for removing some of the epidermis in lieu of dermabrasion[21].

LASIK, laser *in situ* keratomileusis, is the reformation of the cornea by the use of laser pulses. Also known as laser surgery, it can correct for myopia (nearsightedness), hyperopia (farsightedness), and astigmatism (nonsymmetrical imaging). The earliest work of this sort was done by a Spanish ophthalmologist in Bogota, Colombia, named José Barraquer (1916–1998), but this was not *laser* keratotomy. In 1980, Rangaswamy Srinivasan (1929–) showed that one could ablate living tissue (among other things) with an ultraviolet laser with no ill effects[22]. Five years later, in 1985, Stephen Trokel (1934–) showed that one could ablate and shape the cornea of an eye[23]. The procedure is to carefully measure the shape of the cornea and calculate the required amounts and places of ablation. The process then is to set up the laser under computer control and zap the eye. It takes only a minute or two. The work and expense is in the preparation.

Port wine stains and **tattoos**, as well as unwanted **hair**, can be removed with laser treatment, especially if they are dark. The stain, ink, or hair absorbs the energy of the laser and ablates. Care must be taken to not burn the skin, and so this treatment is better for Caucasians and blondes than others. Tattoo removal was first reported by Leon Goldman (1906–1997) in 1963[24]. A port wine stain, also called *nevus flammeus*, is a vascular anomaly, perhaps most evident on the head of Mikhail Gorbachev. Yellow lasers are used for wine stains because the red absorbs it best.

Laparoscopy is the technique of using a small opening, about one centimeter, and a digital camera to perform the surgery. The first such operation performed on a human was by Hans Christian Jacobaeus (1879–1937) in 1910[25]. Although the term laparoscopy is sometimes used for incisions in the stomach, I use the term for all such non-invasive surgery.

The fabrication of a new hand may certainly be called treatment. This has been done at a modest cost of $2000, and it is speculated it could be less in the future. The process is 3D printing (see Displays, chapter 7), and it is likely it will make all sorts of other organs[26].

20.3 Telemedicine

Depending upon your interpretation, **telemedicine** is as old as when means existed for transferring information over reasonably long distances. Perhaps the first, and rather primitive, use was heliograph signals that warned of bubonic plague in the Middle Ages. During the American Civil War, casualty lists and requests for supplies were sent by telegraph. In 1905, William Einthoven (1860–1927) transferred

[20] Humphreys T *et al* 1998 *Arch. Dermatol.* **134** 1247.
[21] Rubach B 1997 *Arch. Otolaryncol. Head Neck Surg.* **123** 929.
[22] American Institute of Physics, Prize for the industrial use of lasers, online.
[23] Cotliar A *et al* 1985 *Ophthalmology* **92** 206.
[24] Goldman L 1963 *J. Invest. Dermatol.* **40** 121.
[25] Hartzinger M *et al* 2006 *J. Endourol.* **20** 848.
[26] Winter J 2014 *Everything that's Fit to Print, Parade.*

electrocardiograms by facsimile[27]. Then, information was transferred by telephone and by radio in the 1900s and later. During the Korean and Vietnam Wars, medical teams were dispatched by radio but medicine was not administered nor diagnoses made over significant distances. These first occurred during the space program in which astronauts were wired for a variety of tests. The first interactive video link was established between the Nebraska Psychiatric Institute in Omaha and the Norfolk State Hospital over 100 miles away. Doctors in one location could tell doctors in the other what their diagnosis was and what to do about it[28]. The first application of remote anesthesia was accomplished in 1974 by Joachim Stefan Gravenstein (1925–2009)[29]. In 1989, a remote defibrillator was used[30]. It was operated from the base station. It seems clear that, in modern times, these various ways of sharing information remotely rely on high capacity fiber optics, or satellite links and television capabilities, and can be used for all sorts of applications—files, x-rays, dermatological images, facial expressions, etc. The most modern telemedicine applications are remote surgeries. Perhaps the earliest was in 2001 by Jacques Marescaux (1948–), who removed the gallbladder of a patient in Strasbourg, France, from his operating room in New York[31]. The best known system for doing this is the so-called Da Vinci system. It has four robotic arms, one of which is an endoscopic camera; the others have tools. It has been used mostly with the surgeon in the room at the console. One slight disadvantage of remoteness would be the slight delay of communication, about 5 μs per mile. One unusual advantage is that the robotic system smooths any tremor in the surgeon's hand by short term averaging of its position. A quite new telemedicine application is the virtual house call. Using a smartphone or Skype, a patient can show his doctor what is wrong. The rash or lesion can be seen while the doc is at the office and the patient is home[32].

There will be much miniaturization in the future. The endo pill camera will become a multi spectral device that can detect cancer by its spectrum. Very small robots will deliver meds and even isotopes to any place in the veins or arteries. Telemedicine will increase and get better. Patients will have diagnostic tools on their smartphones.

[27] Bashur R and Shannon G 2009 *The History of Telemedicine* (New Rochelle, NY: Mary Ann Liebert).
[28] Zundel K 1996 *Bull. Am. Libr. Assoc.* **84** 71.
[29] Gravenstein J *et al* 1974 *Anesth. Analg.* **53** 606.
[30] Kuhrik K 1992 *Am. J. Nurs.* **92** 28.
[31] 1ST Media collection, online.
[32] *The Arizona Daily Star*, via Associated Press, May 2014.

IOP Publishing

Rays, Waves and Photons
A compendium of foundations and emerging technologies of pure and applied optics
William L Wolfe

Chapter 21

Microscopes—a little of this and a little of that

Although it can be claimed that magnifying glasses might be considered as micro-scopes, I don't think so. I think you really need two lenses to get reasonable magnification from a microscope—that is, a compound microscope. Perhaps magnifying glasses should be called miniscopes. The principle of the compound microscope is shown in figure 21.1. The specimen is focused by the objective lens to a real image, which is inside the focus of the eyepiece lens. It, in turn, generates a virtual im0age as viewed from the sensor, which could be an eye. Microscopes come in a great variety of guises; only the optical ones are considered here. After all, this is a history of optics, not electrons or x-rays.

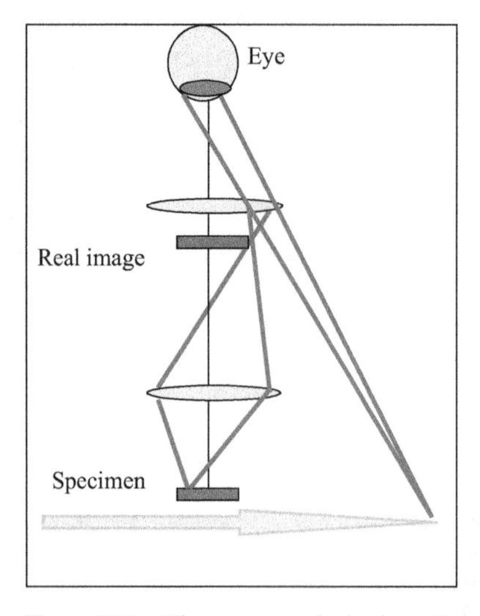

Figure 21.1. Microscope optical schematic

doi:10.1088/978-0-7503-2612-4ch21

The **first (compound) microscope** is said to have been discovered or invented by Hans Lippershey (1570–1619) or his children[1]. Galileo Galilei (1564–1642), who learned of Lippershey's work and made his famous telescope, also made a microscope and called it the little eye, *occhiolino.* But he was much more interested in the heavens. Zacharias Janssen (1585–1632) is also claimed to have invented the microscope, the claim made mostly by his son.

Giovanni Faber (1574–1629) coined the name *microscope*, or μικροσκοπειν—small to look at—shortly thereafter.

Antonie Philips van Leeuwenhoek (1632–1723) learned how to make small lenses with very steep curvatures and therefore obtained large magnification. He did not grind, but used a rod melting technique. They had magnifications of about 270×. Recall that the magnification is the image distance divided by the object distance, and if the focal length is very short, obtained with greater curvatures on the surfaces, the ratio is large. He is often called the father of the microscope because he was so active very early in its use (figure 21.2)[2].

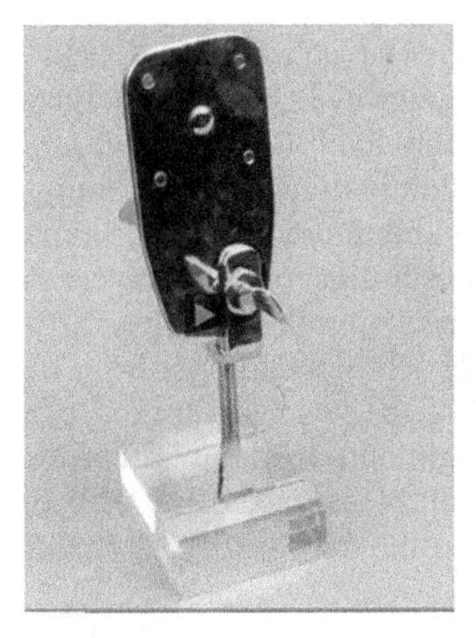

Figure 21.2. Replica of a Leeuwenhoek.

Bright field illumination is the standard method for illuminating the sample. Light is shown from the bottom (with the viewer looking down); the sample absorbs some of it. In some situations, the sample needs to be stained. Typically such illumination is what is called critical illumination; the sample and the source images are in the same plane. Then the image of the source interferes with the image of the sample. Various methods were tried to alleviate this, including diffuser plates and diffuse bulbs, but they all reduced the available light.

In 1893, August Köhler (1866–1948) invented a technique that provided very uniform illumination of the sample, now called **Köhler illumination**[3] (figure 21.3). It focuses the source and the sample in different planes; the light is collimated as it

[1] Van Helden A *et al* 2010 *Origins of the Telescope* (Amsterdam: KNAW Press).

[2] Hoole S 1800 *The Selected Works of Antony van Leeuwehoeck* (London: G. Sydney) online.

[3] Köhler A 1893 *Zetschrift für wissenschaftliche Mikroskopie und für mikroskopische Technik* **10** 433.

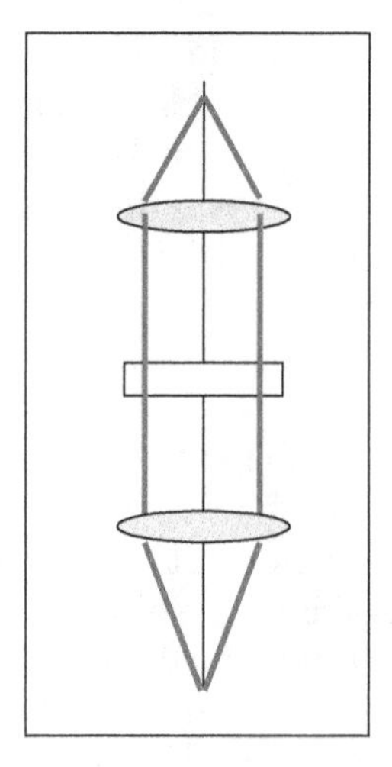

Figure 21.3. Köhler illumination.

passes through the sample. The collector and field lenses send the light to a condenser lens, which collimates it. It then passes through (or off) the sample. An objective lens then sends the light to an eyepiece. This is considerably more complicated than critical illumination, but the sample is illuminated uniformly. And the diagram is considerably more simplified than real systems.

Dark field illumination uses a disk to block the central portion of the beam. The only light that gets to the objective is that which is diffracted by the sample; the rest of the field is dark. Figure 21.4 shows that light from the source is collimated by the collector. The central part of the beam is blocked by the blocker. The condenser focuses the hollow cone of light onto the sample. The direct illumination on the sides is then blocked by the bottom blocker. The scattered or diffracted light goes to the objective and is focused onto the dark field. It is the only bright part.

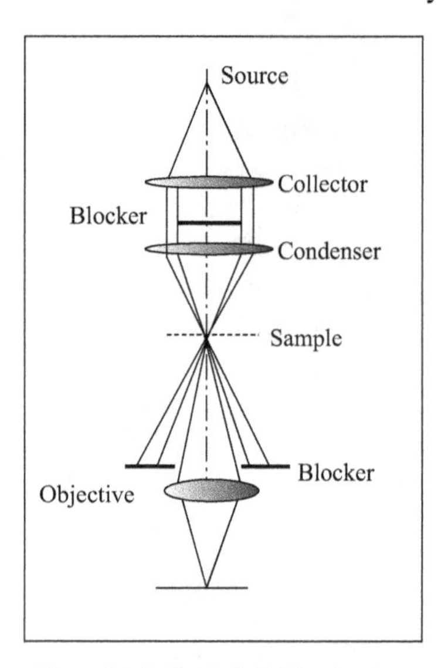

Figure 21.4. Dark field illumination.

Frits Zernike (1888–1966) invented the **phase contrast microscope** in 1953[4]. It makes use of the change in phase of a light field due to its passage through a thicker material or one of different refractive index.

Georges Nomarski (1919–1997) conceived the **differential interference contrast** microscope, also known by his name, in 1952[5]. It involves the interference of two sheared beams of light that then reveal the path length gradient in the sample. It is closely related to the phase contrast method.

The **confocal microscope** was invented by Marvin Minsky (1927–2016) in 1957[6]. It is a technique by which light is focused onto a spot on the sample, and that spot is used for examination. The spot is scanned across the image in some sort of raster pattern. This provides a two-dimensional image. The third dimension can be obtained by a change in focus and even the use of computed tomography. This technique is especially useful in fluorescent microscopy.

Confocal laser scanning microscopy was developed by Thomas Cremer (1945–) and Christoph Cremer (1944–) in 1978[7]. This is a point-by-point scanning device that can measure the vertical profile of the sample. Of course, it uses a laser, and it provides a three-dimensional representation of the sample.

The **resolving power** of a lens, as defined in the field of microscopy, is the distance between two objects that can just be resolved (similar to the Rayleigh and

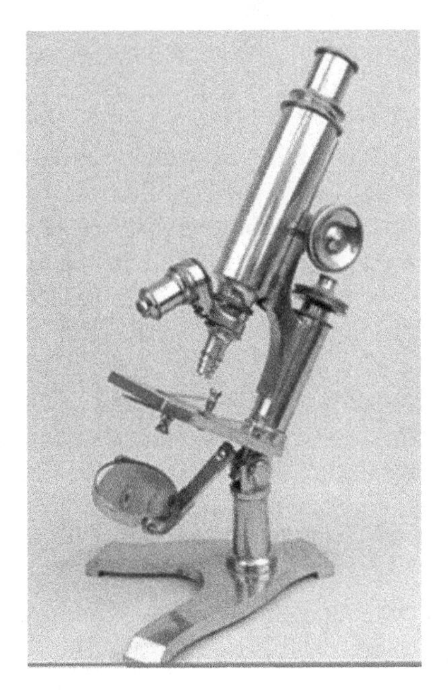

Figure 21.5. Representative microscope.

[4] Zernike F 1942 *Physica* **9** 686.
[5] Nomarski G 1952 *French Patent* 1059123.
[6] Minsky M 1957 *US Patent* 3013467.
[7] Abbe E 1873 *Arch. Mikrosc. Anat.* **9** 413.

Sparrow criteria for telescopes) (figure 21.5). It was derived by Ernst Abbe in 1873[8]. It is given as $\delta = \lambda/(2\, n \sin \alpha)$, where δ is the distance, λ is the wavelength of the light, n is the refractive index of the medium, and a is the field angle. So, if the refractive index is increased, the resolving power is increased—that is, two objects closer together can be resolved; 8 is smaller. Air has an index of about 1, while most oils have an index of about 1.5. Voila! Immerse the sample in a substance of higher refractive index to improve the resolution, an **immersion microscope**. Water could be used, but it is not viscous enough. The first mention of an oil immersion microscope seems to be that of Robert Hooke (1635–1703) in 1678. He apparently did it without the theory of Abbe.

An **ultraviolet microscope** was invented[9] in 1904 by August Köhler and Moritz von Rohr (1868–1940). They reasoned that light of shorter wave-lengths could obtain better diffraction resolution. Check the formula above; λ is in the numerator.

The **fluorescent microscope** was based on the ultraviolet one. It simply used appropriate fluorophores to provide proper excitation for the UV scope. It was the invention of Oskar Heimstädt (1879–1944) in 1911, shortly after the arrival of the UV scope. It is used to investigate the fluorescence of various specimens. This can often be an identifier of specific pathogens, antibodies, or other substances. The design must use either reflecting optics or lenses made of special materials, like quartz (see Ultraviolet, chapter 43).

The **stereoscopic microscope** was also invented by Oskar Heimstädt in 1923[10]. The patent was actually filed two years earlier (figure 21.6).

Richard Zsigmondy (1865–1929) and Henry Siedentopf (1872–1940) invented the so-called **ultramicroscope** in 1903.[11] It was a technique for apparently overcoming the diffraction limit of Abbe. It did its job but did not prove that Abbe was wrong. The microscope used a beam of light that was perpendicular to an array of falling particles. The particles scattered the light, and their size could be determined by counting the number in a given volume. This accomplished the detection of single particles much smaller than the diffraction limit, but not the resolution of two closely spaced particles.

The **digital microscope** is a modern adaptation of the various kinds already described. Apparently the first was by a Japanese company, Hirox, in 1986. It uses a different, but equivalent, lens design and a CCD array. The resolution is determined by the size of an individual pixel, p, and the focal length, f, of the lens $\delta = p/f$. Of course, it cannot be better than the diffraction limit. It *can* be better than p/f if the array is moved a fraction of a pixel width and the two images are combined, a technique called **sub-pixel imaging**. This technique was apparently invented by Terry L Benzschawel and Webster E Howard in 1988[11], but Steve Wozniak (1950–) claims

[8] Encyclopedia Britannica, online.

[9] Heimstadt O 1923 *US Patent* 1,470,670.

[10] Zsigmondy R and Siedentopf H 1902 *Ann. Phys.* **315** 1.

[11] US patent 5341153 (1999).

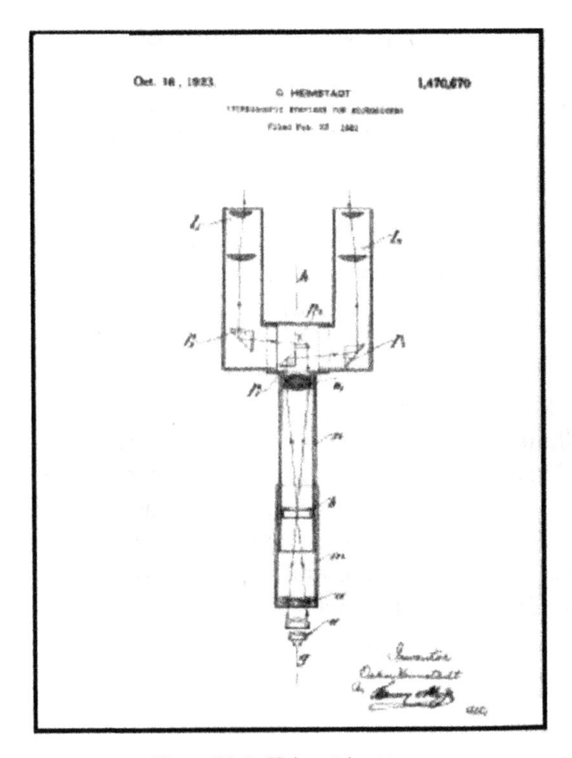

Figure 21.6. Heimstädt patent.

to have used it in 1976[12]. I think it is a fairly easy concept to come by and believe that many have thought about it independently.

Two-photon microscopy uses the upfrequency excitation of light similar to frequency doubling. A laser is used for illuminating the sample with enough photons that two of them combine to cause excitation of light of a higher frequency. This technique can penetrate further into the body, mainly the brain, than simple illumination, ultraviolet light, or infrared. It was pioneered by Winfried Denk (1957–)[13].

The **holographic microscope** was invented by Daniel Courjon and J Bulabois in 1979[14]. It incorporates a standard microscope, the sample of which is illuminated by a laser. The original laser beam is split by a beam splitter so that a reference beam is available that does not go through the microscope. The two are combined in front of the image area to generate the hologram as described in the chapter on holography.

Structured illumination microscopy, SIM, involves using structured light illumination using a pattern. Then the Moiré pattern generated by the interference of the pattern and the sample is analyzed, using Fourier transformations. Resolution better

[12] GRC, The origins of sub-pixel font rendering, online.

[13] Denk W and Svoboda K 1997 *Neuron* **18** 351.

[14] Courjon D and Bulabois J 1979 *J. Opt.* **10** 3.

than diffraction limit is claimed. It appears that the first report on this was in 1993[15]. Some adaptations can view the fields in real time[16].

A related technique for overcoming the diffraction limit is **spatially modulated illumination**[17] in which the sample is moved through another patterned illumination field.

Still another variation on microscopes that claim to exceed the classical diffraction limit is the **stimulated emission depletion, STED**, version. Two lasers are used, one to excite fluorophores and the second to de-excite them. This creates a zero-intensity spot that is scanned over the field[18].

Maybe the latest improvement in microscopy is the FPM, or **Fourier ptychographic microscope**[19]. It promises to exceed the classical resolution limit and have a large depth of field, a larger field of view, and digital refocusing. The general procedure is to illuminate the sample with a large LED array from a number of different angles. This is somewhat analogous to the radar technique of aperture synthesis. Phase information is retrieved and the phases used in effect to create a larger aperture for the microscope. It is also related to tomography in which several images from several angles are used to build up a three-dimensional image. Ptycho is a derivative of the Greek word for layer.

[15] Bailey B *et al* 1993 *Nature* **366** 44.

[16] Photonics showcase, March 2014.

[17] Heintzman R and Cremer C 1999 *Proc. SPIE* **3568** 185.

[18] Fernandes-Suares M and Ting A 2008 *Nat. Rev. Mol. Cell Biol.* **9** 929.

[19] Zheng G *et al* 2014 *Optics and Photonics special issue* (Washington, DC: Optical Society of America); Zheng G *et al* 2004 *Phys. Rev. Lett.* **93** 023903.

IOP Publishing

Rays, Waves and Photons
A compendium of foundations and emerging technologies of pure and applied optics
William L Wolfe

Chapter 22

Military optics—homing in

The military uses optics in essentially three ways: for reconnaissance, communication, and weaponry. There are other minor uses, such as transportation, medicine, and everyday activities, but these are the major ones. Many of the devices the military uses are used in other spheres and are covered there. Accordingly, there are many cross references in this chapter.

22.1 Reconnaissance

Binoculars are the first that come to mind. They are covered in chapter 3.

The periscope is covered in chapter 36 on scopes. It is still basically the same device invented years ago, although I can anticipate the same kind of change that happened with endoscopes: the complex of lenses will be replaced by a camera at the top and a display at the eyepiece.

The sniperscope is a snooperscope with a rifle attached—or a rifle with a snooperscope attached (see chapter 36). It has lately been adapted for use with smartphones[1]. The metascope, described in the same chapter, is a similar variant. All provide night vision.

Drones have been developed lately for both recon and for strikes. They are also called UAV's, Unmanned Aerial Vehicles, and come in sizes ranging from hummingbird to 250 foot wingspans. Boeing's Phantom flies at 65 000 feet and has a payload of 450 pounds. A decent-size optical system would give a resolution of the ground of a few centimeters, plenty good enough for military aerial recon. A hummingbird-sized UAV can flit around near the ground among the buildings of an urban area of conflict. Alfred Nobel would be happy to know that these capabilities are also useful for civilian applications, notably search and rescue.

[1] SOLOSmark riflescope adapter smartphone—Amazon.

doi:10.1088/978-0-7503-2612-4ch22

Laser-based obstacle warning systems are surely a form of recon[2]. These are mostly for helicopters and scan laser beams looking for reflections. This work went on as early as 1966[3].

22.2 Weaponry

The history of **searchlights** is discussed in chapter 36 on scopes. Their use in warfare probably began in Egypt and was made famous in the Battle of Britain. They were the first target designators.

Laser designators are a specific military application of, well, lasers. They are a special kind of target designator. Laser designators are highly collimated so that they pinpoint a target well, and they can be modulated with a variety of codes to differentiate them from the background and countermeasures. Some also act as laser range finders, using time-of-flight techniques. The Army started its development in 1962, but the first laser-guided bomb seems to have been in 1965 with combat testing in 1968 of the BOLT 117, or PAVE 1, developed by Texas instruments[4].

Midas, the Military Defense Alarm System, was a sensor in geosynchronous orbit that detected the launch of missiles. It is discussed in the chapter on infrared. It has been succeeded by several improved systems as part of **SDI**, the Strategic Defense Initiative. Those are also discussed there. SDI included both these systems that detected launches and sensors that detected and intercepted the payloads in midcourse.

Laser weapons were conceived almost as soon as the laser was conceived and demonstrated. Some call it the ultimate weapon. It hits the target almost instantaneously. It takes about five microseconds to hit a target a mile away. It does not require a supply of shells, although it does have to have a good supply of energy. There is no windage or gravity effect on the beam; you shoot where it is.

The Airborne Laser Lab was probably the first of the demonstration programs by the Air Force Weapons Laboratory at Kirtland Air Force Base in Albuquerque. They first made some successful ground tests using 100 kW (kilowatt) carbon dioxide lasers in 1972. In 1993, it shot down several Sidewinders and a drone that mimicked a Russian cruise missile in tests in Albuquerque[5].

Northrop Grumman Corporation announced in 2009 that it had successfully tested a 100 kW electric laser[6]. In 2011, the Navy announced it had used a solid state laser to do extensive damage to a fast moving boat at a distance of some miles[7].

In addition to the test mentioned above, the Navy shot down several unmanned aircraft in 2012 and has been deployed to the Persian Gulf[8].

---

[2] Büchtemann W and Eibert M 1995 *Laser Based Obstacle Warning System for Helicopters*, AGARD-CP-563; Sabatini R *et al* 2014 *Metrology for Aerospace*.

[3] Personal development.

[4] Wikipedia, Paveway online.

[5] Kyrasazis D 2013 *Opt. Eng.* **52** 070143-1.

[6] Northrop Grumman.

[7] Kaplan J 2011 *Navy Shows off Powerful New Laser Weapon* Foxnews.com.

[8] *All Systems Go: Navy's Laser Weapon Ready for Deployment*, America's Navy, online.

A Boeing demonstration vehicle about the size of a garbage truck has a 10 kW laser that has destroyed about 150 targets in all sorts of weather. They are planning a more powerful laser[9]. They probably use some sort of adaptive optics to accomplish this (see Telecopes, chapter 43).

Figure 22.1. The Norden bombsight.

The **Norden Bombsight** is legendary in both its reputation for accuracy and its secrecy during World War II (figure 22.1). The optics in it were a simple telescope and a prism. The 'magic' was in the fact that wind speed, aircraft speed, altitude and direction, and the angle to the target determined by the sighting telescope were all calculated by what today would be considered an archaic, mechanical computer. A rotating prism kept the target fixed in the sighting telescope. The entire system was on a gyroscopically stabilized platform. It had a theoretical CEP (circular error probable, the diameter of the circle into which half of the hits would be) of 75 feet, although in combat conditions it was many times this. Still, it was by far the best we had. The secrecy was compromised even before World War II in 1938 by Herman W Lang, who was an employee of the Carl L. Norden Company. One of my German–American employees at Honeywell in 1967 told me he had one on his desk throughout the entire war.

There exist several optical techniques for the guidance of air-to-ground missiles. These include inertial, celestial, and map-matching systems.

Inertial guidance is based on knowing where the target is and directing the missile to it. The location of the missile is determined by gyroscopes and accelerometers. Early versions used iron gyros, but these have been replaced by laser ring gyros and fiber optic gyros (see Optics Olio, chapter 26). An unfortunate example of this is the V2 rockets of World War II.

Celestial guidance is based on looking at a star. It is used to update an inertial system by star sighting (see star trackers in chapter 26). It has been used independently of inertial guidance for the Apollo Lunar Excursion Module and other spacecraft[10].

Terrestrial guidance uses a combination of tracking the altitude pattern of the intended path and using map matching. Altitude maps are stored in the missile

[9] *Gizmag* Sept. 2014, online.

[10] Personal participation.

memory, and a radar altimeter makes the measurements to compare the data. Then, usually a map from the memory is compared to the image obtained with a camera.

22.3 Others

The Navy has fitted several ships with fiber optic cables for their communication needs. The cables are lighter than copper, have broader bandwidths, and are more secure. That is lightening with light.

Tanks now have night driving and aiming capabilities using infrared cameras.

The Navy has instrumented some ships with LED lighting to reduce the electrical noise and eliminate the potential hazard, as well as to save energy and extend the times between replacement.

IOP Publishing

Rays, Waves and Photons
A compendium of foundations and emerging technologies of pure and applied optics
William L Wolfe

Chapter 23

Mirrors—through one darkly

Undoubtedly the very first mirrors were placid ponds and lakes. We know the story of Narcissus admiring himself in such a mirror. Early manufactured versions consisted of obsidian glass from volcanoes. They were recovered from graves in Anatolia, in what is now Turkey, dating to about 6000 BC[1]. The ancient Egyptians and Mesopotamians polished copper to make mirrors, and the Indians of Central and South America even polished stones[2]. Probably the first metal-backed glass mirror was made in Sidon (now Lebanon) in the first century, but only later in the third century did they become at all common. Some say that Justus Freiherr von Liebig (1803–1873) was the first to make a metal backed mirror in 1835, but Pliny the Elder (23–79) mentions glass mirrors in the first century[3].

Most early mirrors were plane mirrors. However, Diocles (240–180 BC) described a parabolic mirror[4], and Claudius Ptolemy (c. 100–170) considered both convex and concave spherical mirrors[5].

A note about these words: the geometric figures parabola, hyperbola, ellipse, and sphere are all two-dimensional figures. Their three-dimensional counterparts are paraboloid, hyperboloid, ellipsoid, and sphere. Many optikers use these interchangeably—that is, they describe three-dimensional shapes by their two dimensional counterparts. When we say a mirror is a parabola, we know we really mean paraboloid!

Archimedes (287–212 BC) is said to have saved Syracuse during its siege in 214 BC by the use of shields. His soldiers lined up and all redirected sunlight to the ships that were laying siege[6]. This account is subject to argument. Some argue that the

[1] Albenda P 1985 *Notes Hist. Sci.* **4** 2.
[2] *Mirror History: Invention of Mirrors and its Origins*, online.
[3] Castro J 2013 Who invented the mirror *Live Science* online.
[4] Toomer G 1976 *Diocles, On Burning Mirrors* (Berlin: Springer).
[5] C Ptolemy, *Almagest*, online.
[6] Several online references, particularly the UnMuseum.

soldiers using parabolic mirrors would not work because the parallel light from the Sun would come to a different focus for each of them. But maybe they were plane mirrors. Two recent experiments were tried; one burnt, but the other did not. No matter, it is a good story.

The mirror for the Palomar telescope was the first that was coated by a vacuum technique due to John Donovan Strong (1905–1992).

A very useful development was that of so-called zero expansion materials for mirror blanks. These took trade names like **Zerodur** and **ULE glass**. Whereas Pyrex and other glasses have thermal expansion coefficients of about one part in a million per degree, these special formulations of crystals and glasses have coefficients of about one part in a billion. These low values help both the grinding and polishing and stability in use. Zerodur is a glass-ceramic of aluminum, silicon, oxygen, and lithium that was developed and then produced in 1968[7]. ULE glass was developed by Corning in the early 1960s[8].

Mirrors developed for anti-ballistic missile systems were sometimes made of **beryllium** since it is light and has a low atomic number; it is therefore relatively immune to the products of an atomic explosion. The problem was to coat them properly with a highly reflective material with a similar expansion coefficient. Their production was also problematic since beryllium vapor is highly toxic. Other candidates for this application were silicon carbide and various epoxies[9].

In 2000, Roger Angel (1941–) first generated a very **large glass mirror** under the football stands at the University of Arizona (figure 23.1). He did it by spinning borosilicate crown glass at a temperature a little above 1000 °C so that it could flow into a honeycomb structure[10]—at about the consistency of honey! He overcame the problem of thermal expansion of the finished blank by controlling the temperature with a flow of fluid in the cells. The large circular cell was filled with cylinders of hexagonal cross section. The glass chunks, about a foot in size, filled the cell. When they melted, they flowed between the cylinders and covered the top. These are now

Figure 23.1. Arizona 8-meter mirror blank.

[7] Wikipedia, online.
[8] Ibid.
[9] W Wolfe, personal experience.
[10] *The Arizona Daily Wildcat*, March 25, 2013.

the mirrors used in almost all the large telescopes. The degree of curvature of the paraboloid generated by the spin is determined by the angular velocity. After they have been spun and cooled, the final figure has to be ground and polished since some of them were off-center paraboloids[11].

One-way mirrors are not mirrors at all. They are really windows. They work as a mirror if the space on one side is brilliantly lit and the other is dark. Those who use these devices to observe patients, felons, or others, sit in a dark room while the subject sits in a brightly lit one. Some one-way mirrors have light coatings of reflective material like silver or aluminum. Sometimes they are called half-silvered mirrors, but they are more accurately called partially reflective-coated mirrors. A true one-way mirror, one that transmits light in one direction but not in the other, violates the second law of thermodynamics. I was reminded of this in a patent case, which is briefly discussed in chapter 7 on displays.

A very special one-way mirror system has been proposed for a Jewish **mechitzah** (figure 23.2), a wall that separates the men from the women in a Jewish temple[12]. One interesting design is shown in the figure. The so-called one-way mirrors are shown as the diagonals. The tops are brightly colored panels. The bottoms are black. So the men see the reflection of the bright patterns, which obscure the women, while the women see the men and the black background. It has been reported that such a device was planned to be installed at the Kotel, the Western Wall, in Jerusalem.

Figure 23.2. The one-way mirror system for a Jewish mechitza.

All astronomical mirrors are front-surface mirrors, while most vanity mirrors have a reflective coating on the back. This is because it is easy to make back-surfaced mirrors, and they remain protected. But they generate two images, one from the coating and one from the front of the glass. This is intolerable in scientific telescopes. So they suffer contamination and have to be re-coated every so often.

[11] Burge J 2013 presentation College of Optical Sciences, University of Arizona.
[12] Matzav.com, August 23, 2010.

IOP Publishing

Rays, Waves and Photons
A compendium of foundations and emerging technologies of pure and applied optics
William L Wolfe

Chapter 24

Nonlinear optics—outside the lines

This is the field of optics in which materials do not have responses proportional to the light that is incident upon, but rather they respond to the square or a higher power of its intensity. This gives rise to a number of useful phenomena, the most important of which is frequency doubling. Frequency tripling has also been observed, but I know of no important application of it.

The origin of nonlinear optics was at the University of Michigan in Ann Arbor, MI. Peter Franken (1928–1999) and his associates detected ultraviolet light at a wavelength of 347.1 nm coming from a quartz crystal illuminated by a ruby laser, radiating at 694.2 nm—half the wavelength, but twice the frequency[1]. Just a year later and about ten miles away in the Ford Scientific Research Laboratory in Dearborn, MI, Robert Terhune (1926–2014) demonstrated frequency tripling[2].

I think a reasonable way of thinking about this is with photons. Two photons combine, pooling their energy, adding their energy. Since the photon energy is $h\nu$, Planck's constant times the frequency, the sum of two photons has twice the energy and twice the frequency, since the constant is constant.

The main application of nonlinear optics today is doubling a laser frequency in order to get a different color—wavelength. Since it has become rather commonplace we shall see it in many applications in which higher frequency laser light is employed, perhaps with photolithography.

[1] Franken P *et al* 1961 *Phys. Rev. Lett.* **7** 118.
[2] Terhune R *et al* 1962 *Phys. Rev. Lett.* **8** 404.

doi:10.1088/978-0-7503-2612-4ch24

IOP Publishing

Rays, Waves and Photons

A compendium of foundations and emerging technologies of pure and applied optics

William L Wolfe

Chapter 25

Optical design—the rays on d'etre

Although the term 'optical design' can be understood to mean the design of any and all kinds of optical instruments, including for instance lasers, it also has the more restrictive meaning of designing lens and mirror systems. It is also called lens design, although it includes mirrors. Optical engineering involves taking an optical design and integrating it into an optical instrument like binoculars. The context of the present chapter is lens (and mirror) design. It must begin with the use of the laws of reflection and refraction and of geometry to obtain useful results. It also includes an understanding of the aberrations of optical systems since every good designer uses these in his design process, and the specification of what good means.

The origin of optical design is clearly unclear. Lippershey and Galileo had telescopes and microscopes, but they certainly did not do ray traces to design them. The early Venetian spectacle makers probably did about the same. Trial and error. The telescopes of Gregory, Cassegrain, and Newton were based on their knowledge of geometry and not lens design by ray tracing. Optical design may have started with Gauss, who understood focal planes and lengths, but maybe with Abbe and his German contemporaries, who understood optical aberrations. The characteristic functions and eikonals of Hamilton and Bruns were certainly instrumental in the advances. At about the same time that the Germans were starting real optical design, the English were doing the same largely through the efforts of Alexander Eugen Conrady (1866–1944), who emigrated to England from Germany. It was a laborious procedure until the advent of the computer in the 1950s. Then, many optikers used their knowledge of geometry and optics to create their own design programs. Systems have constantly improved whereby one can optimize just about anything in a design and even find the global optimum, although that is still a subject of investigation. These advances have enabled the design of many exotic systems that are multi-focal, asymmetric, anamorphic, anastigmatic, optimized, toleranced, tolerable, cheaper and harder to make—and easier to make.

doi:10.1088/978-0-7503-2612-4ch25

The detailed history of optical design is especially hard to trace. (Harder than rays.) Certainly Isaac Newton (1643–1727) knew some of it. He pointed out that a better lens was appropriate for refractive telescopes—a double meniscus[1]. James Gregory (1638–1675) and Laurent Cassegrain (1629–1693) both understood the geometry of conics, and they used this knowledge to combine a spherical or paraboloidal primary and an ellipsoidal[2] and hyperboloidal[3] secondary, respectively, in their telescope designs. Gregory's design was first built by Robert Hooke (1635–1703) in 1673. Newton's reflecting telescope with a parabolic primary and flat secondary came a little after Gregory's design but before Hooke's construction.

Carl Friedrich Gauss (1777–1855) was the one who in 1841 introduced the ideas of focal planes and focal lengths[4]. We still use the Gaussian, or paraxial, approximations for beginning designs—when the angles are small. James Clerk Maxwell (1831–1879), of Maxwell equations fame, and Ernst Abbe (1840–1905) showed[5] that aberrations were inherent in all optical systems. (They did not show that optical designers are aberrant!) William Rowan Hamilton (1805–1865) introduced his **characteristic function**, which was, essentially, the optical path. Heinrich Bruns (1848–1919) introduced the **eikonal**, which was, essentially, the same thing[6]. (A function by any other name....) Hamilton's path is based on Cartesian coordinates, Bruns on direction cosines. From either of these one can get the **Seidel or third order aberrations** of Philip Ludwig von Seidel (1821–1896), enunciated in 1856[7]. Another way to look at it is that the sines used in the geometric calculations can be replaced by their series. If only the first term is used, one obtains paraxial optics, with small angles and no aberrations. When the second term is used, one obtains the third-order aberrations (since the sine series starts as $x + x^3/3! + x^5/5! + \cdots$). Logically, there are fifth and higher order aberrations. The five third-order monochromatic aberrations are spherical aberration or sphericity, coma, curvature of field, distortion, and astigmatism. These are the monochromatic aberrations, occurring for a single wavelength of light. The first order chromatic ones are lateral and longitudinal chromatic aberration. They are due to the fact that the refractive index is a function of wavelength[8]. An alternative to the Seidel aberrations is the polynomial representation[9] of Frits Zernike (1888–1966). Seidel expanded the aberration function in a power series; Zernike did it with circle polynomials. The **Zernike polynomials** are especially useful with laser applications.

---

[1] Newton I 1704 *Opticks.*

[2] Gregory J 1663 *Optica Promota.*

[3] Cassegrain L 1672 *Journal de scavans.*

[4] Gauss C 1841 *Dioptrische Untersuchung* (Gottingen: Dieterich).

[5] Maxwell J 1856 *Philos. Mag.*; Abbe E 1893 *Grundzuge der Theorie der optischen instrumente nach ABBE* ed S Czapski and O Eppenstein (Leipzig: Nabu Press).

[6] Hamilton W 1828 *Trans. R. Irish Acad.* **15** 69; Bruns H 1895 *Abh. Kgl. Sachs. Ges. Wiss. Math-Phys.* **21** 323; Bruns H 1895 *Leipzig Phys. Math. Ber.* **21** 325.

[7] Seidel L 1856 *Astr. Nachr.* **43** 381.

[8] Buchdahl H 1968 *Optical Aberration Coefficients* (New York: Dover).

[9] Zernike F 1934 *Physica* **1** 689.

Optical design consists of arranging the lenses or mirrors so that they have proper curvatures and separations to accomplish good imaging, but good imaging has to be defined; there are a number of such criteria that describe whether a given optical system is good or not. Perhaps the first was the **Rayleigh criterion** of John William Strutt, the third Baron Rayleigh (1842–1919). The Rayleigh criterion states that two point sources are resolved if the maximum of the Airy pattern of one falls on the first minimum of the pattern of the other. It arises from the **Airy pattern**, which was first observed by John Frederick William Herschel (1792–1871) in 1828[10] and subsequently described theoretically[11] by George Biddell Airy (1801–1892). For a circular aperture, the pattern is a Bessel function of the first kind divided by its argument. Its first zero is at 3.83 and the resulting distance is $3.83\lambda/\pi d = 1.22\lambda/d$ where λ is the wavelength and d is the diameter of the aperture. For a rectangular aperture, the diffraction pattern is the sine divided by its argument and the result is simply λ/d. Airy patterns for a square aperture and a circular one are shown in the figure on the left with solid and dashed lines, respectively. The one in the middle shows the Airy patterns of two, point sources that are just resolved, and the one on the right shows their combined pattern (figure 25.1).

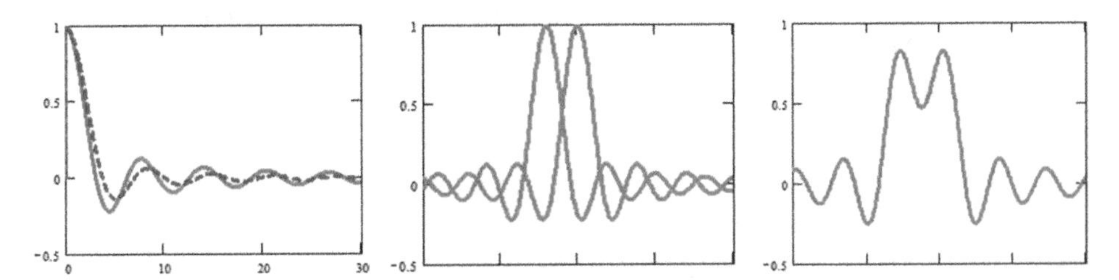

Figure 25.1. Airy patterns.

The **Strehl ratio** is based on the diffraction pattern of a perfectly corrected optical system. It was defined by Karl Strehl (1864–1940) as a measure of the performance of a telescope or other instrument imaging point sources[12]. It is the ratio of the actual height of the point source image to that of the height of the central maximum of the (ideal) diffraction pattern.

Still another performance measure is the **optical transfer function**. It is an optical analog to the electrical transfer functions and responses of audio systems, a plot of the ratio of the output to the input as a function of frequency. In optics, the frequency is the **spatial frequency**, usually specified in lines per inch. The ratio is usually the modulation, defined as the difference in amplitudes divided by their sum. The origin of this is not completely clear as it seems to have arisen in several places at

[10] Herschel J 1821 *Philos. Trans.* **111** 222.
[11] Airy G 1835 *Trans. Camb. Philos. Soc.* **5** 283.
[12] Strehl K 1895 *Z. Intrum.* **15** 362.

the same time. One source was Pierre Michel Duffieux (1891–1976)[13] in 1946. Its use was greatly furthered by Harold Horace Hopkins (1918–1994), Edward L O'Neill, Adolf Lohmann (1926–2013), and Joseph W Goodman in their works and books[14]. Fourier optics and the use of transfer functions was not universally accepted at first, although it is standard now (figure 25.2). One noted investigator, Giuliano Toraldo di Francia (1916–2011), stated, 'the eye is not the ear, and any interpretation of optics in terms of frequency is too farfetched'[15]. But we have now fetched it far. The figure shows two hypothetical transfer function curves. The blue one cuts off at a spatial frequency of 7 (pick your units); the other at 10; it still has an MTF of 50% at 7. This means that for the optics represented by the blue curve there is just a smear at 7 but there is still 50% modulation for the system represented by the red curve.

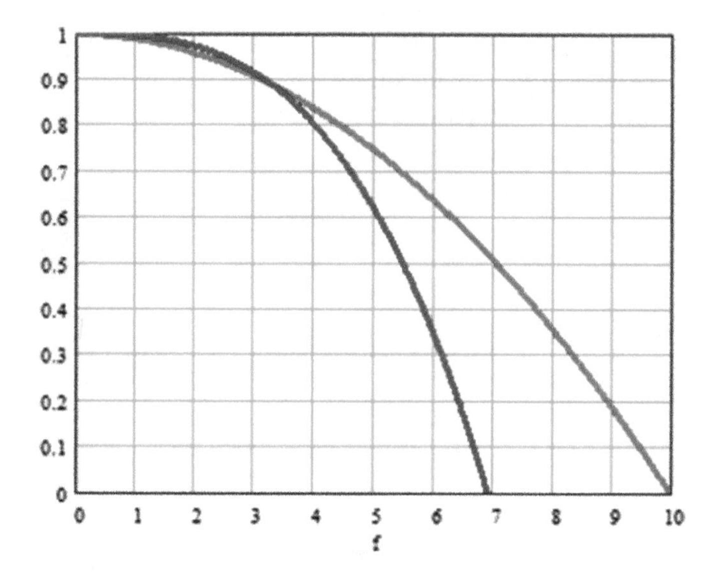

Figure 25.2. Representative MTF curves.

I learned how to trace rays in 1953 from a textbook by Jenkins and White[16]. It provided a table into which you could put entries from your calculations, and the calculations were logs and antilogs and logsines and arcsines, etc. The table below shows the procedure for tracing one ray through one surface! It took about 20 min for one ray and one surface. The entire table is shown here, but without any entries.

[13] Duffieux P 1946 *L'Integral de Fourier et ses Applicationa l'Optique* (Besançon: Faculté des Sciences).
[14] Oneill E 1968 *Optics*; Goodman J 1968 *Introduction to Fourier Optics* (New York: McGraw Hill); Lohmann A 2006 *Optical Information Processing* (Ilmenau: Universitätsverlag Ilmenau); Hopkins H 1962 *Proc. Phys. Soc.* **79** 889.
[15] DiFrancia G 1955 *Opt. Acta* **2** 51.
[16] Jenkins F and White H 1950 *Fundamentals of Optics* (New York: McGraw-Hill).

The message is that it was extremely laborious to calculate the position of even one ray through one surface! The fifteenth row of the fifth column shows the position of the ray after this refraction.

	$h = 1.5$	$h = 1.0$	$h = 0.5$	$h = 0$
log h				
colog r_1				
logsin φ_1				
log n				
colog n'				
logsin φ'				
φ_1'				
φ_1				
θ'				
colog sin θ'				
logsin φ_1'				
log r_1				
log $(r - r_1')$				
$r_1 - r_1'$				
r				
s				

A short time later in the '50's, some of my colleagues, at least those at Kodak and the Institute of Optics, were using Marchant calculators to carry out these tasks. It was still a laborious process, and it made the understanding of aberrations that much more important. They actually trained themselves to enter data with their left hands so they could pull the lever with their right and be faster[17] (like Bob Shannon, figure 25.3). Then came real computers[18]. A discussion of the development of the use of computers in lens design at the Institute of Optics is online and in a history of that institution[19].

[17] R Shannon, who did it.

[18] Feder D 1963 *J. Opt. Soc. Am.* **2** 1209; Wynne G 1963 *Appl. Opt.* 2 1223.

[19] Shannon R 2004 Three influential decades at mid-century - the Institute of Optics online and *A Jewel in the Crown, 75th Anniversary Essays, The Institute of Optics, The University of Rochester* ed C Stroud (Rochester, NY: Meliora Press).

Figure 25.3. Bob Shannon.

The computers of those days were central computers, like UNIVAC and ENIAC. One had to buy time on them and get in line, but it was well worth it. Thousands of rays could be calculated in a relatively short time. On some, the input was coded on IBM cards that were stacked on input trays[20]. As of this writing, one can have a program on a desktop or laptop computer that will do the job for you. It will trace thousands of rays; it will calculate the performance in terms of spot diagrams and MTF's and the like. It will even optimize surfaces and spacings based on given criteria. It makes the understanding of aberrations and their causes almost irrelevant—almost. There are over thirty such design and optimization programs from which to choose[21].

Three of the most popular programs are CODE V, OSLO, and ZEMAX. They have been reviewed[22] by Michael Hatcher back in 2003, and many changes have occurred since then, but that is a good starting point to compare them. ZEMAX was written by Ken Moore in 1990 while he was still a student at what is now the James C. Wyant College of Optical Sciences. It was written specifically to be used via the Windows operating system. It is therefore quite user friendly. The name was first just MAX, after his dog, but had to be changed because of patent or copyright infringements. CODE V was first released in 1975 by the people of Optical Research Associates, in particular Frank Harris. ORA was recently bought by Synopsis Optical Solutions Group. OSLO, Optical System for Layout and Optimization, was written by Doug Sinclair when he was a student at The Institute of Optics in the 1970s. It was a product of Sinclair Optics but is now owned by Lambda Research Corporation. Each of these systems is continually undergoing revision, improvements, and expansion. The visuals are improved, as are the optimization techniques, ease of use, and compatibility with CAD programs.

[20] Shannon R 2004 *A Jewel in the Crown* ed C Stroud (Rochester, NY: Meliora Press).

[21] Links-OpTaliX: Optical Design Software, on the net.

[22] Hatcher M *Design Software which Package do you Need?* Optics.org.

Automatically finding the global optimum is still a challenge. In some cases, the program leads you to a local optimum that is not the best design; the best is the global optimum.

Although optical design is very mature, we shall see better, more comprehensive global maxima programs and those which incorporate nano structures and freeform designs. As noted elsewhere, we can now design with nano structures many of the lenses and similar devices endemic to radar, like Luneburg lenses, for instance.

IOP Publishing

Rays, Waves and Photons
A compendium of foundations and emerging technologies of pure and applied optics
William L Wolfe

Chapter 26

Optics olio—mishmash

Olio, not oleo, a miscellaneous collection of unrelated things, a mishmash like my desktop. This chapter includes those optical things that just did not seem to fit anyplace else.

26.1 The greenhouse effect

Global warming has been in the news for the last several years. The greenhouse effect, that which goes on in a real greenhouse, is the input of solar energy in the visible where the glass is transparent and the trapping of the infrared radiation from the plants by the glass that is opaque in that region of the spectrum, beyond about 2 μm. The atmospheric **greenhouse effect** is similar but depends upon the transparency of the atmosphere in the visible and opacity in the infrared. The effect is mostly in the region from about 8 μm to about 14 μm, and it is mostly due to absorption by carbon dioxide and water vapor. The bottom curve in figure 26.1 shows the transmission of the entire atmosphere. The curves above it show the contributions of each of the constituents. Clearly, carbon dioxide and water vapor are the major players. Ozone occurs only in the upper atmosphere. Methane and carbon monoxide are minor players.

It, the atmospheric greenhouse effect, was first discussed by Jean-Baptiste Joseph Fourier (1768–1830) in 1824[1], who suggested that Earth would be far cooler if it did not have an atmosphere. That is true, and we would not know about it! John Tyndall (1820–1893) speculated about the changes in climate and determined that there were gasses in the atmosphere that could affect the climate, namely carbon dioxide and water vapor[2]. Svante Arrhenius (1859–1927) speculated in 1896 that this could give rise to global warming and it might be due to the creation of carbon

[1] Cowie J 2007 *Climate Change: Biological and Human Aspects* (Cambridge: Cambridge University Press); Fourier J 1827 *Memoire sur las temperatur globe terrere et des espaces planetaires* and a translation, online.
[2] Tyndall J 1873 *Heat Considered as a Mode of Motion* (New York: Appleton).

doi:10.1088/978-0-7503-2612-4ch26

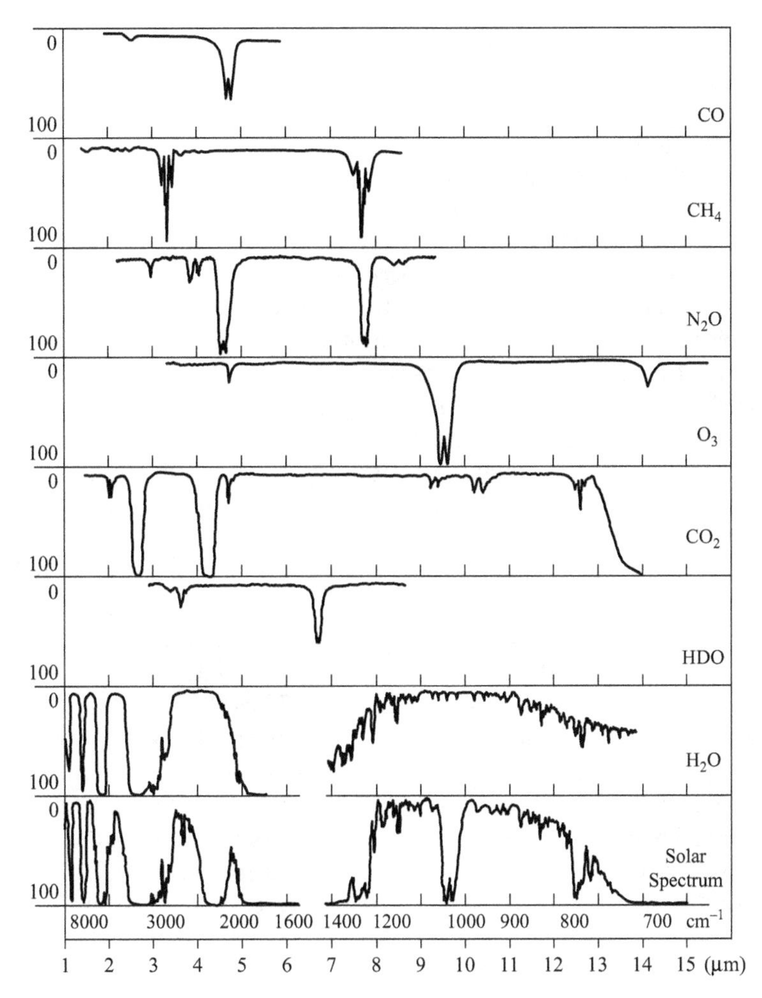

Figure 26.1. Transmissions of atmospheric gasses.

dioxide by humans[3]. E O Hulburt (1890–1982) did the same some years later[4]. There have been many studies and many opinions, too many to cite here. However, it seems clear that we have seen a recent increase in the global temperature, but it is not clear exactly how much is of human origin and how much is of a cyclic nature. Several studies show four cycles, the highs of which are just as high as they are now, and this was over many years, long before the Industrial Revolution[5].

The **Doppler effect**[6] is known as the red shift in optics and is invaluable in astronomy. It was first enunciated by Christian Andreas Doppler (1803–1853) in 1843. The relative frequency shift is approximately equal to the ratio of the source

[3] Arhennius S 1896 *Philos. Mag.* **41** 8.
[4] Hulburt E 1931 *Phys. Rev.* **38** 1876.
[5] Harris C and Mann R *Global Temperature Trends from 2000 BC to 2040 AD*; C3 Headlines, online.
[6] Doppler C 1843 *Gesellschaft der Wissenschaften* **2** 465.

(or receiver) to the speed of light, v/c, but it can get more complicated when relativistic effects need to be taken into account. The Doppler effect applies more broadly to other waves, notably acoustic ones including the change in pitch one hears from an approaching and receding train whistle. I find it curious that his publication is related to double stars—Doppler on Doppelsterne.

Searchlights are generally large parabolic mirrors with carbon arc sources at their foci. They have been used in warfare as early as 1882 by the British in their war with Egypt[7] and as part of anti-aircraft procedures. The Battle of Britain is a prime example of their use: the searchlights spotted the Nazi planes, and the anti-aircraft batteries took them out. Lighthouses are another example in which searchlights are combined with Fresnel lenses (see Lenses, chapter 17). Their use in fairs to attract people is well known and apparent in the night sky.

Retroreflectors are optical devices that return an optical beam back from whence it came no matter what their angular position is. A plane mirror is a retroreflector only if it is perpendicular to the incident beam. Retroreflectors that can handle beams at an angle come in two forms: cube corners and cat's eyes.

Cube corners, which are often called corner cubes, are literally the reflectorized corners of cubes (figure 26.2). They are often cubes with corners. So both appellations are appropriate. Their action may be understood by the fact that an incident beam undergoes three reflections, each of which reverses its direction. Cartesian coordinates (x,y,z) become $(-x,-y,-z)$. Thus the beam has been reversed in all three coordinates and therefore returns upon itself. One analogy I like is when you hit a racquetball into a lower corner of the court, it comes right back at you—and hopefully hits the floor before your opponent gets to it. The diagram shows just one dimension. They are used in surveying[8]; NASA put several on the Moon[9]; since then, many others have ranged against the Moon using retroreflectors. They have been proposed as modulating devices for free space communication[10].

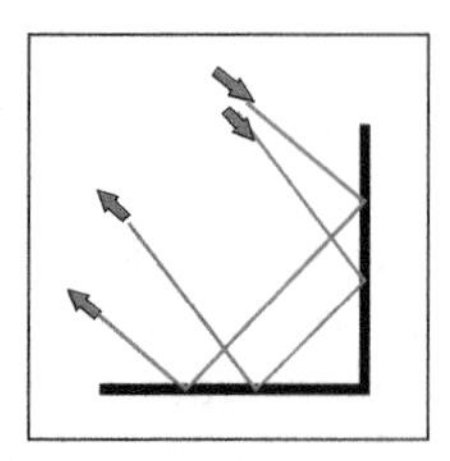

Figure 26.2. Cube corner.

Cat's eyes are refractive spheres such that the entering beam is directed by refraction to the focal point of the rear spherical surface, which is sometimes reflectorized. It then acts as a collimating mirror no matter which direction the

[7] Sterling C 2008 *Military Communications* (Santa Barbara, CA: ABC-CLIO).
[8] *Scientific American*, October 1990.
[9] Faller J *et al* 1969 *Science* **166** 99.
[10] Wikipedia.

incident beam comes from. If the refractive index is correct, a simple sphere will do, but otherwise it may have to be a cylinder with spherical ends. The cat's eye was certainly first invented by a cat—or its maker. The artificial one was invented by Percy Shaw (1890–1976) in 1934[11]. Figure 26.3 shows an incident horizontal beam and an oblique one from above. Recall that the focal point of a mirror is half the radius. Notice that both collimated beams enter the sphere and exit in the exact opposite direction. The use of these kinds of retros, beads, in traffic signs began in the 1920s but really made their mark in the 1930s when there was more traffic and smaller beads. Potters Inc.[12] used these in movie screens and road signs. A few years later, in 1937, 3M got into the act. By 1947, they had learned to cover the beads with a flat, transparent coating to protect them from the elements. In 1963 the Rowland brothers of Rowland Products introduced sheets of cube corners formed of plastic, each about 0.2 mm in size.

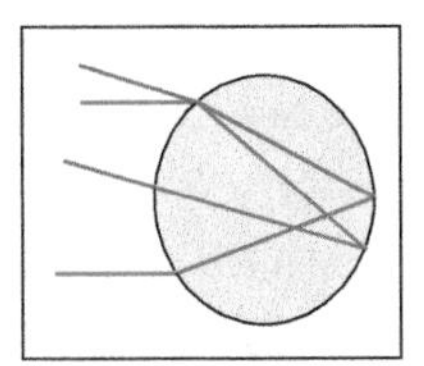

Figure 26.3. Cat's eye.

Horizon sensors are used on spacecraft as part of the navigation and stabilization system. You have to know which way is up! Most detect the emission of carbon dioxide radiation from the high-altitude atmosphere[13]. An astronaut viewing the Earth may see the ground or he may see high altitude clouds by virtue of their solar reflection. A simple, visible detector will have the same problem. But carbon dioxide has a constant mixing ratio in the atmosphere and extends relatively high in the atmosphere. It radiates copiously at about 15 μm. So investigations have been made, as have horizon sensors to use this technique[14]. The first horizon sensor was on the first Discoverer satellite, launched in 1959. Most of these sensors performed a dither scan around the periphery of the Earth disk that was generated by a scanning flat mirror (figure 26.4). Detection was accomplished by a thermistor bolometer, some with immersion lenses (see Infrared, chapter 13). A few used just a few points of crossing of the horizon. Some use an Earth balancing technique. In this case, the entire Earth is imaged on two sensors and their output is balanced—actually four to get both directions[15]. An interesting technique that was never used is the so-called inside-out horizon sensor. It uses a conical mirror to invert the image of Earth and then scans its inside edge (figure 26.5)[16]. A very different application of a horizon sensor is in an unmanned aircraft. In this, essentially the same techniques are

[11] British patent 436290 (1935).
[12] Lloyd J A *Brief History of Retroreffective Sign Face Sheet Materials*, online.
[13] DeWaard R and Weiner S 1967 *Appl. Opt.* **6** 1327.
[14] Duncan J *et al* 1965 Infrared horizon sensors *IRIA State of the Art Report* University of Michigan.
[15] Thomas J and Wolfe W 1969 *Spacecraft Earth Horizon Sensors*, NASA SP 8033; available online.
[16] Kilpatrick J 1962 *Appl. Opt.* **1** 147.

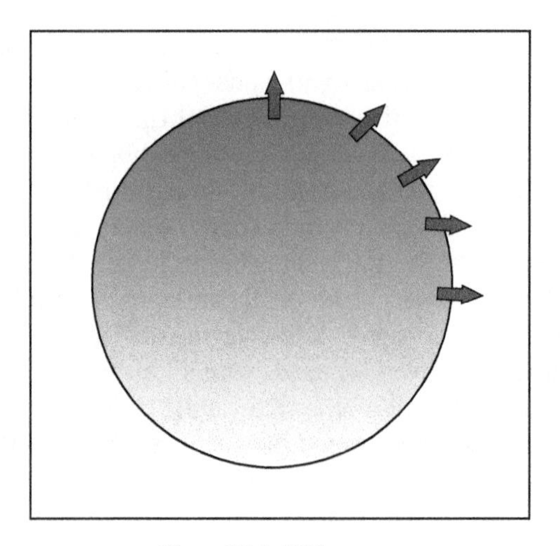

Figure 26.4. Dither scan.

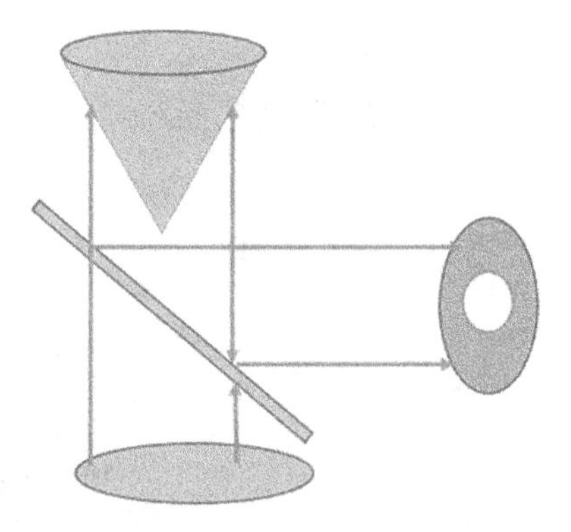

Figure 26.5. Inside-out horizon sensor.

used—the 15 μm band and balancing. But the idea is to keep the unmanned vehicle stabilized[17]. I guess that is what horizon sensors on satellites do, too.

The **optical mouse** is now used to control computers. A mouse in my day was a creature that you trapped, and spam was something you ate. But they both have their computer counterparts. We don't want spam, but we use the mouse to control our computers. The first versions were mechanical devices with roller balls that sensed where they were on a mouse pad. Then came the optical mouse. At first, I thought the optics would be used to communicate with the computer, but it is used for orientation on the mouse pad or other surface. They all use an optical source, usually an LED, to illuminate the surface and photodetectors to sense the reflected return. The first ones were demonstrated in 1980 by Steven Kirsch (1957–) and

[17] Taylor B 2003 Horizon sensor attitude stabilisation: a VMC auopilot *18th Int. UAV Systems Conf.* (Bristol, UK) online.

Richard Francis Lyon (1952–). The Kirsch design used a grid on the pad; the Lyon design used a rudimentary imager and tracked dots reflected from the pad[18]. Modern devices use imagers with about 400 pixels, frame skipping, automated learning, and image correlation techniques[19]. That is a very smart mouse! The light source is often an LED, but may also be a diode laser. The sensors are silicon. The laser provides somewhat better resolution.

For couch potatoes, the **TV optical remote** was a blessing—nay, a necessity. In fact, you could not be very good couch potato without it. Remote control was not new; it occurred, for instance, with model airplanes. The earliest remote control device was patented[20] by Nikola Tesla (1856–1943) in 1898 and guided a boat in a pool during an exhibition in Madison Square Garden. It is interesting that, although his patent deals essentially with radio waves, he was (perhaps) prescient enough to include other emanations which might include light. The first optical remote control for TV was developed by Eugene Polley (1916–2012) in 1955 called the Flash-Matic. It consisted of a flashlight in the remote and four photocells on the corners of the TV. It replaced a 'remote' that was a long wire and was followed by an ultrasonic system called a clicker—it clicked on bars of different lengths to make inaudible sounds of different, ultrasonic frequencies. This system was developed by Robert Adler (1913–2007); it was replaced by the infrared remote in the 1980s.

Every time we go to the grocery store, or almost any store, we encounter the point-of-sale **optical scanner**. In short, it is an optical device that uses a beam of light to scan over a barcode to record a sale and to update the inventory. There are several types: flatbed, wand, two-dimensional, and multiple detector. The essence is that the scanner reads the black-and-white pattern, which has information in it about the cost and type of the product or about a location. The first of these, in a somewhat different form, were colored plates on the sides of railroad cars to identify ownership and car type. These were used from about 1968 to 1978[21]. Although he did not invent the bar code, Alan B Haberman (1929–2011) is largely responsible for its adoption over dots and bullseyes and other patterns. It was first used commercially on a pack of Wrigley chewing gum in 1974[22]. In 1948, a graduate student at Drexel University by the name of Bernard Silver (1924–1963) and his friend Norman Joseph Woodland (1921–2012) heard a plea for automatic checkouts from the president of a large food chain. They pondered the issue for a while and extended Morse code dots to lines and used a 500-watt incandescent bulb and a phototube to scan the lines.

Photocopiers became popular and ubiquitous with the advent of the Xerograph system. Before that, there were other means of copying documents. Carbon paper and mimeographing were not optical. Photostats were copies made with a camera. These photostats were first used by George C Beidler (1878–1940) in 1906, and the

[18] Lyon R 1981 *VLSI systems and Computations* ed H Kung *et al* (Rockville, MD: Computer Science).

[19] Gordon G *et al* 2002 *US Patent* 6433780.

[20] Tesla N 1898 *US Patent* 613809.

[21] Cranstone I *A Guide to ACI (Automatic Car Identification)/KarTrak*, on the net.

[22] *New York Times*, June 15, 2011.

name photostat became generic[23]. He moved his operation to Rochester and was eventually purchased by the Haloid company. That company purchased the patent rights from Chester Floyd Carlson (1906–1968) for his electrophotography process. Carlson demonstrated his process in 1938, but it was not until 1942 that he got a patent[24]. The Haloid Company, later named Xerox, introduced this process commercially in 1959. A plate or drum with a photoconductive surface is electrostatically charged and then illuminated by the light reflected from the item to be copied. Since the surface is a photoconductor, the white areas conduct away the charge and the black areas retain it. The toner, essentially black carbon, is spread over the paper to be copied, and it sticks to the charged (black) areas. The last step is heating to fuse the carbon particles. I used an early version in the mid-50s. We had to flip the tray back and forth to get the toner to spread over the paper. They are much more automated now. Xerography (dry writing) has become a generic word like Frigidaire. The first electrostatic color copier was sold by Canon in 1973.

Digital copiers are gradually replacing those using the Xerox process. They use a combination of a laser scanner to obtain the digital version of an image and a laser printer to make the copy.

The **laser printer** is an adaptation of the Xerox copier. A laser scanner is used instead of the image and its illumination. It was developed by Gary Keith Starkweather (1938–) in 1969[25]. IBM came out with a kluge in 1976, the IBM 3800; a more compact version was produced by Xerox in 1981. Of course these early models were expensive, but economy of scale and Moore's law has reduced the costs considerably—about $100. Their main competition is the inkjet printer, which is not an optical device.

The 3D printer is an adaptation of these digital printers and thoughts that are associated with tomography in which layer by layer imagery is developed. Layer after layer of the image of an object is laid out until a true three-dimensional object is reproduced—over the airwaves in some instances.

Laser levels are a boon to both construction workers and do-it-yourselfers. They provide a level line all the way around a room or other site. One version is mounted on a tripod, and the laser shines on a mirror that spins. The straight-line version is useful for surveyors and even those in mines[26]. There must be an independent means for making sure the laser beam traverses a horizontal plane. A rotating laser level of the type described here was patented by Joseph F Rando (1921–2002) *et al* in 1977[27] and Steve Orosz in 1996[28]. It is also applied to construction and other uses. Mark Hunter's company claims that he invented it in 1987[29].

Laser gyroscopes are based on the Sagnac interferometer and are often called ring laser gyroscopes, or RLG's (see Interferometers, chapter 15). They were first

[23] Beidler G 1906 *US Patent* 810388.
[24] Carlson C 1942 *US Patent* 2297691.
[25] Reilly E 2003 *Milestones in Computer Science and Information Technology* (Westport, CT:Greenwood).
[26] Brod L 1976 *MS Thesis* University of Arizona.
[27] Rando J *et al* 1977 *US Patent* 4062634.
[28] Orosz S 1998 *US Patent* 5836081.
[29] *Spot on*, on the net.

described by Warren M Macek (1932–2017) and D T M Davis Jr in 1963[30]. They have remarkable performance numbers: 0.01 percent per hour bias uncertainty and 60 000 h mean time to failure. This is partly due to the fact they have no moving parts (except photons). The main competition for RLG's is the fiber optic gyroscope; the mechanical rotators of Elmer Ambrose Sperry (1860–1930) are now passé. The FOG, or **fiber optic gyroscope**, is a close cousin to the RLG. Instead of several mirrors forming the Sagnac interferometer, coils of fiber optics do. The more coils the better. The idea was first proposed by V Vail and R W Shorthill in 1976[31].

Photoelasticity is the phenomenon based on birefringence, stress birefringence. When a transparent body is put under stress, the crystal structure is changed in a way that produces birefringence. There are different refractive indices in different directions and different polarizations. These produce patterns of color by interference that reveal different areas and amounts of stress. This was first described by David Brewster (1781–1868) in 1815[32]. The technique is used mostly for visualizing stress patterns in various materials under a variety of different loads. The patterns you may see with sunglasses in your car's side windows is a result of intentionally induced stress in the glass. The glass is stressed by tempering it so that if it breaks it will go into zillions of small pieces and no (dangerous) shards.

[30] Macek W and Davis D 1963 *Appl. Phys. Lett.* **2** 67.

[31] Vail V and Shorthill R 1976 *Appl. Opt.* **15** 1099.

[32] Brewster D 1815 *Philos. Trans.* **299**.

Chapter 27

Optical societies—high society

Technical societies have done much to advance the field of optics by inviting the interchange of information, mentorships, and encouragement. The three main ones, in terms of membership, are OSA, the Optical Society of America; SPIE, the International Society of Optical Engineers; and ICO, the International Commission on Optics. The European Optical Society and the Infrared Information Symposia are also worthy of note.

27.1 OSA

The Optical Society of America was formed in 1916 in Rochester, NY, by thirty charter members with Perley G Nutting (1873–1949) as its first president. W Lewis 'Lem' Hyde (1919–2003) writes[1] that he believes that Charles Edward Kenneth Mees (1882–1960), Nutting's boss at Kodak, had a strong influence on its founding, believing in the free transfer of information. It had its first meeting at Columbia University under the auspices of the American Association for the Advancement of Science. Its journal, the *Journal of the Optical Society of America*, was first printed in 1918, when its membership had risen to 200; in 1922, OSA had its first exhibit. It bought a building in Washington, DC, that became its headquarters in 1960. It published *Optics—an Action Program* in 1963, a contributing progenitor of the College of Optical Sciences at the University of Arizona. Society membership

[1] Hyde W L 2006 *Opt. Photon. News* **19**.

doi:10.1088/978-0-7503-2612-4ch27

reached 4500 as OSA approached its golden anniversary in 1966. It launched its third publication, called *Optics News*, in 1975 and its fourth in 1977, called *Optics Letters. Optics News*, later in 1989 to be called *Optics and Photonics News*, became a monthly publication about various topics in optics written on a relatively non-technical basis. *Optics Letters* was established to publish short notes with relatively short time to publication. Membership rose to 10 000 in 1988. Two new 'publications' were initiated in 2012: the *Optics ImageBank* and *OSA Digital Archive.* Today, the society boasts almost 20 000 members in more than 100 countries all over the world, and accordingly changed its name to the Optical Society in 2008. It issues in print and online nine different publications.

OSA has had one executive secretary, one executive director, and one CEO to date: Mary Warga (1904–1991) from 1960 to 1969, Jarus Quinn (1930–2012) from 1969 to 1993, and Elizabeth Rogan since his retirement. There have been almost 100 different presidents over the years.

The society has over twenty different awards recognizing excellence in the various fields of optics. The primary award is the Frederik Ives Medal.

27.2 SPIE

SPIE. The international society for optics and photonics

SPIE was formed as the Society of Photographic Instrumentation Engineers in 1955. It has maintained this set of initials, although its optical reach has broadened considerably. It was originally a group of engineers who designed and operated the scientific photographic instruments, mostly for range instrumentation[2]. The first meeting was in the old Carolina Pines Restaurant in Hollywood, CA, with 74 in attendance. In 1964, 9 years after its founding, SPIE changed its official name to reflect the change in the optics industry; it became the Society of Photo-optical Instrumentation Engineers and expanded further into a national society. It has recently changed its name once again, this time to SPIE, the International Society for Optics and Photonics. By the 1980s, the society had taken in many members from outside the United States; it was on its way to becoming an international society. Robert R 'Bob' Shannon (1932–), the president in 1980 stated, 'Clearly there was a need that was met, and SPIE became an agreeable partner to many of the groups in Europe.' The first meeting in Europe was in 1983 in Geneva, Switzerland, with 1300 attendees from 26 different countries. This internationalization has had its positives and negatives. The word has spread; more people from all over interact. But at one point the US government almost shut down a meeting since it believed that too much classified information was being disclosed. In the end, only one paper was barred and the information exchange was deemed more beneficial than any disclosures.

[2] Anon 2005 *From Photography to Photonics, 50 Years of SPIE* (Bellingham, WA: SPIE Press).

The society now has numerous awards and scholarships. Its premier award, the Gold Medal, was inaugurated in 1977 and has been awarded every year since then.

The first executive secretary was E J Carr. Joseph Yaver (c. 1945–) became Executive Director in 1969 and lasted 24 years, until 1993 when Jim Pearson took his place. Joe presided over a great increase of membership, a move of the headquarters to Bellingham, WA, and expansion into Europe to make it a truly international society. Jim had been a vice president of United Technologies and head of its optical facility in Florida. He lasted just a few years until his wife's homesickness caused him to move back. In 1999, Eugene Arthurs took the reins as Executive Director of SPIE until 2019 when Kent Rochford became the Executive Director.

In 1998 and the years before and after, SPIE and OSA considered joining forces, becoming one society. There were many pros and cons, and even more individual opinions. OSA was a more or less traditional scholarly society with a longer history and established way of doing things. Some said stodgy. SPIE was more of an industrial society, more technical, more practical. But OSA had its *Applied Optics* journal that was almost an engineering journal, and SPIE certainly had many theoretical and scientific presentations. It would be more economical for practitioners to go to just one set of meetings, some argued. Others said that the separation of the disciplines was essential. Coordination was argued; finances were discussed; eventual directorships and officers were bandied about. Would the twain ever meet? The answer was no, and it was by a very narrow margin. As I recall, it was one vote. Today, both societies exist side by side and have cooperated on a variety of endeavors. They are both prospering, and optical engineers and scientists are the better for the split, I think. Many of us are members of both societies, and benefit from both. One of my books was sponsored by both societies.

SPIE has had an evolution and expansion of its publications. It is now widely known for the 'Yellow Books,' the unrefereed proceedings of its meetings. Although they are not reviewed, they have been a source of plenty of information on a timely basis. But reader beware. The first regular publication was cleverly titled 'SPIE-GLASS.' Just say it! Most Americans would say *spyglass*, but most Europeans would pronounce it *spee-glass*. Think American for the pun to work. The now monthly (and somewhat voluminous) publication of technical papers, the journal *Optical Engineering*, started in 1962. The SPIE Optical Engineering Press was established in 1988 in order to publish texts, monographs, and other scholarly works not contained in the journals or proceedings. This has turned out to be quite successful.

27.3 ICO

The International Commission on Optics was formed in 1948 by Abraham Cornelius Sebastian van Heel (1899–1966) for the purposes of diffusing the knowledge of optics throughout the world. It is an affiliate commission of the International Union of Pure and Applied Physics (IUPAP) and a scientific associate of the International Council for Science. It arranges and sponsors periodic meetings at member countries on an irregular basis. It also sponsors lectures and travel, and publishes a newsletter. Its focus is more scientific than engineering. It maintains member countries, but not individual members. It does not publish a regular journal, nor does it have an established headquarters.

27.4 EOS

The European Optical Society was founded in 1991 after a series of agreements and trials. The European Optical Committee existed before 1984, but it was not a true society. It became the optics division of the European Physical Society (EPS) in 1984, but was still not considered a scholarly society. In 1986, several countries formed the European Federation for Applied Optics, which they called Europtica. As part of SPIE's expansion internationally, it agreed with Europtica and EPS to hold one major optical meeting in Europe once a year. Then in 1990, the EPS agreed with Europtica to form the European Optical Society, EOS. And it happened in 1991. This was in part so it would not be subsumed by SPIE, and partly by European pride. The EOS today has more than 6500 societal, corporate, individual, associate, and student members from all over Europe and worldwide. It publishes an online journal, the *Journal of the European Optical Society: Rapid Publications.*

27.5 IRIS

The Infrared Information Symposia, IRIS, is a series of classified meetings that deal with infrared science and technology. It grew out of an organization called GMIR, or Guided Missile Infrared. Don't blame me if these acronyms sound funny. It was a group on the west coast that got together periodically to share information about infrared technology and mostly about infrared guidance systems. It was later sponsored by the Office of Naval Research (ONR). There was no official membership, and no set of officers. It was run by ONR, and most infrared workers in the industrial military complex attended meetings at one time or another. Venues varied. Attendance was typically 100–200, and specialty meetings on particular subjects like optical materials or detectors were often held. A proceedings was published of the papers presented. The director of IRIA (Infrared Information and Analysis Center) was the editor.

In the interest of full disclosure, I need to reveal that I was president of SPIE during its discussions with OSA for unification, and I was director of IRIA and editor of the publication. I was on the board of SPIE during many discussions with EOS. That is why there are no footnotes.

IOP Publishing

Chapter 28

Polarization—some circular reasoning

Light is polarized. Today we know that light is a transverse electric and magnetic field so that if light is traveling in the *z* direction, the electric and magnetic components are in the *x* and *y* directions—perpendicular or transverse to the direction of motion. Light can be plane polarized, circularly polarized, or elliptically polarized in general. That is, the electric field can be oriented in the *y* direction, for instance, or in the *x* direction, or at an angle to either; or it can describe a spiral in the *x*–*y* plane as the wave moves along. By convention the orientation of the electric field determines the state of polarization. The figures show light traveling from left to right with the electric and magnetic fields for two different polarizations (figure 28.1).

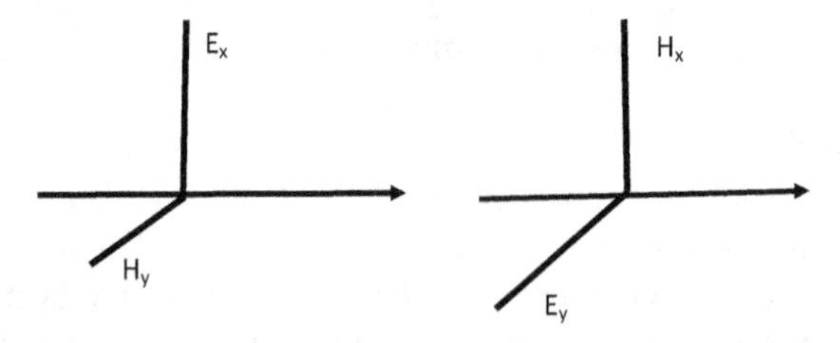

Figure 28.1. Representation of vertical and horizontal polarization.

One theory supposes that the Vikings may have discovered it by viewing things through Iceland spar (a crystal of calcium carbonate) that was abundant in the northern climes.

The first established recognition of optical polarization was by Rasmus Bartholin (1625–1698) in 1669. He wrote about his observations of two images being generated by propagation through the same crystal. They are polarized perpendicularly to each other[1].

[1] Bartholin E 1669 *Experimenta crystalli islandici disdiaclastici quibus mira & insolita refractio detegitur* (Copenhagen ('Hafniæ'), Denmark: Daniel Paulli); Cheshire F 1923 *J. R. Microsc. Soc.* **262**.

doi:10.1088/978-0-7503-2612-4ch28

Isaac Newton (1642–1727), in his famous book on optics in 1704, noted that his optical corpuscles have sides[2]. (For his theory, see Light, chapter 18).

Étienne-Louis Malus (1775–1812) also experimented with Iceland spar in 1808. Where would we be without that crystal? He noted that the reflection of the sun from a nearby window changed in intensity when he rotated the crystal through which he was looking. He also determined that polarization could be introduced by other non-metallic materials upon reflection, and he derived the law that predicted the change in transmission with position[3].

David Brewster (1781–1868) in 1815 repeated some of the experiments of Malus and finally obtained what is now called the **Brewster relationship:** the angle of complete polarization is the inverse tangent of the refractive index, that is arctan(n) or $\tan^{-1}(n)$[4].

Dominique François Jean Arago (1786–1853) first noticed the colors that can be generated by the interference of polarized light in 1811[5], and Jean-Baptiste Biot (1774–1862) published about this and the fact that the sky light is polarized[6]. He also noted **circular polarization**. One of the problems with polarization and the transverse nature of light at that time was how to explain unpolarized light. In 1821, Augustin-Jean Fresnel (1788–1827) realized that **unpolarized light** was light in which the direction of polarization changed very fast; it was randomly polarized[7].

Fresnel derived in 1827 the equations for the transmission and reflection of polarized light, the famous **Fresnel equations**. They are far too complex to present here; the interested reader is referred to appendix E. I was surprised to find that he did this long before the electromagnetic theory described by James Clerk Maxwell (1831–1879). They can be used, with proper simplifications, to derive Snell's law, Brewster's angle, the critical angle, and the degrees of polarization of the reflected and transmitted beams from both dielectrics and metals[8].

Optical activity, or optical rotation, the rotation of the direction of polarization, was observed by Biot and Louis Pasteur (1822–1895)[9]. Rotary polarization was also observed by Arago and Herschel.

George Gabriel Stokes (1819–1903) developed the technique of describing **polarized light by a vector representation** in 1852[10]. The four parameters are obtained by experimentation and are usually designated by S with subscripts 1–4, or I, Q, U, and V. S_0 is I_0 where I_0 is the total intensity of light. S_1 is $I_1 - I_0$, where I_1 is the intensity of horizontally polarized light. S_2 is $I - I_0$ where I_2 is the 45° polarization. Finally, S_3 is $I_3 - I_0$, where I_3 is the left-handed circularly polarized light. Some multiply these by two, but who cares right now. Suffice it to say that four simple

[2] Newton I 1704 *Opticks*.

[3] Malus E 1809 *Journal l'Ecole de Polytechnique* **8** 219.

[4] Brewster D 1815 *Philos. Trans. R. Soc.* **105** 125–59

[5] Hellemans A and Bunch B 1988 *The Timetables of Science* (New York: Simon and Schuster).

[6] History—Polarized Light, online.

[7] Lowry M 1935 *Optical Rotatory Power* (New York: Dover) and online.

[8] Born M and Wolf E 1959 *Principles of Optics* (Oxford: Pergamon).

[9] Hecht E 1998 *Optics* (Reading, MA: Addison-Wesley).

[10] Chandrasekhar S 1960 *Radiative Transfer* (New York: Dover).

measurements of the polarization states of light provide the information to set up the Stokes vector.

Jules Henri Poincaré (1854–1912) devised a way to represent the state of polarization on a sphere, now called the **Poincaré sphere**[11]. Any position on the sphere represents a particular and unique state of polarization.

Hans Mueller (1900–1965) at MIT developed a technique in 1943 by which one could calculate the degree and type of polarization generated by various optical elements using the Stokes vector and what is now called the **Mueller calculus** using the Mueller matrix[12]. He did this as a possible method of using secret code for the military. The Mueller matrix is a 4 × 4 element matrix that operates on an input Stokes vector to obtain an output Stokes vector. The elements determine the operation of the polarizer.

At the same time, Robert Clark Jones (1916–2004), then working at the Polaroid Corporation, developed an alternate method[13]. He was motivated by Edwin Land (1909–1991), who thought that it would be important for the entire field of polarized light. And Polaroid and Land were leaders in that field. The **Jones matrix** is a 2 × 2 element matrix with complex numbers.

It is interesting to note that the two of them worked completely independently—both in Cambridge, MA—at the same time until Jones attended a lecture by Mueller and discovered that they were both working on the same problem and came up with equivalent but different solutions. Mueller's calculus uses 4 × 4 matrices of real numbers, whereas Jones' uses 2 × 2 matrices of complex numbers. The two are indeed comparable[14]. I believe that you determine the elements of the Mueller matrix by trial and error, by trying, for instance, the elements that give an input vector an appropriate output vector—like the input is unpolarized and the output is vertically linearly polarized. Jones told me that you find the elements in his matrices by inspection; they are obvious. Maybe to him!

Polarimetry is the determination of the state of polarization of a light beam.

One important application of polarized light is the area of ellipsometry, so named since it uses elliptically polarized light. It was first recognized by Paul Karl Ludwig Drude (1863–1906) in 1888[15]. The process is to shine light of a known polarization onto a sample, usually a thin film, and then measure the polarization state of the reflected or transmitted beam. The change in state is a measure of the properties of the material and can be analyzed by either the Mueller or the Jones matrices. The real work comes after these have been obtained, determining the properties of the material that give rise to these polarization properties. Ellipsometry can be done at a single wavelength or over a spectral region.

---

[11] Poincaré H 1892 Theorie Mathematique de la Lumière (Paris: G. Carré).

[12] Mueller H 1943 Memorandum on the polarization optics of the photoeleastic shutter *Report of the OSRD Project OEMsr 576*; Parke N 1948 Matrix optics *Thesis* MIT.

[13] Jones R 1941 *J. Opt. Soc. Am.* **31** 488.

[14] Wolfe W (ed) 2001 *Optical Engineer 's Desk Reference* (Bellingham, WA: SPIE).

[15] Drude P 1889 Über die Gesetze der Reflexion Dissertation University of Göttingern.

IOP Publishing

Chapter 29

Prisms—an acute discussion, not obtuse

Optical prisms are mainly of two different types, dispersive and non-dispersive. The former disperse light into its component colors; the latter are specifically designed to avoid doing this. The dispersive prisms are used in spectrometers and are discussed in that chapter. Non-dispersive prisms are used to invert, revert, and otherwise redirect the path of a beam of light. That is, they either flip the image top to bottom or they switch the image from left to right. (They are not like the old auto ad that the key works either upside down or downside up; there is no either in that.) They are notably used in many binoculars, and that application is discussed in that chapter. They are also used as polarizers and beam steerers.

Contrary to popular belief, a prism is not necessarily a three-sided figure. Prisms can be quadrangular, as well as many other interesting shapes. A prism has three or more flat sides that extend in one direction. The ends usually dictate the name. Shown below are three prisms (figure 29.1). The one on the left is a familiar right-angle triangular prism. The next is a square, quadrangular prism, and the last is an arbitrary one I made up just to show how wild they can be. Later in this chapter all sorts of geometries and combinations of prisms will be shown that were all meant to accomplish some sort of optical function.

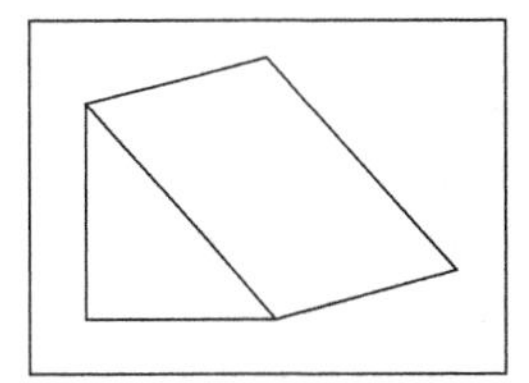
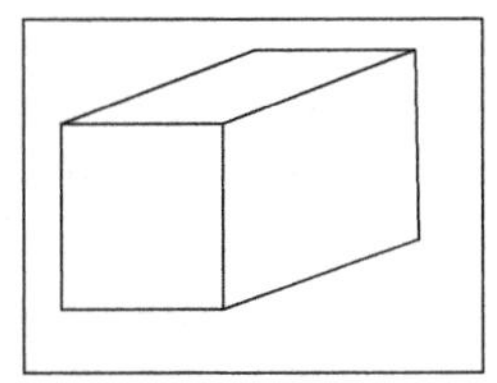
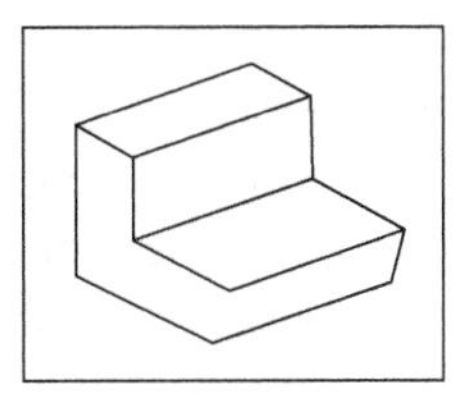

Figure 29.1. Three examples of prisms.

doi:10.1088/978-0-7503-2612-4ch29

29.1 Dispersive prisms

The **Abbe constant deviation prism** is a 30–60–90 prism, as shown in figure 29.2. It deviates a selected wavelength of light at exactly 60°. It was first devised and reported on in 1874[1] by Ernst Abbe (1840–1905). The angles are, clockwise, 90, 60, and 30. The light comes in from the left, is refracted, then internally reflected, and finally refracted out. The different colors are not shown. There is a band of them surrounding the selected one, all at constant deviation.

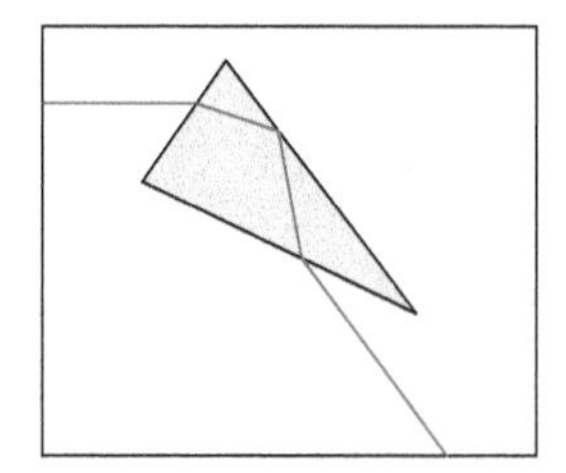

Figure 29.2. Abbe prism.

The **Pellin–Broca prism** is also a constant deviation prism, this time 90° (figure 29.3). It was invented by François Philibert Pellin (1847–1923) and Elie André Broca (1863–1925) in 1899[2]. It is not a triangular prism, but quadrangular and quadrilateral, as shown. It is a 90, 75, 135, 60 degree four-sided prism. It could be just triangular, but it would be a very loooooooong prism. There will be a similar band of colors about the design wavelength.

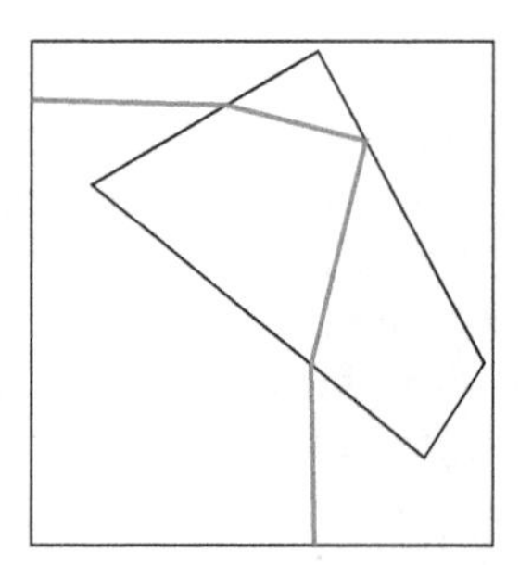

Figure 29.3. Pellin–Broca prism.

The **Amici prism** (figure 29.4), not to be confused with the Amici *roof* prism, consists of two prisms in contact. The front prism is usually made of crown glass, the second of flint. It is designed so that the central wavelength beam is undeviated but displaced, so that it is a direct view spectral disperser. Giovanni Battista Amici (1786–1863) realized a little later, in 1860, that he could 'double down.' He could use

[1] Abbe E 1874 *Jenaische Z. Naturwiss.* **8** 96.
[2] Pellin P and Broca A 1899 *Astrophys. J.* **10** 377.

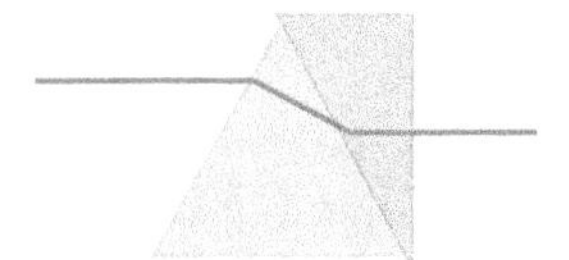

Figure 29.4. Amici prism.

two of these back to back, thereby returning the input beam parallel to itself but displaced[3].

Compound prisms consist of things like double Amici's, triplet Amici's and other variations on the theme[4].

A **grism** is a combination of a grating and a prism. The grating is ruled or etched on one face of the prism. They have been used in the NICMOS (Near Infrared Camera Multi-Object Spectrometer) instrument in the Hubble Telescope for imaging and determining spectra of celestial objects—prior to its correction[5]. Grisms are normally dispersive non-deviating devices.

29.2 Non-dispersive prisms

There are many, many non-dispersive prisms designed to do different things to different beams of light. They can deviate, displace, invert, revert, and even send the beam back to exactly from whence it came. The secret to keeping a prism from dispersing the light is to have the rays impinge on every interface perpendicularly or reflect internally. Probably the most common of these is the **Porro prism** after Ignazio Porro (1801–1875) (figure 29.5). It is a simple right-angle triangular prism that inverts and reverts the incoming image.

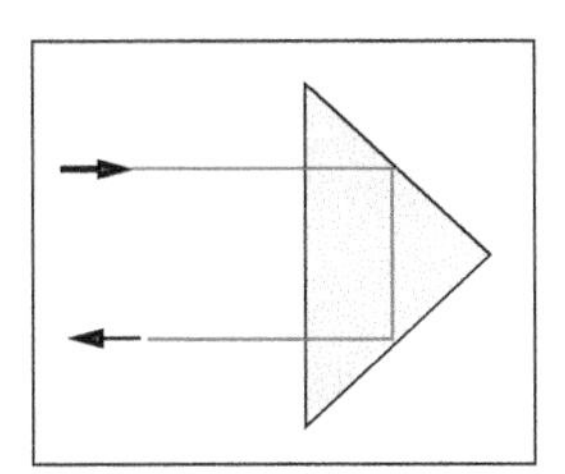

Figure 29.5. Porro prism.

[3] Donati G 1862 *Nuovo Cimento* **15** 292.
[4] Browning J 1871 *Mon. Not. R. Acad. Sci.* **31** 203.
[5] Storrs A and Bergeron L *Instrument Science Report*, NICMOS 97-027; Storrs A and Bergeron L *Hubble Space Telesope NICMOS Grisms* (Boulder, CO: Space Science Institute).

29.3 Porro prism

The **Amici roof prism** after Amici may be considered a simple right-angle prism with an extra reflection formed by a roof-like back surface. This extra reflection on the roof maintains the left-rightedness, but still inverts the image. Inverted, but not reverted (figure 29.6).

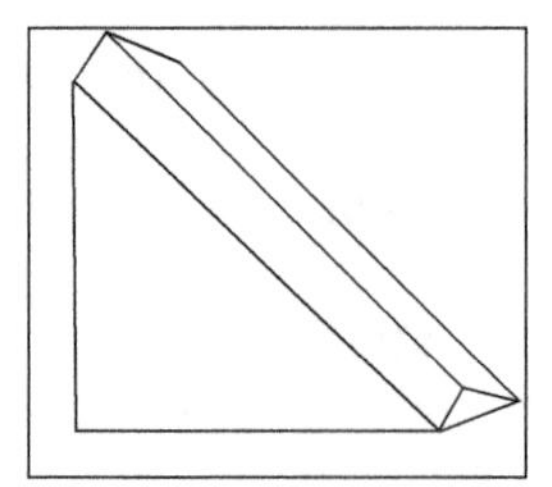

Figure 29.6. Amici roof prism.

The **pentaprism** uses just three of its sides for reflection, and those sides need to be reflectorized since the critical angle for internal reflection is not attained. It does not invert or revert, but it does rotate the image 90°. There is also a version with a roof, and a mirror analog (figure 29.7).

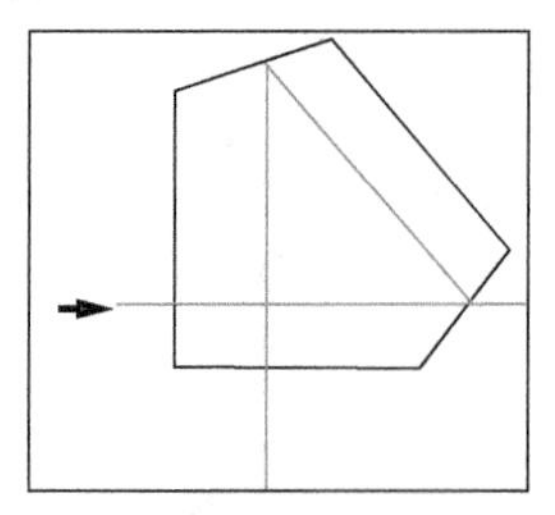

Figure 29.7. Pentaprism.

The **Dove prism** inverts and reverts, as well as rotating the image at twice its own rotational rate. It was described by Heinrich Wilhelm Dove (1803–1879) in 1851[6] but patented in 1838 by Delaborne (figure 29.8)[7].

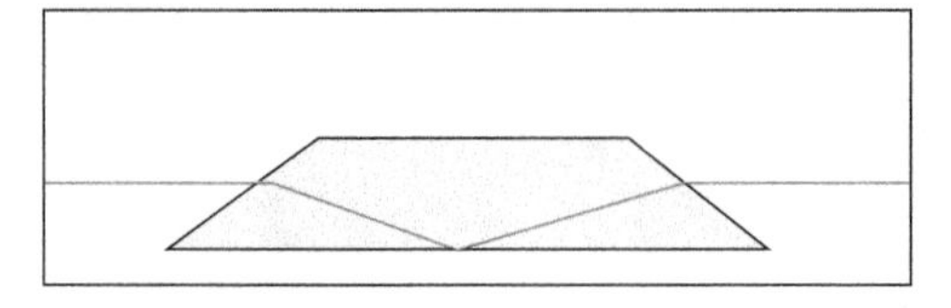

Figure 29.8. Dove prism.

[6] Dove H 1853 *Pogg. Ann.* **83** 189.
[7] Delaborne 1838 *French Patent* 5941.

29.4 Polarizing prisms

The **Nicol prism** was the first of the polarizing prisms, invented by William Nicol (1770–1851) in 1828. It is made of Iceland spar, a form of calcite that is birefringent (has different refractive indices for different polarizations of light) (figure 29.9). Light enters the rhomboid from the left. (This is a real quirk of optical people; light always travels from left to right!) It is then divided into the two polarizations, the ordinary O-ray and the extraordinary E-ray. The O-ray is reflected at the interface by total internal reflection, while the E-ray passes through. The two sections are joined by an appropriate glue, like Canada balsam, shown as the dashed line. The ordinary ray has a refractive index of 1.658 and the glue of 1.55. So the O-ray must impinge on the interface at an angle of 69°; the E-ray index is 1.486, so it will not be reflected. (I have not shown the refractions at the surfaces).

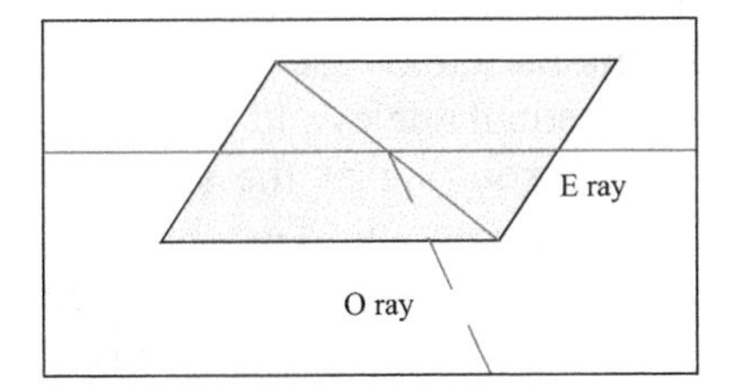

Figure 29.9. Nicol prism.

The **Wollaston prism** was invented by William Hyde Wollaston (1766–1828) (figure 29.10). It also consists of two calcite prisms cemented together, originally with Canada balsam. In this, however, both the O-ray and the E-ray exit the prism, but in different directions. The light is shown entering from the left (by convention). The exiting ray shown are horizontally polarized; the other vertically.

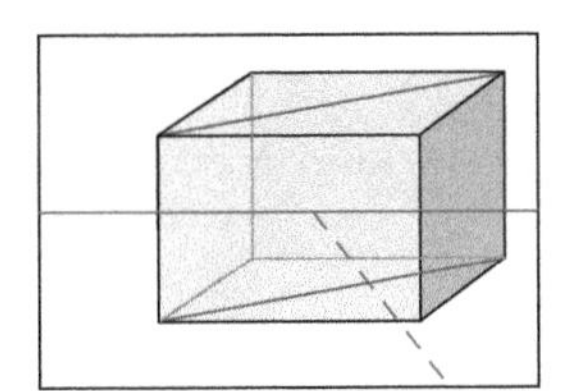

Figure 29.10. Wollaston prism.

A number of other polarizing prisms are variations of the Nicol, with different geometries, arrangement of the polarizing axes, and an air space rather than glue. These have largely replaced the Nicol prisms and include the **Glan–Thompson**, **Glan–Foucault**, **Rochon**, **Sénarmont**, and **Nomarski**. They were invented by P Glan in 1880 as an air-spaced design[8] and Silvanus Phillips Thompson (1851–1916) with a

[8] Glan P 1880 *Carl's Repert.* **16** 570.

cemented version a year later[9]. The Foucault variety of the Glan prism was invented by Jean Bernard Léon Foucault (1819–1868) in 1857[10]; it uses an air space. Alexis-Marie de Rochon (1741–1817) invented the prism now named after him in 1783[11]. The Sénarmont prism, after Henri Hureau de Sénarmont (1808–1862) is very similar to the Rochon[12]. The Nomarski prism after Georges Jerzy Nomarski (1919–1997) is a variant of the Wollaston prism.

There are many other varieties that were all cataloged in a publication by the staff at Frankford Arsenal and in the *Handbook of Optics*[13].

Risley prisms have been used in two ways: as variable attenuators and as optical scanning elements. As variable attenuators, two prisms with modest absorption are slid across each other to generate different total thicknesses and a resultant change in transmission. They have also been used for over 125 years for scanning and beam steering applications[14], although their origin is unclear. The two prisms are rotated at different rates and in different directions (clockwise and counterclockwise) to generate a variety of different patterns. The first pattern, a simple circle, shown below is for both prisms having equal angles and rates; the phase doesn't matter. The second is for two equal prisms rotating at the same rate in opposite directions, starting with their apices both horizontal. The third is the same, but with them starting vertically. The next pattern, a rosette, is for two equal prisms but the second rotating at ten times the first and starting horizontally. If they start at any other place, the pattern is rotated. The next pattern results from one prism having 0.3 times the deviation of the other and rotating at 20 times it. Again, a phase shift rotates the pattern (figure 29.11).

Figure 29.11. Risley prism scan patterns.

[9] Thompson S 1881 *Philos. Mag.* **5** 349.
[10] Foucault L 1857 *C. R.* **45** 238.
[11] West C and Jones R 1951 *J. Opt. Sci. Am.* **41** 976.
[12] Bennett J 1995 Polarizers *Handbook of Optics vol II* ed M Bass (New York: McGraw-Hill) ch 3.
[13] Arsenal F 1952 *Design of Fire Control Optics* (Washington, DC: US Govt Printing Office); Wolfe W 1996 Non-dispersing prisms *Handbook of Optics vol II* ed M Bass (New York: McGraw-Hill) ch 4.
[14] Carter W and Carter M 2006 *Eos, Trans. Am. Geophys. Union* **87** 273; Wolfe W and Zissis G (ed) 1978 *The Infrared Handbook* (Washington, DC: US Govt Printing Office).

IOP Publishing

Rays, Waves and Photons

A compendium of foundations and emerging technologies of pure and applied optics

William L Wolfe

Chapter 30

Radiometry and photometry—a precise and accurate treatment

These two topics are closely related. Radiometry is the measurement and calculation of the amount and distribution of radiation whether it be visible, infrared, ultraviolet, or even radio waves. Photometry deals with the same subjects but only for visible quantities. Thus, most of the concepts are applicable for both, but a special language has been constructed for photometry. Photometry was the forerunner of radiometry because people could see the light.

30.1 In the beginning

Perhaps the beginning was Hipparchus of Nicaea (~190–120 BC), who made a list of almost 900 stars according to their brightness and position in about 129 BC, but these were measurements by eye. The eye is a notoriously poor absolute measuring apparatus, but it makes comparisons very well. So this was a list of comparative measurements. Claudius Ptolemy (~100–170) noted Hipparchus' observations and followed him[1]. Perhaps the foundation of photometry was with **Pierre Bouguer** (1698–1758) who in 1725 operated a comparison photometer that was described later[2]. A little later, **Johann Heinrich Lambert** (1728–1977) enunciated some basic laws of photometry[3]: the inverse square law, cosine law, irradiance addition law, etc. **Count Rumford**, Sir Benjamin Thompson (1753–1814), developed a photometer of similar type that was also a comparison instrument[4]. It was not until later that a semi-quantitative measurement was made by way of the visual-comparison photometer by Carl August von Steinheil (1801–1870). Using a

[1] C Ptolemy, *Almagest*, (170).

[2] Bouguer P 1729 *Traité d optique sur la gradation de la lumiére* (Paris: Jombert).

[3] Lambert J 1892 Photomethasive de mensura et gradibus luminus colorum et unbrae *Klassiker d. exacten Wissenschaften* ed E Anding (Leipzig: W. Engelmann).

[4] Rumford C 1794 *Philos. Trans.* **84** 67.

doi:10.1088/978-0-7503-2612-4ch30

split optical system, he could make adjustments until the (star) image in question appeared to be exactly equal in brightness to a reference[5]. Friedrich Zöllner (1834–1882) improved this type of photometer by using a kerosene lamp as a standard and a pair of polarizers to adjust its intensity on the screen next to the image by cross-polarizing them an appropriate amount[6]. Perhaps the first objective or absolute, rather than comparison, measurement was made by Wilhelm August Eberhard Lampadius (1772–1842) in 1814 when he used a series of semi-opaque sheets to attenuate the image until it disappeared[7]. Then a certain image might be declared three sheets to the…! A number of different comparison-type photometers were developed in the late 1800s and were adapted to use with a reference source of light, which was initially the standard candle introduced by Count Rumford in about 1795[8].

30.2 Standards

The first photometric standard candle was literally a **sperm candle** of a given size, burning at a given rate, but it was still not entirely reproducible or consistent. It was later improved many times by the **Carcel lamp**, the **pentane lamp**, and the **Hefner lamp**. The Carcel lamp[9] burnt a specific oil with a wick fed at a controlled rate. The pentane lamp[10] burnt (yes) pentane without a wick but with a controlled rate of feed of the gas. The Hefner lamp[11] again used a controlled feed but of a volatile fluid, along with a lens. The source, using a solid wick and burning amyl acetate, was invented by Friedrich Franz von Hefner-Alteneck (1845–1904) in 1884 and used with other fuels by several others[12]. It was understood, and it is easy to understand, that flame sources are inherently unstable due to changes in feed rate, humidity, air turbulence, and the like. Accordingly, there was a search for an incandescent metal standard. The first was molten platinum—in fact, melting platinum. The temperature of solidifying or melting (pure) platinum is well known, so the uncertainty comes largely from its composition, its purity. This standard became known as the ***bougie décimate***. The modern standard is not much different from this. In 1946, the CIE, the International Commission on Illumination (or *Commission internationale de l'éclairage)*, defined the ***new candle*** as radiation from a blackbody of a certain size at the temperature of freezing (solidifying) platinum and is realized by a blackbody simulator cavity. Then the *new candle* became the ***candela*** in 1948.

The standard candle is maintained by NBS (now NIST) by law. Its realization is a long, thin cylinder truncated in a cone and surrounded by melting platinum, which is

[5] Steinheil C V 1836 Elemente der Heiligkeitsmessimgen am Sternhimmel: Denschrift. der Königliche Bayer *Akademie der Wissenschaft, Mathematische-physik, Classe II, München.*

[6] Zöllner J 1861 Grundzüge einer allgemeinem Photometrie des Himmels (Berlin: Mitscher & Röstell).

[7] Lampadius W 1814 *J. Chem. Phys.* **11** 361.

[8] American Academy of Arts and Sciences 1876 *The Complete Works of Count Rumford* vol 5 (London: Macmillan).

[9] Carcel B 1802 *Ann. Des Arts Manuf.* **6** 269.

[10] *NBS Bulletin* 10, 391 (1914).

[11] Hefner F 1883 *ETZ* **4** 445.

[12] *Centalblatt für das deutsche Reich* 20, 124 (1893); *J. Gas Lighting*, 64 (1894); etc.

surrounded by fused thoria that is in turn surrounded by unfused thoria, the latter two for insulation. As far as I know, it was last used in 1931 to calibrate a set of secondary tungsten standards. The first use of such a standard was that of Coblentz in 1915[13]. The standard candela was replaced in 2019 by the General Conference on Weights and Measures as a definition:

> The candela [...] is defined by taking the fixed numerical value of the luminous efficacy of monochromatic radiation of frequency 540×10^{12} Hz, K_{cd}, to be 683 when expressed in the unit lm W^{-1}, which is equal to cd sr W^{-1}, or cd sr kg^{-1} m^{-2} s^3, where the kilogram, metre and second are defined in terms of $\Delta\nu_{Cs}$.

The gold point standard is a similar cavity operated at a lower temperature and useful in the infrared[14]. An alternative to standards based of radiation is a standard detector. Jon Geist developed such a one based on a silicon detector with a determinable responsivity[15].

30.3 Luminous efficacy (the eyeball curve)

The terms in photometry are based in part by the relative luminous efficiency or luminosity function of the human eye, dubbed the luminous efficacy. It is shown in figure 30.1 and points out that we are most sensitive to yellow light at about 550 nm. A lumen[16] might be considered a light watt. Related units are the candela, lux, and nits. The luminous efficacy curve is therefore important in photometry. The human eye has a different spectral response in dim light than it does in bright light—the scotopic (dim light) and photopic (bright light) responses, respectively—and different

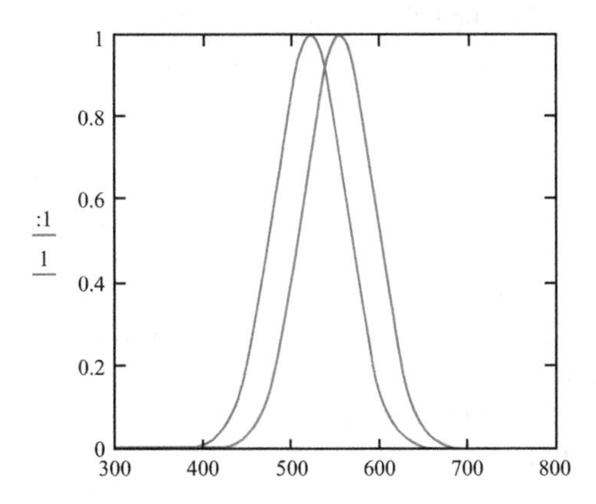

Figure 30.1. Gaussian representations of the photopic and scotopic efficacies

[13] Coblentz W 1915 *Bull. NBS* **11** 87.
[14] Lee R 1966 *Metrologia* **2** 150.
[15] Geist J 1979 *Appl. Opt.* **18** 760.
[16] Moon P 1942 *J. Opt. Soc. Am.* **32** 348.

people have different characteristics, different responses. Therefore, a standard eye had to be defined or determined. Some say it was the eye of Perley Nutting, a famous worker at the NBS. Actually, it is the average of many test personnel[17]. Both the photopic and scotopic responses can be represented quite well by Gaussian curves with maxima at 555 nm and 521 nm, respectively, and with standard deviations of 0.034 and 0.030, also respectively. I have used these in many calculations.

30.4 Radiometry

The advent of radiometry may be considered as the first quantitative measurement since the field of radiometry—the measurement of the amount of radiation—includes the visible as well as the infrared, ultraviolet, and other regions. So it certainly involves the first (non-eye) detectors of radiation. However, radiometry still depends upon a comparison to a standard. So I consider photometry the measurement of radiation in the visible spectrum and with respect to the eye, while radiometry is the more general measurement and evaluation of the quantity of radiation using an electric detector. Radiometry involves not only the measurement of the amount of radiation and how it is transferred, but also the radiometric properties of materials, like transmission, reflection, scattering, and emissivity.

30.5 Radiation transfer

Radiative transfer involves the calculation of the transfer of radiation in free space, in optical systems, in our atmosphere, in the ocean, and in other worlds. Some of the earliest ideas about radiation transfer assumed that light went from the eyes to the object and did not include evaluations of how much. Count Rumford refuted the idea of caloric[18], the substance that transferred heat, which was introduced[19] by Antoine Lavoisier (1743–1794) in 1783. Rumford found that continued boring of a cannon did not reduce the amount of heat produced; caloric was not used up; it was not conserved as it would have to be for the theory to hold up.

Calculations of radiant transfer, which is one part of radiometry, starts with the Poynting vector from electrodynamics, but it soon reduces to the consideration of ray bundles. The Poynting vector[20], invented by John Henry Poynting (1852–1914) in 1884, is $\mathbf{S} = \mathbf{EXH}$, the curl of the electric and magnetic fields. He stated it as the electromotive times the magnetic intensity times the sine of the included angle in equation form. If one solves the wave equation for each in spherical coordinates, the term $1/r$ appears. Multiplied together, it results in the inverse square law. This is the basis for propagation of energy from a point source. However, if the source is not a point, but a circular disc, integration over its surface must be taken into account. Then the equation for the irradiance, E, in W m^{-2} is $E = I/R^2\,[1 - (r/R)^2 + (r/R)^4 - \cdots]$. This shows that when the dimensions

[17] Nutting P 1914 *Trans. Illumin. Eng. Soc.* **9** 633.

[18] Rumford C 1798 *An Experimental Enquiry Concerning the Source of Heat which is Excited by Friction.*

[19] Lavoisier A 1783 *Reflexions sur le phlogostique.*

[20] Poynting J 1884 *Philos. Trans. R. Soc. Lond.* **175** 343.

of the source, indicated by r, are on a par with the dimensions of the distance, R, the inverse square law needs to be amended (where I is the radiant intensity).

What is often referred to as the cosine law of radiation is based on the fact that the radiant emittance has a cosine distribution, but it is better stated, I think, as the law of constant radiance: the radiance is independent of angle. Such radiation is also known as Lambertian, after Johann Lambert, cited above.

It has been shown that radiance is the fundamental property in radiation transfer, and that radiance, L, is conserved throughout an optical system of uniform refractive index and that reduced radiance, L/n^2, is conserved in regions of differing refractive index[21].

30.6 Radiometric material properties

The measurement of reflection, transmission, scattering, and related radiometric properties of materials is also an aspect of radiometry. There are many different properties to measure and many ways to do it.

Scattering from surfaces is discussed both theoretically and experimentally in the chapter on that subject.

Reflectometers measure the amount of light that is reflected in the specular direction. Several types have been designed. The simplest is undoubtedly substitution and has probably been used by many anonymous investigators: shine the light on a detector in position A without the mirror under test. Then insert a mirror, M, at 45° and move the detector to position B; measure the ratio (figure 30.2).

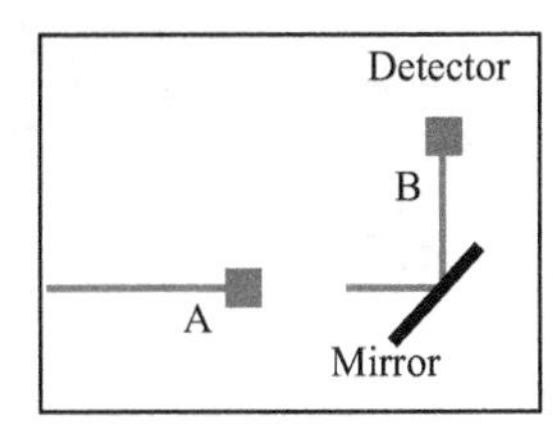

Figure 30.2. Substitution method of reflectivity measurement.

The Strong VW device[22] and the Bennett–Koehler instrument[23] are representative of more complicated systems. Each of these uses a different technique for eliminating the reflectivity of the instrument mirror from the measurement. The Strong system shines the light off an instrumental mirror and to a detector. It then interposes the sample mirror and reflects off it twice and the instrumental mirror once. So the first measurement by the VW technique yields $\rho_1 = V_1$ and the second gives $\rho\rho_1\rho = V_2$. Here ρ is the unknown reflectivity, ρ_1 is the instrument reflectivity and V_1 and V_2 are the measured values. A small amount of math gives that the reflectivity is given by the square root of the ratio of the voltages (figure 30.3).

[21] Wolfe W 1998 *Introduction to Radiometry* (Bellingham, WA: SPIE Press).

[22] Strong J 1938 *Procedures in Experimental Physics* (Englewood Cliffs, NJ: Prentice Hall).

[23] Bennett H and Koehler W 1960 *J. Opt. Soc. Am.* **50** 1.

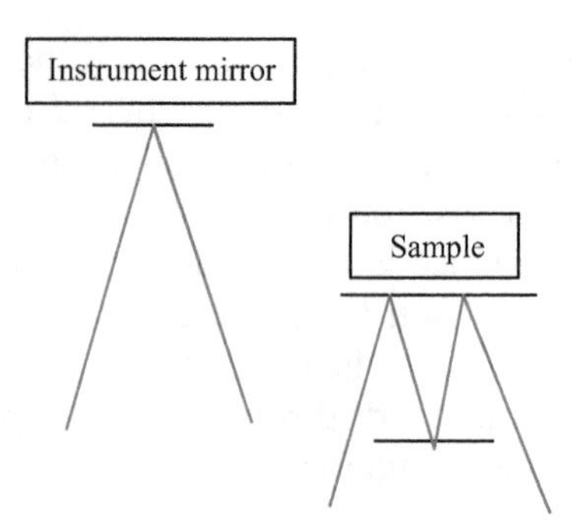

Figure 30.3. Strong VW reflectivity measurement schematic

The B–K arrangement is much more complicated. The first six mirrors shown in the figure are for convenience; the light is shone off the sample S and reflected back off it, just as in the Strong technique. Then the sample is removed and another measurement made. The auxiliary mirrors shown as dashed lines are to ensure that mirror 11 has the same reflectivity as mirror 9. If not, the difference is taken into account. The calculation of reflectivity is much more complicated with many more terms, but still just algebra (figure 30.4).

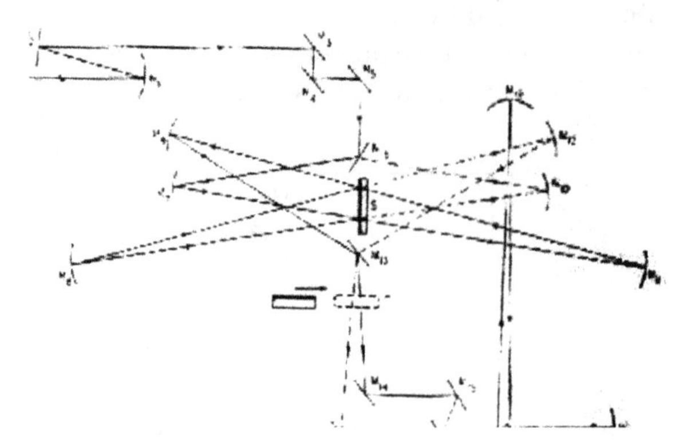

Figure 30.4. Bennet–Kohler reflectivity measurement arrangement

Directional-hemispherical reflection has been measured by several different methods, including the use of a Coblentz hemisphere[24], an integrating sphere and a pair of paraboloids. This is reflection of light from a given direction that reflects into an entire hemisphere.

In the hemisphere of William Weber Coblentz (1873–1962), the sample and the detector are placed side by side near the focus. Light enters from a hole in the overlying surface, reflects off the sample, and then back from the hemisphere to the detector. The measurement can be improved if the hemisphere is ellipsoidal[25].

[24] Coblentz W 1905 *Investigations of Infrared Spectra* (Washington, DC: Carnegie Institute of Washington).
[25] Dunn S, Richmond J and Parmer J 1966 *NBS Spec. Publ.* **300** 7.

It can also be done with an integrating sphere, an entire sphere[26]. In this case, the flat sample reflects light that enters from an opening in the sphere, reflects to all portions of the hemisphere above it, and back down to a nearby detector. Another technique uses two facing paraboloids, like clamshells, with the detector at the focus of one and the sample at that of the other. Gier, Dunkle, and Bevans have also carried out the measurement with a hohlraum (hole or cavity space—a complete enclosure, usually spherical) that is just a box[27] (figure 30.5).

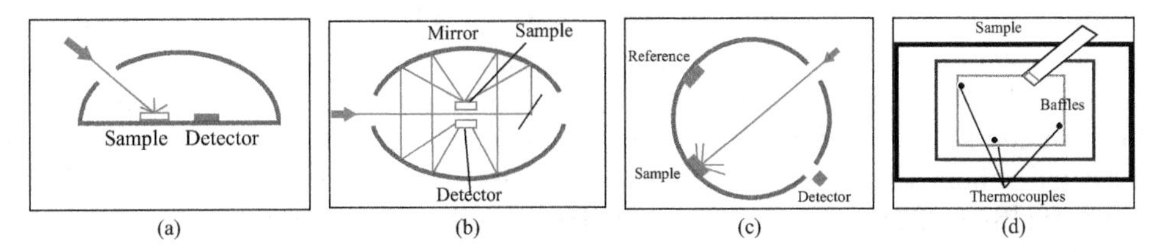

Figure 30.5. (a) Coblentz hemisphere, (b) Coblentz hemisphere with paraboloids, (c) integrating sphere, (d) Hohlraum.

Total hemispherical emissivity has been measured[28] by Lou Drummeter and Emmanuel 'Manny' Goldstein by suspending an aluminum sphere in a hohlraum on a thin cord and heating it (figure 30.6). The spectral directional emissivity has been measured by the NBS. The technique[29] is to use a spectrometer as a receiver and compare the output of the sample to that of a blackbody simulator at the same

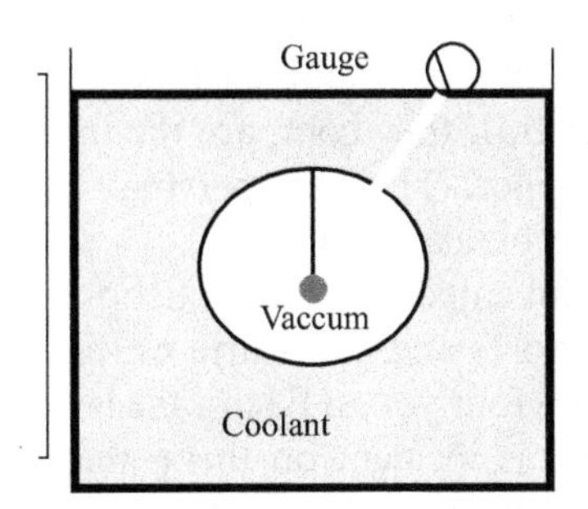

Figure 30.6. Drummeter–Goldstein arrangement for emissivity measurement.

temperature.

Gustav Robert Kirchhoff (1824–1887) was able to equate absorptivity and emissivity (under equivalent conditions)[30]. Thus measurements of either give values

[26] Jacquez J and Kuppenheim H 1955 *J. Opt. Soc. Am.* **45** 460.
[27] Gier J, Dunkle R and Bevans J 1954 *J. Opt. Soc. Am.* **44** 58.
[28] Drummeter L and Goldstein E 1960 Vanguard emittance studies at NRL *Surface Effects on Spacecraft Materials* ed F Claus (New York: Wiley).
[29] Richmond J 1971 *NBS Spec. Publ.* **300** 182.
[30] Kirchhoff G 1860 *Ann. Chem. Phys.* **109** 275; Kirchhoff G 1860 *Philos. Mag.* **20** 1.

for the other. As a result, several other techniques have been used to measure either, or both, of these quantities.

One of these was used to measure the absorption coefficient of a material. It was done by Roy Potter (1919–2019) and Don Stierwalt to measure the absorption coefficient. They compared the output of the sample with that of a blackbody at the same temperature (using a cold surround). Then, knowing the emissivity, they knew the absorptivity, and, knowing the thickness, they knew the absorption coefficient[31].

The **integrating sphere** has been an important part of radiometry since its invention in 1892. It is essential in measuring things like diffuse and hemispherical reflection. It consists of a sphere whose insides are covered with a matte surface and has several holes in it. One hole is used for a source, another for a detector, and a third for the sample under test. Of course, there are variations on this theme. The first such sphere was considered by W E Sumpner, but it was Friedrich Richard Ulbricht (1849–1923) who put it into use. He named it a Kugelphotometer, or sphere photometer[32].

30.7 Radiometers

There are basically two kinds of radiometers, which are devices that measure the amount of radiation—not the number of radios. There are those that compare the sample to an internal blackbody cavity and those that use an equivalent electrical input. The latter are alternately called **electrical substitution radiometers** or **absolute radiometers**. They use a cavity to collect radiation from a source and then electrically heat the cavity to obtain the same output[33]. Probably the first of these were instruments by Guild[34]. These have been used to measure the Stefan–Boltzmann constant σ with an uncertainty of about 0.01 percent. The other type has been described by Zissis[35]; in general, they compare the input radiation with an internal reference by way of a chopper. The chopper may reflect the radiation from an internal cavity, or it may be blackened itself.

The **Crookes radiometer**, invented by William Crookes (1832–1919) in 1873[36], is the one sold in many scientific stores and planetariums (figure 30.7). It has an evacuated glass globe inside of which are four vanes. Each vane is reflective on one side and black on the other. When sunlight is incident on them, they turn—up to 3000 rpm. The standard explanation is that the black sides absorb more energy from the Sun and energize more molecules, which bombard the surface. So the vanes spin with the black side trailing. But, if there is sufficient vacuum, the vanes go the other way around. In this case, a photon absorbed by the black side receives one photon momentum, but it reflects off the reflective side, thereby yielding two photon momenta—and the vanes

[31] Stierwalt D and Potter R 1962 *Proc. of the Int. Conf. on the Physics of Semiconductors (London)*.

[32] Wilks S 2014 *Optics and Photonics News*.

[33] Blevin W and Brown W 1971 *Metrologia* **7** 15; Geist J and Blevin W 1973 *Appl. Opt.* **12** 2532; Quinn T and Martin J 1984 *Metrologia* **20**.

[34] Guild J 1937 *Proc. R. Soc. Lond.* **161** 1.

[35] Zissis G 1978 Radiometry *The Infrared Handbook* ed W Wolfe and G Zissis (Ann Arbor, MI: University of Michigan).

[36] Crookes W 1873 *US Patent* 182172.

will go with the reflective side trailing. This effect is too small to be seen in the Crookes radiometer, but it was demonstrated in the **Nichols radiometer**[37], which is similar but has its mirrors mounted on the end of a quartz fiber in the manner of a torsion balance. It is thereby much more sensitive. It was invented by Ernest Fox Nichols (1869–1924) and Gordon Ferrie Hull (1870–1956) in 1901.

Figure 30.7. Crookes radiometer.

30.8 Nomenclature

No treatment of the history of photometry and radiometry would be complete without a discussion of the nomenclature. The nomenclature of photometry is chiefly that of the units. A lumen is a light watt—that is, the light energy per unit time, the flux of light energy. A lux is a lumen per unit area. A candela (once called a candle) is a unit of flux per solid angle. A nit is a candela per square meter. The lumen was introduced in 1864 by André Blondel (1863–1938), as well as the phot and stilb[38]. Since then, the CIE has routinely upgraded and improved the nomenclature. Some consider this photometric unit business as nitpicking!

The nomenclature of radiometry is more closely related to the names of the different quantities and various systems for naming them. The basic quantities are the flux, flux density, flux per unit solid angle, and flux per unit area and solid angle. These were called flux, irradiance, radiant emittance, intensity and radiance, respectively—with two for incoming and outgoing flux density. But they could also be reflected, transmitted, emitted, spectral total, etc. So Jon C Geist and Ed Zalewski introduced the Chinese Restaurant System[39]—pick one of these, and one of those, and this one, too. So, for instance one could have reflected spectral radiance or total transmitted irradiance. The word 'intensity' had at least three different meanings, so Fred E Nicodemus (1911–1997) started with pointance and areance and also created

[37] Nichols E and Hull G 1901 *Phys. Rev.* **13** 307.
[38] Moon P 1942 *J. Opt. Soc. Am.* **32** 355.
[39] Geist J and Zalewski E 1973 *Appl. Opt.* **12** 435.

exitance and incidance[40]. There are three different quantities involved: the flux of energy, the flux of photons, and the flux of light to which the eye is sensitive. It has been recommended that the symbols be subscripted u, q, and v to denote this. But Jones recognized the generality of it and introduced what he called fluometry, the flux of things[41]. He then introduced stearance to go with exitance and incidance. He retained intensity, declining the suggestion of Nicodemus. By the way, the three meanings of intensity are watts per unit area as Emil Wolf would have it, Watts per unit area and per unit solid angle as Neville Woolf would have it, and watts per unit solid angle as I would have it. So intensity means whatever W(o)olf(e) says it means!

The practice of radiometry has long since emerged from the laboratory and become part of remote sensing. Many satellites now have multispectral sensors which must be calibrated radiometrically. The humble practice of radiometry is now a space science.

[40] F Nicodemus, personal friendly argument!

[41] Jones R 1963 *J. Opt. Soc. Am.* **53** 1314.

IOP Publishing

Rays, Waves and Photons
A compendium of foundations and emerging technologies of pure and applied optics
William L Wolfe

Chapter 31

Reflection—noitcelfeR

The earliest consciousness of the reflection of light was surely from such things as puddles of water, placid lakes, ponds and the like. Narcissus is well known for falling in love with his reflection. The ancients used containers of water as mirrors. Aristophanes, (~450–388 BC) in his *Clouds*, describes the melting of a wax by reflection of the Sun's rays on it. Euclid (325–265 BC), the famous geometer, noted that rays travel in straight lines. Archimedes (~290–212 BC) is said to have used the shields of many warriors to reflect the solar radiation to burn enemy ships during the 213 BC siege of Syracuse. His reputed technique has been replicated twice with different results. Hero of Alexandria (~10–70) is believed to be the first to enunciate the law of reflection, although those mentioned above seemed to have known it instinctively. He did it in terms of the shortest path, which is known today as Fermat's principle. Al Hazen (~965–1040), a famous Arab contributor to optics also known as Ibn al-Haytham, discussed reflections from curved surfaces. Roger Bacon (1220–1292) is known in part for his studies of reflection and refraction. It would seem that many of these researchers have experimented with and discovered the simple law of geometric reflection—that the angle of reflection is equal to the angle of incidence and is in the same plane.

There are several types of reflection with which we are all familiar: the specular reflection, as from a mirror discussed above; diffuse reflection, in which light goes in all directions; spectral, total, and reflection in a band of wavelengths of light. Light can also be polarized by reflection, and there can be total internal reflection as light emerges from dense media. Finally, there is bidirectional reflectance in which the angles of both incidence and reflection are arbitrary.

31.1 Nomenclature

Several words are used in this discipline—reflection, reflectance reflectivity, diffuse, specular, spectral, hemispheric, and so on. Reflection is the process. Reflectance is the ratio of reflected light to incident light for a substance. Reflectivity is the same

doi:10.1088/978-0-7503-2612-4ch31

for an ideal substance with no impurities, diffuse means reflection in all directions. Specular means in the direction of the reflection angle—equal to the incident angle. Spectral means as a function of wavelength, and hemispheric means into the overlying hemisphere.

Brewster's angle is the angle for which there is 100% polarized reflected light. David Brewster (1781–1868) first observed this, although Étienne-Louis Malus (1775–1812) tried. The equations of Fresnel, which were presented somewhat later in 1815, predict this. The equation that determines this polarizing angle is $\theta_B = \arctan(n_2/n_1)$, where θ_B is Brewster's angle, n_2 is the index of the denser material, and n_1 is that of the less dense material, often air.

Total internal reflection occurs when light goes from a medium of higher refractive index to one with a lower value. The angle at which this occurs is called the critical angle; it can be obtained from Snell's law. The critical angle is given by $\theta_c = \arcsin(n_2/n_1)$, where the n's have the same meaning as before. Probably the earliest demonstration of guided light by internal reflection was in 1841 by Jean-Daniel Colladon (1802–1893). So he understood total internal reflection. A version of Snell's law was discovered by Ibn Sahl (~940–1000) in 984. It is not clear whether any of these investigators really put two and two together.

Specular reflection from a dielectric is dictated by the Fresnel equations, obtained by Augustin-Jean Fresnel (1788–1827) in about 1820[1]. They are discussed in detail and mathematically in appendix E, and methods of measurement are discussed in the chapter on radiometry.

Diffuse reflection, like that from a piece of paper, is discussed in the radiometry chapter, along with several historic methods of measurement.

The laws of reflection (and refraction) can be obtained by a solution of the Maxwell wave equations with proper boundary conditions. See appendix E for a summary of this procedure.

[1] Whittaker E 1910 *A History of the Theories of Aether and Electricity* (Dublin: Dublin University Press).

IOP Publishing

Rays, Waves and Photons
A compendium of foundations and emerging technologies of pure and applied optics
William L Wolfe

Chapter 32

Refraction—as the light is bent

Refraction is the bending of a ray of light as it traverses from a medium of one density to another, a redirection of the waves of light as they go from one medium to another (of different density and refractive index). This is caused because light has different speeds in different media. The ratio of the speed of light in vacuo to that in a medium is the refractive index, $n = c/v$, where c is the speed of light and v is the phase velocity of the light in the medium through which it traverses. See appendix B for phase and group velocities.

It seems that the first understanding of refraction, although limited[1], was by Ptolemy (~85–165) in 140. He studied the refractive effects of water and the atmosphere.

Later, Ibn Sahl (~940–1000) in Arabia came up in 984 with something equivalent to **Snell's law** and used it for the design of lenses[2].

Then, in 1602, Thomas Harriot (1560–1621) discovered what is now known as Snell's law[3].

And in 1621, Willebrord Snellius (1580–1626) enunciated his law, the law of refraction that today we call Snell's law. Rene Descartes (1596–1650) did the same shortly thereafter, so it is sometimes called Descartes' law (especially in France, although I have never heard it called that).

Newton stated in 1704 the axiom that the angle of reflection was equal to the angle of incidence, that the refracted angle is smaller than the incident angle, that the two angles are accurately or very nearly in a given ratio as the sines, and that different colors refract different amounts[4]. It would seem that he really did understand how waves of light refracted, but his law required that light traveled faster in

[1] C Ptolemy, the *Almagest*.

[2] Rashed R 1990 *ISIS* **81** 464.

[3] The Galileo Project, Wikipedia, online.

[4] Newton I 1704 *Optiks*.

doi:10.1088/978-0-7503-2612-4ch32

denser media than in air or vacuum. This was based on his corpuscular theory of light.

Apparently, the term **index of refraction** did not arise until Thomas Young (1773–1829) used it in 1807[5]; he did not use the symbol n. Others[6] used n, m, and μ, but n finally prevailed and is used for refractive index almost universally today. In many areas, n is sufficient for defining the refractive index, but in some the complex index of refraction is used, in which $\tilde{n} = n - ik$. The complex index incorporates the real part of the refractive index, n, and the imaginary part, k, where k is the extinction coefficient that designates the amount of absorption in a material.

Ernst Abbe (1840–1905) developed what is now, and has been for many years, called the **Abbe refractometer** in 1869[7]. It originally measured liquid samples between two prisms at the wavelength of the sodium D lines by measuring the critical angle. Other lines are now used in modern versions, and certain plastics and other solids can be measured. The **Pulfrich refractometer**[8], invented by Carl Pulfrich (1858–1927), is strictly one that uses grazing angle incidence and measures the resulting critical angle. However, many other variations are also called by this name. Most references to refractometers are for those that measure liquids or are used in the chemistry or biology lab. One such is the **V block refractometer**, a V that is cut in a solid and forms a prism, designed by J V Hughes in 1941. The liquid is poured into the V and the light is bent by refraction.

But the word refractometer also includes very sophisticated instruments that measure the refractive index of glasses and crystals that are used in optical instruments.

There exist many ways of measuring the refractive index of solids. Some are adaptations of the gas and liquid refractometers. Simply put a solid in one of the arms. Others use one of the many phenomena that involve the refractive index; these include focal length of lenses, interference, critical angle, and polarization. Perhaps the earliest device used to measure the refractive index of solids was that of Michel Ferdinand d'Albert d'Ailly (1714–1769). He measured the focus of a microscope (or other optical device) with and without the sample interposed. The difference is the difference in optical path, nd. Measure the thickness, d, and calculate the index[9]. The normal (perpendicular) reflection from a dielectric gives the value of the refractive index; the reflection, ignoring multiple passes, is just $r = (n - 1)^2/(n + 1)^2$. This can be solved by trial and error, but the multiple passes and the absorption can get in the way. The refractive index can be measured interferometrically since the optical path difference is nd. These are not very precise measures, generally with an uncertainty of about one percent.

---

[5] Young T 1807 *A Course of Lectures on Natural Philosophy and Mechanical Arts* (London: Joseph Johnson).
[6] Fraunhofer J 1817 *Denkschriften der Königlichen Akademie der Wissenschaften zu München* **5** 208; Brewster D 1815 *Philos. Mag.* **45** 126; Herschel J 1828 *On the Theory of Light* (London).
[7] Abbe E 1874 *Neue Apparate zur Bestimmung des Brechungs-Zerstreuimgsvermögens und fester Körper und flüssiger* (Jena: Mauke's Verlag).
[8] Pulfrich C 1888 *Z. Instrum.* **8** 47.
[9] Ally M 1767 *Histoire des Academie Royale des Sciences*; Miller A 1968 *J. Opt. Soc. Am.* **58** 428.

A classical method for the measurement of refractive index is the measurement of the angle of **minimum deviation**. This technique was used extensively by the US National Bureau of Standards (now National Institute of Science and Technology)[10]. The prism is rotated, and the refracted beam goes one way and then backs up. The turning point indicates minimum deviation, and the beam is symmetrical and perpendicular to the prism bisector. The equations to get the refractive index are easy to solve (figure 32.1).

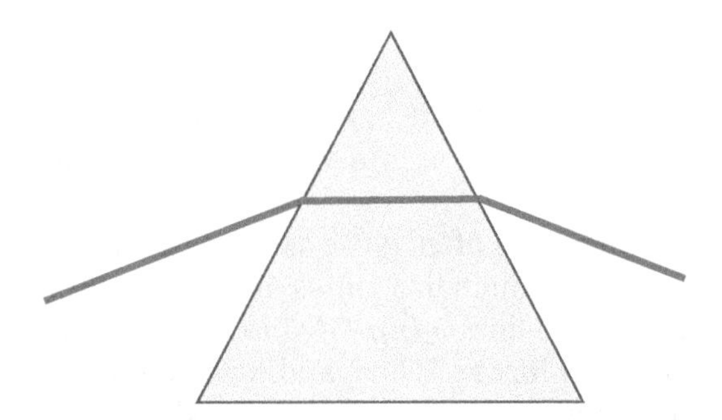

Figure 32.1. Minimum deviation.

A technique for measuring the refractive index of exotic materials in the infrared and at low and high temperatures is the method of **normal incidence**. Whereas the minimum deviation technique requires observation of the minimum angle, the normal incidence technique requires aiming the incident beam perpendicularly to the front surface. The instrument, developed by Ben Platt, was able to measure refractive indices from about 0.4 μm to 25 μm and at temperatures from about 20 K to about 500 K using an appropriate spectrometer and dewars[11]. In the normal incidence method, light is made to impinge on the prism at a right angle to its face by autocollimation. There is then only one refraction at the other surface, and the refractive index can be calculated from that angle (figure 32.2).

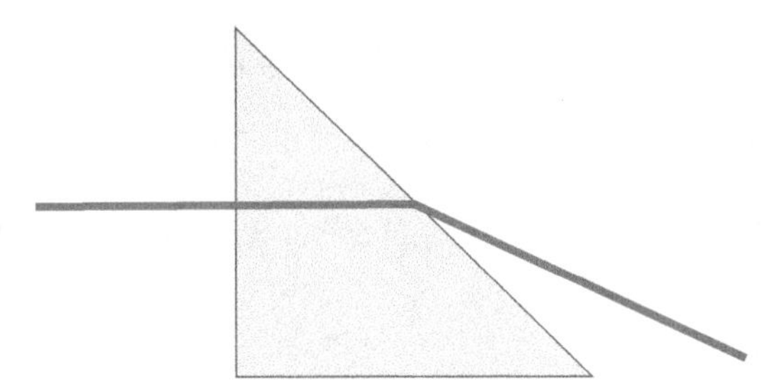

Figure 32.2. Normal incidence.

[10] Malitson I 1965 *J. Opt. Soc. Am.* **55** 1205.

[11] Platt B 1976 Instruments for measuring properties of infrared transmitting materials *Dissertation* University of Arizona.

Another useful and different technique is that of **Kramers and Kronig** due to Ralph Kronig (1904–1995) and Hendrik Anthony Kramers (1894–1952). The mathematical basis is the relation between the real and imaginary parts of a complex analytical function. It was derived independently by Kronig in 1926[12] and Kramers in 1927[13]. It is useful for measuring n and k over a wide spectrum with an uncertainty of about one percent.

The change of the refractive index with wavelength is called dispersion, although dispersion can be applied to other phenomena. The theory is based on the damped harmonic oscillations of the absorbers in the material. The incoming light causes the constituents of the material to vibrate. They vibrate most when their characteristic frequencies match the frequency of the light—a resonance, but one that is damped by surrounding forces. The equation is discussed in appendix B, along with the approximations described below.

In lens design, the dispersion is often given as the Abbe number, which is the ratio of the refractive index at the sodium line minus one to that of the difference between the indexes at the blue and red lines = $(n_D - 1)/(n_F - n_C)$. The wavelengths are 587.6 nm for the D line, 486.1 nm for the F line, and 656.3 nm for the C line.

Several different investigators have also developed equations that describe the variation—that is, $n(\lambda)$. These include the Sellmeier, Herzberger, and Cauchy. These are used to fit the data and find intermediate values by interpolation.

The **Cauchy formula**, described by Augustin-Louis Cauchy (1789–1857) in 1836[14], is

$$n(\lambda) = A + B/\lambda^2 + C/\lambda^4 + \cdots$$

where as many terms are used as is necessary, but often just two suffice.

The **Sellmeier equation**[15] was developed by Wilhelm Sellmeier (1859–1928) in 1871 as an improvement on Cauchy's; it is

$$n^2(\lambda) = 1 + A_i\lambda^2/(\lambda - a_1)^2 + A_2\lambda^2/(\lambda - a_2)^2 + A_3\lambda^2/(\lambda - a_3)^2$$

where the A's and the a's are constants to be fitted.

The **Herzberger dispersion equation**[16], enunciated by Maximilian Herzberger (1899–1982) is

$$n = a + b\lambda^2 + c/(\lambda^2 - 0.035)^2 + d/(\lambda^2 - 0.035)^2$$

where a, b, c, and d are adjustable constants.

It seems to me that Cauchy was simply curve fitting, that Sellemeir was at least giving the nod to the theory, and that Herzberger assumed that in the region of

[12] Kronig R 1926 *J. Opt. Soc. Am.* **12** 547.

[13] Kramers H A 1927 *Atti del Congresso Internationale dei Fisici, (Transactions of Volta Centenary Congress) Como* vol 2 pp 545–57.

[14] Cauchy A 1836 *Mémoire sur las Dispersion de la Lunière in Nouveaux Exercices de Mathématiques* (Prague: Calve).

[15] Sellmeier W 1871 *Ann. Phys. Chem.* **219** 272.

[16] Herzberger M 1942 *J. Opt. Soc. Am.* **32** 70.

interest there was just one resonance at 0.035 μm. These and the theory are described in more detail in appendix B.

The **complex index of refraction** is written several ways to include both the refractive index and the extinction—or absorption—coefficient. It is usually written as $n - ik$, where n is the real refractive index and k is the absorption or extinction coefficient. It is also written in other forms: $n + ik$, $n(1 - ik)$, $n(1 + ik)$.

Birefringence occurs in anisotropic crystals in which the refractive index is different in different directions. It was first observed, or at least recorded, by David Brewster in 1815[17]. However, **double refraction** was observed by Rasmus Bartholin (1625–1698) in 1669[18]. It is the basis of photoelasticity and some polarizing prisms.

Refraction is refraction. However, refractometry of the eye, phorometry, will soon be done automatically with adjustable lenses or adaptive mirrors making the measurement without the trial and error of the optometrist.

[17] Brewster D 1815 *Philos. Trans.* **299**.

[18] Bartholin R 1669 *Experimenta crystalii islandic disdiaclastici quibus mira & insolita refracto detegitur* (Denmark).

IOP Publishing

Rays, Waves and Photons
A compendium of foundations and emerging technologies of pure and applied optics
William L Wolfe

Chapter 33

Relativity—not uncle Louie and aunt Sadie

33.1 The role of optics

Relativity is intimately involved with optics. The speed of light is a constant in all frames of reference. A frame of reference is where you are—on Earth, in a bus, on a plane, in a train, or even on a meteor. Nothing goes faster than light. Information is carried on a light beam. The speed of light is the same in all directions and whether a body is moving or not. The classic experiment that led to the demise of the luminiferous ether and the development of the special theory was the Michelson–Morley experiment with an optical interferometer. The modern concepts of relativity seem strange and unreal. The idea that time changes in moving frames of reference or in more intense gravity does not agree with our experiences and seems bizarre. However, the initial idea of relativity does seem logical, that the laws of physics do not change from one frame of reference to another. That is, for instance, that gravity has an equal attraction whether you are in a moving train or sitting by the TV at home. Galileo[1] was the first to enunciate this idea. He asserted that the laws of physics were equally true for a crew of men in the hold of a ship as they were for those on land. The men in the ship did not know they were in a moving vehicle. They were not the oarsmen, just passengers. And the ship was on a very smooth sea. Modern examples invoke trains that are on perfectly smooth tracks with no curves and no changes in velocity. As the two approach, which is moving and which is still, or are they both moving? It cannot be determined, so it is all relative. Einstein[2] took up Galileo's idea and added that the speed of light is always the same, in every direction and whether the source is moving or not, and nothing goes faster than it.

33.2 A Gedanken experiment

Einstein's gedanken experiment went something like this: imagine someone standing on the railroad tracks observing light stream by at 186 000 miles per second. Then

[1] Galilei G 1638 *Discorsi, e Dimostrazioni Matematiche, á due nuoue scienze* (Leiden: Elzevier).
[2] Einstein A 1905 *Ann. Phys.* **17** 891.

imagine a woman in a train traveling in the same direction at 2000 miles per second (in a very fast train). Does the woman observe the light traveling at 184 000 mps? If so, the physics is different in the two different frames of reference. This is not allowed by either Galileo's or Einstein's theory of relativity. So either relativity is wrong or the simple law of the propagation of light is wrong. It was on a beautiful day in Bern, while he was walking with a friend, that he suddenly realized that time was not a constant[3]. This theory, now law, has been verified experimentally with atomic clocks on transcontinental planes[4]. That is the easiest one to understand, but there have been about 100 different tests, every one of which supported the theory[5]. That is why I call it a law or principle and not a theory.

33.3 Scientific background

The scientific background of special relativity starts with Galileo Galilei (1564–1642) as noted above. His was the observation that there is no preferred frame of reference. As noted in the chapter on light, Maxwell derived the equations that predicted light as an electromagnetic wave motion, but the question still was how did it propagate. The postulation was in a luminiferous ether[6]. This stuff had remarkable properties that were conjured up just to make it work. It had to permeate all space. It had to be ethereal enough that feathers were not affected by it, or planets, but it had to be stiff enough that it supported these light vibrations. In a classical experiment, it was proved to be a figment of the scientists' imagination. The Michelson–Morley experiment[7], by Albert Abraham Michelson (1852–1931) and Edward Williams Morley (1838–1923) in 1887, showed that there was no ether drift. Then Hendrik Antoon Lorentz (1853–1928) postulated an immobile ether. He was working on extending James Clerk Maxwell's (1831–1879) work to moving bodies—in reference frames that move. In order to make this description, he introduced the concept of local time but did not carefully describe it[8]. He also introduced the idea that physical lengths are shortened in the direction of motion. His reference-frame transformations were improved by Jules Henri Poincare (1854–912) in a letter to Lorentz in 1905; they are now known as the Lorentz transformations (poor Henri!). In 1889, George Francis FitzGerald (1851–901) published his idea of physical contraction to help explain the Michelson Morley results, although FitzGerald still framed it in terms of the ether[9]. Although Walter Isaacson in his wonderful biography of Einstein does not include these events, I believe they were on Einstein's mind when he pondered all this. It is also questionable whether the Michelson–Morley experiment influenced Einstein, although it did put the ether to rest. Knocking out ether rather than having it knock you out is an interesting concept.

[3] Isaacson W 2007 *Einstein* (New York: Simon and Schuster).
[4] Gwinner G 2005 *Mod. Phys. Lett.* **1** 20.
[5] Roberts T What is the experimental basis of Special Relativity, online.
[6] Fizeau H 1851 *Philos. Mag.* **2** 568; Newton I 1704 *Opticks.*
[7] Michelson A and Morley E 1887 *Am. J. Sci.* **34** 333.
[8] Lorentz H 1904 *Proc. R. Netherlands Arts Sci.* **6** 809 available online, but tough to read.
[9] Fitzgerald G 1899 *Science* **13** 390.

The idea that an object is shortened in the direction in which it moves is not so absurd. One can envision it being squished a little by that motion. The idea that time is changed is another matter; it just does not happen in our experience. But these changes are minute. They are dictated by a factor that involves the speed of light and the velocity in question. The factor involved, but not the whole story, is $\sqrt{(1 - v^2/c^2)}$. One example is the Supersonic Transport airplane at about 2000 mph. The factor is 0.999 998 507. This is just not noticeable in our regular lives, but it is significant in science. We would have to go at about five million miles an hour before there would be a one percent change!

33.4 General relativity

No such thought experiment was available for considering the theory of general relativity that holds in all frames of reference, inertial or not. For this, Einstein spent two years studying under Marcel Grossman, his friend, former classmate, and then chairman of the math department at the ETH, the Swiss Eidgenössische Technische Hochschule, to become proficient in tensor analysis. This enabled him to formulate the multidimensional procedures for coordinate transformation in four-dimensional space, which he announced in 1915[10]. One result was that time slows down in regions of higher gravity. Einstein proposed three ways to test the general theory: the precession of the perihelion of the orbit of Mercury (the point nearest the Sun), bending of light by the Sun[11], and the gravitational redshift (a change in the frequency of light as it passes through a gravitational gradient).

The **precession of the perihelion of Mercury** is only about 1.5° per century, but it had been observed by astronomers. That is about 54 arcsec per year, a quantity easily measured—with patience. It was originally thought to be the result of some nearby object, but the general theory explained it.

The classical experiment that proved general relativity was during an **eclipse of the Sun** in 1919. The expedition was headed by the theoretical physicist Arthur Eddington (1882–1944). By observing a star during this event, they could determine that the gravity of the Sun attracted and bent the rays of light from the star, Einstein's second proposal[12].

The **gravitational redshift** was not conclusively shown until 1959[13], long after the general relativity became established as a law. These results were confirmed much later by an independent means, using a maser shot vertically into space[14]. There is a vertical gravitational gradient, and the maser would experience a frequency shift.

I find it fascinating that the general theory of relativity was proven by explaining a historic anomaly and then by experiments that showed the phenomena that it predicted. That is the way science should work.

[10] Einstein A 1916 *Ann. Phys.* **49** 769.
[11] Einstein A 1936 *Science* **84** 506.
[12] Dyson F, Eddington A and Davidson C 1920 *Philos. Trans. R. Soc. Lond.* **220** 291.
[13] Pound R and Rebka G 1959 *Phys. Rev. Lett.* **3** 439.
[14] Vessot R *et al* 1980 *Phys. Rev. Lett.* **45** 2081.

Although general relativity has little effect on our daily lives, there are some exceptions. The GPS, Global Positioning System, requires relativistic corrections for both its position in our gravitational field and the speed of the satellites[15]. Another, but less direct effect, is the existence of black holes and even the correct understanding of the expansion of the Universe.

I sit in the top row of our church in the round so that the sermon does not last as long, but it only saves me about one quadrillionth of a second each Sunday.

[15] Ashby N 2003 Relativity in the global positioning system *Living Rev. Relativity* **6** 1.

IOP Publishing

Rays, Waves and Photons
A compendium of foundations and emerging technologies of pure and applied optics
William L Wolfe

Chapter 34

Remote sensing—admiration from afar

In the broadest sense of the phrase, remote sensing means sensing from a distance, any distance, looking at things with which you are not in contact. One way of looking at it is if you are looking at it, it is remote sensing. Cameras, telescopes, microscopes, all sorts of optical instruments sense things remotely. In a narrower sense, this phrase has come to mean investigating things by means of optical or other techniques from satellites and aircraft, and in many cases in various parts of the spectrum either singly or at the same time. I will use this narrower meaning in this chapter. Remote sensing has been used for military intelligence purposes, for crop assessment, for firefighting, and for other applications.

The first picture taken from aloft was from a tethered hot-air balloon by Gaspard-Félix Tournachon (1820–1910) in 1858. Several investigators used kites and even pigeons. Alfred Bernhard Nobel (1833–1896), the inventor of dynamite and originator of the prize, was the first to take a picture from a rocket in 1897. Wilbur Wright (1867–1912) took the first aerial photograph in Italy in 1909[1]. Sherman M Fairchild (1896–1971) designed a camera with a behind-the-lens shutter in 1917; it was a significant improvement because it was much faster. He was also the one to commercialize the field, adding this application to military ones[2].

Certainly the program of **reconnaissance satellites** for military intelligence purposes qualifies as remote sensing. Satellites were, however, preceded by the U-2 program that used high altitude aircraft [3]. The first U-2 flight was in 1955. The plane was designed and built at the Lockheed Skunk Works by Clarence Kelly Johnson (1910–1990). The U-2's flew at an altitude of 70 000 feet, much higher than any other aircraft at the time. One could generate about 3600 feet of film with a ground resolution of about two to three feet. That is enough to identify plane types,

[1] Professional Photographers Association, online.
[2] *History of Remote Sensing*, Aerial Photography, online.
[3] Bauman P 2009 *History of Remote Sensing, Part II* online.

doi:10.1088/978-0-7503-2612-4ch34

but not individual models, and to count them in air fields. Note that for a resolution of three feet and an altitude of 70 000 feet and yellow light, a diffraction limited aperture would have a diameter of about 15 mm (1.7 in); it was probably a bit larger than that, about 50–100 mm (2–4 in)—probably about like a normal camera. In fact, James Gilbert Baker (1914–2005) designed a 180 inch F/13.85 lens for the camera, which means it was just about 13 inches in diameter with a 13 × 13 inch focal plane. We all know that the U-2 flights ended with the downing of Francis Gary Powers (1929–1977) in 1960. Surely that was incentive for satellite recon.

The satellite program began in 1957 with the W-117 satellites[4]. It was followed by the **Corona** system that provided the intelligence imagery from space from 1960 until 1972[5]. The early systems were photographic and dropped the rolls of film in capsules that were retrieved by C-119s and C-130s, some out of the drink, some out of the air. The satellites of the Corona program were labeled KH-1 through KH-6, probably for KeyHole, a peeking technique. The KH-7 and 8 were labeled **Gambit**, and KH-9 **Hexagon**[6] and **Big Bird**. The first three KH vehicles had a ground resolution of about 40 feet; KH-4 reduced it with a larger aperture to about five feet. It flew at about 700 000 feet altitude so that its aperture diameter had to be about 8.5 cm (about 4 in) using the same diffraction limit assumptions. It was probably bigger. The Hexagon camera was a folded Schmidt (see Telescopes, chapter 42) with a 20 inch aperture and speed of F/3. According to Clause and Miller[5], we are now up to KH-13 with some remarkable imagery. That imagery and recent publications of North Korean ships appears to me to have a resolution of a few inches!

Today's satellite recon systems surely use large apertures and detector arrays, and they send the information to ground stations by telemetry. There is good reason to suggest that the aperture diameters are about 2.4 m and the arrays are as much as 5000 by 5000 pixels—25 megapixels. Raytheon has reported infrared arrays with 4000 by 4000 pixels that are buttable[7]; visible ones are easier. Each pixel is probably about 10 μm on a side, and the vehicles probably come as close as 100 miles. That would give a diffraction limit of about an inch (2.5 cm) at 0.5 μm wavelength, but the F/number would have to be fairly large to match this resolution. With image processing, they might be able to read license plates (per Tom Clancy)—if they were horizontal[8]. Kramer shows that the KH-11B had a ground resolution of about 0.06 m = 6 cm = 2.4 inches in 1992[9]. I have warned the sorority girls to be careful what they wear (or not) on the roofs of their houses!

Remote sensing to many people means viewing crops and cities and oceans and the like with civilian satellites or aircraft. Applications include assessing the health of crops, of global weather, lightning occurrences, mapping wild fires, exploring Mars,

[4] *Aviation Week*, October 14 (1957).

[5] National Reconnaissance Office Report onlineClausen I and Miller E 2012 *Intelligence Revolution 1960: Retrieving the Corona Imagery that Helped Win the Cold War* (Chantilly, VA: CSNR).

[6] Pressel P 2013 *Opt. Photon. News* 28.

[7] D Wolfe, private communication.

[8] Conjecture based on the size of the Hubble telescope and formats of commercial digital cameras.

[9] Kramer H 2002 *Observation of the Earth and its Environment: Survey of Missions and Sensors* (Berlin: Springer).

and more. Crop assessment is done in part by the 0.85 μm band that senses chlorophyll, the 0.8 μn band that senses water content in leaves, and the 1.6 μm band that senses water content in soil. The 10 μn band can be used for mineral detection and evaluation.

The mapping of **wildfires** has a history reaching back to 1971, with hopes and dreams even earlier. The beauty of the technique is that infrared mapping can penetrate a certain amount of the smoke, clearly identify the fire perimeter, and can also identify burning embers left behind. Active optical systems, L1DARS (Light Detection And Ranging), have been used to detennine the altitude of various terrain features and canopy features[10]. Fires are now detected by **AVHRR**, a meteorological instrument, **MODIS**, and **ATSR** (see below for description), but the earliest use of aerial recon with infrared techniques appears to be in Missoula, MN[11]. Many aspects of the fire and its results have been determined by these instruments and those who use them[12].

Landsat, which stands for, you guessed it, land (viewing) satellite, was conceived in 1966 largely as a result of very useful orbital photography from the Mercury and Gemini spacecraft. Landsat 1, also known as **ERTS-1** (Earth Resource Technology Satellite) was launched in 1972 and contained a return beam vidicon (RBV, a type of TV tube) and a multi-spectral sensor, **MSS**. The RBV had limited use or success. The MSS was a whisk-broom scanner. That is, a mirror swept the field of view transverse to the direction of the satellite while the satellite's forward motion generated the other direction of the raster scan. The orbit is Sun synchronous. The bands were 0.5–0.6, 0.6–0.7, 0.7–0.8, and 0.8–1.1 μm; the ground resolution was 68 by 83 m, and the total lateral field was 185 km. The MSS was used in Landsats 1–3, but a longer IR band was added for Landsat 3, a long wave infrared band from 10.4 to 11.6 μm. Landsats 4 and 5 used a different payload, the **Thematic Mapper**. For a change, no acronym. It has four bands in the visible, two in the near IR, and one in the thermal or long wave IR. The thermal IR band has a ground resolution of 120 × 120 m while the others have 30 × 30 m. It has the same swath width. Landsat 6 had an early demise. Landsat 7 had an enhanced Thematic Mapper with a panchromatic (across the visible) band that had 15 km spatial resolution and 60 km spatial resolution in the infrared as well as a stated 5% radiometric accuracy[13]. It was launched in 1999 and is still going. This information is from a NASA online report[14].

The **IKONOS**[15] was a commercial venture by Lockheed in the satellite remote sensing arena. The name derives from the Greek *eikon*, meaning image. The first attempt aborted, but the second in 1999 went into a Sun-synchronous orbit at

[10] LIDAR, Light Detection And Ranging, is a remote sensing method to examine the surface of the Earth, National Geodetic Survey, online.

[11] Wilson R *et al* 1971 *Airborne Infrared forest Fire Detection System: Final Report* (USDA Forest Service).

[12] Lentile L *et al* 2006 *Int. J. Wildland Fire* **15** 319.

[13] From my long experience in radiometry, I would doubt 5% accuracy, but perhaps 5% precision or relative accuracy. See Radiometry (chapter 29) for accuracy and precision.

[14] NASA 2013 *Remote Sensing Tutorial* online.

[15] Broad W 2010 Giant leap for private industry: spies in space *New York Times* available online.

10:30 am at an altitude of about 700 km. The optical assembly, made by KODAK, was a three-mirror anastigmat with a 70 cm aperture (see Telescopes, chapter 42). It operated in the push-broom mode with a linear array of detectors perpendicular to the travel of the vehicle that was pushed along by the motion of the vehicle—like I have to do when I clean the driveway with a push broom. Several linear arrays each with a different spectral response are used. The panchromatic detectors subtend 1.2 μrad, the other bands, 4.8 μrad; the swath width is 0.0162 radians. These resulted in 1 and 4 m ground spots. The spectral bands are 0.45–0.90, 0.445–0.516, 0.506–0.595, 0.632–0.698, 0.757–0.853 μm.

The French satellite **SPOT 1** (Satellite Pour l'Observation de la Terre) was launched with an Ariadne rocket for the first time in 1986. It and SPOT 2 in 1990 and SPOT 3 in 1993 continue to provide imagery with SPOT 6 launched in 2012 and SPOT 7 in 2014. They have evolved to versions 6 and 7, which have the following spectral bands: 450–745, 450–525, 530–590, 625–695 and 760–890 nm. The panchromatic band, the first one, has a ground resolution of 1.5 m; the others, 8 m. The entire footprint is 60 km by 60 km. They fly in a Sun-synchronous orbit at 832 km[16].

Satellites have been used for meteorological, weather, and climatological purposes. The most famous is now the **GOES** (Geostationary Operational Environmental Satellite), but it was preceded by **TIROS** (Television Infrared Observation Satellite).

TIROS was launched in 1960 and was a low orbit satellite—that is, a Sun-synchronous, polar one at an altitude of about 700 km (420 miles). It employed two television cameras and lasted for only 78 days, but it proved the value of such observations. Subsequent versions with various improvements were launched up to 1978. The later versions used the **AVIRR** (Advanced Visible InfraRed Radiometer).

NOAA, the National Oceanic and Atmospheric Administration, employs two satellites in polar orbit to monitor the land and oceans, clouds and aerosols, snow and ice. They both use the **AVIRR** radiometer to make the measurements. They use several spectral bands at 0.6, 0.9, 3.5, 10 μm; one at 1.6 was added later. The first was launched on a TIROS in 1978. The satellites travel in polar orbits and are adjusted so one is at a constant morning time and the other in the evening. It too is a whisk-broom scanner, with the scan going from horizon to horizon. The ground resolution is a bit more than one kilometer, 1.09 km. The swath width is 2500 km. Version 3 has the following spectral bands: 0.58–0.68, 0.75–1.00, 1.58–1.64, 3.55–3.93, 10.30–11.30, and 11.50–12.50 μm. The earlier versions did not have as many[17]. Each radiometer uses an 8-inch diameter Cassegrain telescope.

The **GOES** system is the workhorse of the satellite weather system. In geosynchronous orbit, the instrument is always over the same part of the Earth. The geosynchronous orbit is at an altitude of 22 300 miles, so the ground resolution will not be great, but it need not be for weather purposes. The first launch was in 1975 and placed the payload over the Indian Ocean. It used a **VISSR** sensor system:

[16] SPOT Satellite, Wikipedia, https://en.wikipedia.org/wiki/SPOT_(satellite).

[17] *NOAA Advanced Very High Resolution Radiometer;* online (2013).

Visible Infrared Spin Scan Radiometer. It has eight photomultiplier tubes, PMT's, that operate in the visible, with an S-20 response and two infrared channels using mercury cadmium telluride detectors (see Infrared, chapter 12). The IR channels are filtered to the 10.5 to 12.6 μm band; the detectors subtend an instantaneous field of 0.192 mrad. The detectors are cooled to 95 K (−190 °F) by a radiative cooler. The telescope has a 40 cm aperture, and the instantaneous field is scanned by a flat mirror in front of the primary. The total field is programmable and takes a relatively long time[18]. It maps the temperature variations in the field; cloud tops are cooler than land.

The **VAS** was added to the GOES system in GOES 8, launched in 1994. It is the VISSR Atmospheric Sounder, proposed by Verner Edward Suomi (1915–1995)[19]. An infrared sounder uses several closely spaced infrared bands in the carbon dioxide region, around 15 μn, that 'penetrate' to different levels in the atmosphere due to their different absorption amounts. This allows the measurement of the atmospheric radiation at different altitudes. GOES 8 was 'retired in 2004 and boosted to a graveyard orbit,' according to Wikipedia. There are currently four GOES 4's in orbit, two watching the eastern and western portions of the US and two in reserve[20]. Four more are in various stages of development. They will have significant improvements in resolution, sensitivity, and maneuverability, as well as lightning, solar flare, ultraviolet, and x-ray detection[21]. Details about that are in the category of future history.

Certainly, the first **detection of lightning** was by the human eye. By the way, the myth that it is one second for each mile between the flash and the thunderclap is wrong. It is 5 seconds per mile. Most manufactured lightning detection systems use radio techniques. Lightning is a very impressive part of Mother Nature's armamentarium. There are about 44 000 bolts per day over the surface of Earth. Each bolt has, on average, four stokes, each of which lasts about 30 μs. The power in each bolt can be as much as one billion kilowatts. The thunder is from the supersonic travel of the bolt—supersonic sound.

Lightning is the phenomenon that maintains our electric balance. It is a giant arc coming from the static electricity generated in the clouds. Meteorologists would like to know the overall frequency and occurrence of lightning as part of their modeling of the atmosphere and the climate. One such study to make these measurements was done at the University of Arizona. The technique used a spectral discrimination of lightning against reflected sunlight and a frame-to-frame subtraction[22]. A relatively small instrument, about the size of a breadbox, was designed. Its prototype is shown in figure 34.1. Since there are about 44 000 lightning strikes every day, discrimination and memory are essential. Lightning has a somewhat different spectrum from the Sun since it results from the breakdown of atmospheric constituents and

[18] Centre for Environmental Data Archival, online and personal experience.
[19] Suomi V 1995 *EOS Trans. Am. Geophys. Union* **76** 361.
[20] *Space News*, May 2013, online.
[21] Personal involvement.
[22] Nagler M 1981 Design of a spaceborne lightning sensor *Dissertation* University of Arizona

Figure 34.1. Lightning sensor prototype.

not thermonuclear reactions. The study showed the feasibility, but realization came much later as adjuncts to GOES, since memory technology was not sufficient at the time. In order to reduce the background of solar reflections, frame-to-frame subtraction had to be introduced, and this led to terrabyte memories and processing.

The GOES is complemented by the **POES**, Polar Orbiting Environmental Satellite. POES goes with GOES! POES orbits at a lower altitude than the GOES and travels from north to south, covering Earth with 14 traverses each day at a 520 mile altitude. It uses the **AVHRR** and ATOVS sensors. The AVHRR (Advanced Very High Resolution Radiometer) is a multispectral sensor now with six bands: 0.58–0.68, 0.725–1.0, 1.58–1.64, 3.55–3.98, 10.30–11.30 and 11.50–12.50 μm. The first version had four channels and flew on TIROS-N in 1978, all with a ground resolution of 1.09 km at nadir. The next two versions each added a channel in 1981 and 1998, respectively. The Europeans have introduced **METOP**, their polar orbiting weather satellite, with its first launch in 2012 from Baikonur on a Soyuz rocket. It incorporates the AVHRR and an atmospheric sounder, as well as other instruments. The sounder with the acronym **HIRS**, for High Resolution InfraRed Sounder, uses an imaging radiometer with 19 infrared channels and one visible one. It claims 1 K temperature precision, 10% humidity in 1 km altitude increments. It is a stop-and-stare, whisk-broom scanner with a ground resolution of 10 km and a swath width of ±1092 km.

Relatively current practice has extended the use of multi-spectral imagery to so-called hyperspectral imagery. Hyperspectral imagery is generally considered to be the use of very many bands, rather than a few, that are contiguous. Hence hyper rather than multi. Whereas the multi instruments use relatively few bands that are usually defined by filters, the hypers often use spectrometers of one sort or another[23] (see Spectrometers, chapter 39). One recent example is the **AVIRIS** (Advanced Visible and Infrared Imaging System). It flew in various aircraft and had 224 channels approximately 0.010 μm wide from 0.41 to 2.45 μm. It used four concave grating spectrometers and a whisk-broom scan that was 614 pixels, 1 mrad each for a total swath width of approximately 30°. (This calculates to 35°, but there might have been a little overlap.) The system was first flown in 1987[24]. I have to say it: they

[23] Wolfe W 1997 *Introduction to Imaging Spectrometers* (Bellingham, WA: SPIE Press).

[24] Green R *et al* 1998 *Remote Sens. Environ.* **65** 227.

were very greedy about getting many spectral bands; they were AVIRISious! **HYDICE** (HYperspectral Digital Imagery Collection Experiment) was developed jointly by Hughes Danbury[25] and the Naval Research Lab, and it was flown in The University of Michigan's Environmental Research Institute's aircraft in 1995[26]. It has 210 bands from 0.4 to 2.5 μm. It has a ground resolution from 1 to 4 m, depending upon altitude and a corresponding swath width of 320–1280 m. **PROBE 1** was flown by Earth Sciences Inc. with 128 bands in the same region and a swath width of about 1–6 km with an instantaneous field of view of 2 mrad. **CASI** was flown by ITRES Research with 228 bands from 0.4 to 1.0 μm. **HyMap** by Integrated Spectronics with 200 bands from the visible to the thermal IR. EPS-H by GER flew with 152 channels from 0.43 to 2.5 μm, and two version of DAIS by the same company with about the same bands. Finally, **AISA** by Spectral Imaging flew with 288 bands from 0.43 to 1.0 μm

Additional spaceborne devices have been launched by NASA and the Air Force Research Lab. The first is **Hyperion** on EO-1 with 220 channels from 0.4 to 2.5 μm, the second, **FTHSI**, on MightySat II with 256 bands from 0.35 to 1.05 μm. Much of the information on these imagers is in an online report by Peg Shippert[27].

This technology has even become commercial. You can buy a spectral imaging device. You can hire spectral imaging overflights[28]. You can buy the imagery.

Remote sensing is here to stay and will grow as we launch more satellites, improve multispectral sensing and monitor our crops, seaways, fisheries and atmosphere. It will grow to include drones that cover agricultural fields in a not-so-remote way.

---

[25] This is the result of several takeovers of Perkin Elmer, ITEK, Litton and maybe the Avon lady.

[26] Mitchell P 1995 *Proc. SPIE* **2587** 70.

[27] Shippert P *Introduction to Hyperspectral Analysis* (Research Systems Inc).

[28] Resonon, Merrick, *LLA Instruments, Surface Optics*, on the net.

IOP Publishing

Rays, Waves and Photons

A compendium of foundations and emerging technologies of pure and applied optics

William L Wolfe

Chapter 35

Scattering—scattered thoughts

The blue sky and various hazes were certainly interesting to our ancestors with inquisitive minds. The great Leonardo Da Vinci observed the colors generated by scattering of aerosols: 'I say that the blueness we see in the atmosphere is not intrinsic color…the atmosphere takes on this azure hue by reasons of the particles of moisture…'. John Tyndall (1820–1893) posed two questions to the Royal Society: What causes 'the blue colour of the sky and the polarization of skylight'? And he answered them, 'It rests with the undulatory theory (of light).'

35.1 Particulate scattering

John William Strutt, the third Lord Rayleigh (1842–1919) developed his theory of scattering of light by particles smaller than the wavelength of the light (now called **Rayleigh Scattering)** in a series of papers from 1871 to 1899[1]. He was interested in, and explained, both the color of the blue sky and its polarization. His derivation included the fact that the amount of scattering was inversely proportional to the fourth power of the wavelength—so that blue light scatters about 2^4 or 16 times more than red. This has more recent application to the scattering of light in optical fibers.

Gustav Adolf Feodor Wilhelm Ludwig Mie (1869–1957) developed the equations in 1908 that describe scattering of light by spheres of approximately the same wavelength as the light or somewhat larger. It is now called **Mie Scattering**. He actually solved boundary value problems of electromagnetic radiation[2], but, of course, that applies to light.

Note that the beautiful Arizona sunsets are dependent upon both types of scattering. The clouds, consisting of large droplets of water, scatter in a Mie

[1] Rayleigh L 1871 *Philos. Mag.* **4** 41 107, 274, 447 et seq; Rayleigh L 1881 *Philos. Mag.* **5** 12 81; Rayleigh L 1899 *Philos. Mag.* **5** 47.

[2] Mie G 1908 *Ann. Phys.* **4** 25.

doi:10.1088/978-0-7503-2612-4ch35

fashion, so that the scattering is independent of wavelength. The resultant scattered light travels through the atmosphere where the molecules scatter to the side—the blue more than the red, according to Rayleigh's theory. So only the reds get through.

35.2 Surface scatter

The scattering of light from slightly rough surfaces received attention in the 1960s when optical systems were being used for navigational purposes. One example was the optical sensor for the Lunar Excursion Module of the Apollo mission. It had to perform star tracking and communication with the orbiter in the presence of the Sun[3]. There was even more intensive attention in the 1980s; the performance of infrared systems for the detection and tracking of ballistic missiles could be adversely influenced by the scattering of out-of-field sunlight onto the detector or detector array. Just as in World War II when the Zeros flying out of the Sun were a threat, the ICBM's of the USSR flying near the Sun would also be a threat. Two programs were developed for the calculation of off-axis and out-of-field rejection of nearby sunlight. These were GUERAP and APART, General Unwanted Energy Rejection Analysis Program and Arizona Paraxial Analysis of Radiation Transfer. The first was by Barry Likeness and the second by Bob Breault[4]. GUERAP used a statistical approach to the distribution of scattered radiation[5] at each surface, whereas APART was based on both a paraxial description of the optical system and an empirical surface-scattering function.

During this period, a more precise description of the radiometric characteristics of scatter was introduced by Fred Nicodemus, the BRDF, or bidirectional reflectance distribution function[6]. It is the reflected radiance as a function of the angles of incidence divided by the incident irradiance as a function of its angles. It has since been expanded to BSDF, where S stands for scatter, and the function can describe either reflection or transmission.

Several theories of scattering from what are called microrough surfaces (wherein the rms roughness is less than a wavelength) have been developed. They are based on a perturbation expansion of the solutions to the Fresnel equations for a metallic surface[7]. They are all equivalent, although their forms look different[8]; they include a wavelength factor, wavelength to the minus four as in Rayleigh theory, a geometry factor depending upon the angles involved and the spectrum of the surface height distribution. Others are based on Fourier and radiance representations[9].

---

[3] Personal experience.

[4] Breault R 1979 Suppression of scattered light *Dissertation* University of Arizona.

[5] Likeness B 1977 *Proc. SPIE* **107** 80; Likeness B 1978 *GUERAP III: General Unwanted Energy Rejection Analysis Program User's Manual* (Clearwater, FL: Honeywell).

[6] Nicodemus F 1970 *Appl. Opt.* **9** 1474.

[7] Davies H 1954 *Proc. IEEE* **101** 209; Maradudin A and Mills D 1975 *Phys. Rev. B* **11** 1392; Kroger E and Kretchman E 1970 *Z. Phys.* **237** 1.

[8] Personal calculations.

[9] Harvey J *et al* 1998 *Proc. SPIE* **3426** 51; Harvey J *et al* 1999 *Appl. Opt.* **38** 6469; Harvey J 2012 *Proc. SPIE* **8483** 848304.

The measurements of scattering from surfaces have developed over the years. The geometry is clear: shine light onto a sample from all the different angles in the overlying hemisphere and collect light from all those same angles—independently. Instruments based on these ideas[10] have a fixed sample, but a different kind was developed by Larry Brooks. It had a sample that was fixed on a gimbal so that it could rotate around two axes, a detector that could rotate 360° horizontally, and a set of fixed sources[11].

Léon Brillouin (1889–1969) is usually given credit for the first realization of the scattering of light from structures of varying density and therefore refractive index[12], although Leonid Isaakovich Mandelstam (1879–1944), or Mandelshtam since it is transliterated from Cyrillic, may have preceded him in the laboratory but not in publication[13]. It is the basis of some acousto-optical instruments (see Acousto Optics, chapter 2). The scattering, now called **Brillouin scattering**, is said to be inelastic, involving a change in frequency of the light (and energy). This is caused by the Doppler affect if the density pattern is moving. This may also be considered the interaction of phonons and photons. It is closely related to Bragg scattering, which can be considered its static equivalent in which the variations in density, the refractive index, do not change with time.

Raman scattering is also an inelastic effect. If an electron that has been excited to a state of higher energy relaxes to an energy state different from its origin, it will emit a photon of a different energy, frequency, and wavelength. This effect can be used in some kinds of spectroscopy (see Spectroscopy, chapter 39). Raman scattering was reported in 1928 by Chandrasekhara Venkata Raman (1888–1970) and Kariamanickam Srinivasa Krishnan (1898–1961) in liquids[14] almost simultaneously with Grigory Landsberg (1890–1957) and Mandelstam in solids[15]. Adolph Gustav Stephan Smekal (1895–1959) predicted the inelastic scattering earlier[16], and in much German literature it is referred to as the Smekal–Raman effect. In Russia, they call it the combinatorial effect.

---

[10] Shack R and DeBell M 1974 Surface scatter theory *Final Report* SAMSO TR 74-88, University of Arizona; Harvey J 1974 *Proc. SPIE* **107** 41.

[11] Brooks L 1982 Microprocessor-based instrumentation for BSDF measurements from visible to FIR *Dissertation* University of Arizona.

[12] Brillouin L 1946 *Wave Propagation in Periodic Structures: Electric Filters and Crystal Lattices* (New York: McGraw Hill); Brillouin L 1922 *Ann. Phys.* **17** 88.

[13] Mandelstam L I 1926 *Zh. Russ. Fiz. Chem. Ova.* **58** 381.

[14] Raman C and Krishnan K 1928 *Indian J. Phys.* **2** 387.

[15] Landsberg G and Mandelstam L 1928 *Naturwissenschaften* **16** 557; Landsberg G and Mandelstam L 1928 *J. Russ. Physico Chem. Soc. Phys. Sect.* **60** 335.

[16] Smekal A 1923 *Die Naturwissenschaften* **11** 873.

IOP Publishing

Rays, Waves and Photons
A compendium of foundations and emerging technologies of pure and applied optics
William L Wolfe

Chapter 36

Scopes—scoping it out

Scope has at least two meanings: the range of things and a viewing instrument. This chapter is about a range of different kinds of scopes—the scope of scopes. Telescopes and microscopes deserve their own chapters due to their importance and widespread use and variations. This chapter is about miscellaneous scopes—olioscopes.

The colposcope (cervix viewer) is used to examine the genitalia of women to determine the existence or absence of diseases such as cancer. The earliest such devices were invented by Hans Peter Hinselmann (1884–1959) in 1925 as an improvement over simple lenses[1]. Current designs use a coaxial illuminator and receiving lenses with beam splitters to allow multiple access to the images (figure 36.1). In this way an instructor and student can see the same thing, or the investigator can view and photograph at the same time. A representative schematic

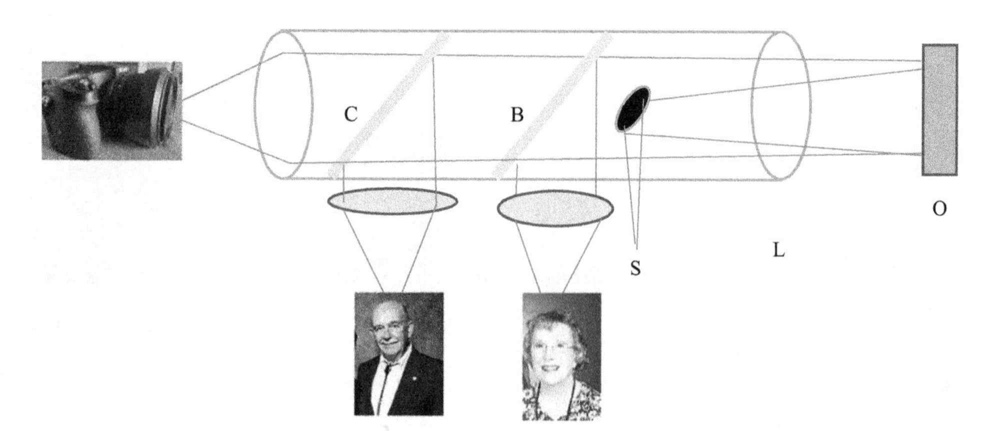

Figure 36.1. Colposcope schematic.

[1] Hinselmann H 1925 *Münch. Med. Wochenschr.* **77** 1733.

is shown here. The light is emitted from the source S and shone onto the object O by way of a collimating lens L and diagonal mirror. The light reflected from the object is collimated by the lens L and partially reflected by the beam splitter B to the first viewing port. The remaining light is reflected by the second beam splitter C to a second port. A third port is possible at the left end of the colposcope. Each of the ports has its own optics—eyeball, camera, etc.

The kaleidoscope (beautiful form viewer) was invented by David Brewster (1781–1868) in about 1815 and patented later[2]. It was an immediate success, selling about 200 000 in just three months in London and Paris[3]. It was later improved by Charles Green Bush[4]. It consists of a series of mirrors at different angles, typically 60°, and loose, colorful objects that provide entertaining viewing, especially as it is moved about.

The teleidoscope (remote beautiful form viewer) is a kaleidoscope with lenses. It was invented by John Lyon Burnside III (1916–2008) who obtained a patent in 1970[5]. It can be thought of as a telescopic kaleidoscope.

The otoscope (ear viewer) was first described and illustrated in France in 1363 by Guy de Chauliac (~1300–1368). In Italy in 1838, Ignaz Gruber (1803–1872) invented the first funnel-shaped speculum (mirror), although he didn't publish his findings. In Germany in 1864, E Siegle invented a pneumatic otoscope, a model that allows the user to administer air pressure. These scopes have been used for inspecting both ears and noses.

The periscope (near viewer) is said by some to have been invented by Johannes Gutenberg (~1395–1458), he of the printing press[6]. It is clear that he 'printed' a mirror that some used at a religious event as half a periscope. Some say it was Hippolyte Marié-Davy (1820–1893) in 1854[7]. Others say it was Thomas N Doughty during the American Civil War[8] or Simon Lake (1866–1945) in 1902[9], and improved by Howard Grubb (1844–1931). The rotary periscope was patented by Rudolf Gundlach (1894–1957) in 1936[10].

The endoscope (inside viewer) started with a rigid body, became semi flexible, then flexible with optical fibers, then with a small camera and a wire, and finally with a pill. It comes under many guises, depending upon which body part is being examined: hysteroscopy, laparoscopy, colonoscopy, etc. And this I love: esophagogastroduodenoscopy! The first, very primitive use was a speculum used by Hippocrates (~460–375 BC) for rectal and vaginal inspections. The first

[2] Brewster D 1817 *British Patent* 4136.

[3] S Talbot, *The Perfectionist Projectionist:* Phillip Carpenter, 24 Regent Street, London, online.

[4] Bush C 1873 *US Patent* 143271.

[5] Burnside J L 1970 *US Patent* 3661439.

[6] Wikipedia.

[7] Roscoe T 1949 *United States Submarine Operations in WWII* (Annapolis, MD: United States Naval Institute); Hoar A 2011 *The Submarine Torpedo Boat, Its Characteristics and Modern Development* (Redditch: Read Books).

[8] San Francisco National Park Association.

[9] Lake S 1930 Submarine, *The Autobiography of Simon Lake* (New York: Appleton).

[10] Wikipedia.

instrument that might really be called an endoscope was used in 1806 by Philipp Bozzini (1773–1809) and called a lichtleiter, a light conductor[11]. It was primitive by today's standards, a rigid device consisting of a light source, lens, and mirrors, but it was used for a fair number of anatomical cavities. The next major advance resulted when Harold Horace Hopkins (1918–1994) invented the fiber optic endoscope, which he called the fibroscope[12]. His invention was accompanied by simultaneous versions by A C S van Heel (1899–1966), Narinder S Kapany (1926–) and B O'Brien Sr (1898–1992). Then came an improved version in which a small camera is at one end of the tube and a receiver at the other. In fact, it seems the receiver can be anyplace nearby. The ultimate seems to be the pill[13]. A CCD camera is placed inside a large pill along with a transmitter. The pill is swallowed and wends its way through the digestive system, but it can be guided by an external magnet. The images are transmitted to an external receiver.

The borescope (hole viewer) is a device that is used to inspect the inside of tubes. They come in both rigid and flexible varieties. The flexible ones use fibers. Either can have an objective and an eyepiece and a folding mirror or prism to reflect the side of the bore to the viewer. Some also have an illumination device. As with endoscopes, there is now a version that uses a video camera at the distal end. One version using fiber optics was invented by Brian O'Brien and Narinder Kapany of American Optical Company and Imperial College, respectively.

The stereoscope (solid viewer) allows us to see two-dimensional things in what seems to be three dimensions. We see regular three-dimensional objects by virtue of the fact that we have two eyes. They each see the object at a slightly different angle, and our mind does the triangulation. We only see in three dimensions at a limited range, a range in which the triangulation angles are sufficiently large. So in 1838 Charles Wheatstone (1802–1875) arranged two views of the same scene at slightly different angles and combined them with mirrors[14]. He also opined that had Leonardo da Vinci (1452–1519) used a cube rather than a sphere in an illustration, he would likely have understood this triangulation effect. Brewster improved Wheatstone's device in 1849 by using lenses rather than mirrors. Other improvements came from Jules Duboscq (1817–1886)[15] and notably by Oliver Wendell Holmes (1809–1894), who made what is undoubtedly the most popular version in 1859.

Three-dimensional movies and television are now with us. They are based on the same principles of triangulation. For most movies, two images of different polarization are presented and viewed with polarized glasses. Avatar may be the most popular of these. The potential for holographic television is with us (see Holography, chapter 12).

[11] Bozzini P 1806 *J. prakt. Arzneyk. Wund. neyk.* **24** 107.

[12] Berg G 1995 *Surg. Endosc.* **9** 6; Hopkins H and Kapany N 1954 *Nature* **173** 39.

[13] Stryker.com for instance.

[14] Wheatstone C 1838 *Philos. Trans. R. Soc. Lond.* **128** 371.

[15] Brenni P 1996 *Bull. Sci. Instrum. Soc.* **51** 7.

The riflescope (rifle or groove viewer) is called by various names, but it is a form of telescope mounted on a rifle. It would not be surprising if shortly after the telescope was invented that someone tried it on a rifle. It provides a magnified view of the target and an overlay of an aiming reticle. The first documented version of a rifle sight or riflescope was by Morgan James in 1836[16]. Zeiss offered a riflescope based on the design of Sigurt-Horstmar Freiherr von Beaulieu-Marconnay (1900–1953) in 1892. Variable power scopes appeared in the late 1940s. Incremental changes were made until Swarovski introduced the laser rangefinder with the riflescope. It seems everybody is the best riflescope maker and all made the first one, according to the internet!

The sniperscope (sniper viewer) also called the snooperscope, is a rifle-mounted aiming device that uses an infrared searchlight and a near-infrared sensitive receiver, called an image converter tube[17]. Some reserve the word snooperscope for a sniperscope without a rifle. The converter tube is a standard photosensitive cathode that has response into the near infrared. The searchlight has a filter that passes near infrared but blocks visible light[18].

Metascopes (beyond viewers) were another version of the image converter tube and also served as snooperscopes and other night vision devices[19]. They employed phosphors that had responses further into the infrared.

The iconoscope (image viewer) was an early television camera invented[20] by Vladimir Kosmich Zworykin (1888–1982).

The oscilloscope (oscillation viewer) is just barely an optical device in which the electrons excite the phosphor to create the trace on the screen. The oscilloscope was preceded by several devices that recorded electrical waveforms, mostly based on the movements of galvanometers. One of these, by William Du Bois Duddell (1872–1917), used the galvanometer to scan across photographic paper[21]. The modern oscilloscope began with the invention of the CRT, Cathode Ray Tube, by Karl Ferdinand Braun (1850–1918) in 1897[22]. The early versions were rather unstable due to variations in the vacuum and the emitter. In 1931, Zworykin developed a high-vacuum, permanently-sealed version with a thermionic emitter that was usable in laboratories[23].

The stethoscope (chest examiner) is not an optical device. In this case, the scope arises from scopos, which means examination.

A horoscope is an entirely different matter, and a horrorscope is only valid at the end of October!

---

[16] Chapman J 1844 *The Improved American Rifle* (New York: Bienfield Publishing).

[17] Pratt T 1947 *J. Sci. Instrum.* **24** 312.

[18] Morton G and Flory L 1946 *Electronics* **19** 192; Morton G and Flory L 1946 *RCA Rev.* **7** 385.

[19] O'Brien B 1946 *J. Opt. Soc. Am.* **36** 369.

[20] Zworykin V and Morton G 1940 *Television: The Electronics of Image Transmission* (New York: Wiley).

[21] Hawkins N 1917 *Hawkins Electrical Guide 6* (New York: Audel).

[22] Keller P 1991 *The Cathoderay Tube: Technology, History, and Applications* (New York: Palisades Press).

[23] Kularatna N 2003 *Digital and Analogue Instrumentation testing and Measurement* (London: Institution of Engineering and Technology).

Chapter 37

Solar energy—the Sun is my doing

The intentional use by man of solar energy has a long history; the use of solar energy on this planet, by man and other species, is as old as Earth itself. The Sun is our ultimate source of energy. All of our fossil fuels have their origin in the energy from the Sun. Our vegetables and our meats and our eggs come ultimately from the energy provided by the Sun. This section, however, deals only with the direct, intentional uses of the Sun by man. Farming and mining do not count! The two main methods of using solar energy are direct heating of some fluid and conversion to electricity by a photovoltaic effect, often called a solar cell.

Probably the oldest direct uses of solar energy were about 700 BC when they used magnifying glasses for starting fires. (And I had to rub two Boy Scouts together!) A few centuries later, Greeks were using mirrors to do the same thing. It is reported that in 212 BC, Archimedes (~287–212 BC) used a collection of shields brandished by soldiers to set fire to the Roman ships laying siege to Syracuse[1]. Roman bathhouses and Anasazi pueblos had south-facing rooms to collect the sunlight. Horace-Bénédict de Saussure (1740–1799) is the first to be credited with using solar energy to cook. He used a stack of boxes with glass covers and attained a temperature of 190 °F (88 °C) in 1767[2]. He later improved the apparatus by using several glass plates and a completely black interior, thereby obtaining a temperature of 230 °F (110 °C). Thus was born the ancestor of the greenhouse. Both John Frederick William Herschel (1792–1871), of infrared fame in the UK, and Samuel Pierpont Langley (1834–1906), of bolometer fame in the US, were intrigued and made their own versions in 1880 and 1881. They confirmed de Saussure's idea that it was the glass that blocked the return of heat and that it was atmospheric constituents in the open air[3].

---

[1] *Time*, November 1973; The history of solar.
[2] Horace de Sausere and his hot boxes of the 1700s, online.
[3] S Butti and J Perlin, A golden thread 2500 years of solar architecture and technology.

doi:10.1088/978-0-7503-2612-4ch37

Shortly thereafter in 1860, Augustin Mouchot (1825–1911) proposed making a steam engine using this effect, one in which solar energy was collected with a concave disc in order to generate steam. In 1908, William J Bailey invented a solar water heater with coils on the top that were heated by the Sun and a storage vat inside the house[4].

In 1839, Alexandre-Edmond Becquerel (1820–1891) discovered the photovoltaic effect in which light is converted to electricity. The first solar cell was invented by William Weber Coblentz (1873–1962) in 1913[5]. In 1953, three guys at Bell Labs developed the first viable photoelectric cell: Gerald Pearson (1905–1987), Daryl Chapin (1906–1995), and Calvin Souther Fuller (1902–1994).

William G Cobb (1918–2011) demonstrated the first solar-powered car, although it was really just a model car, in 1955[6]. Ford has described a solar-powered car of limited performance. It is an electric vehicle that plugs in to the grid but also has photovoltaic panels on its roof. The solar generation provides about 20 miles of travel[7]. Hans Tholstrup (1944–) drove the first solar-powered car between Sydney and Perth, Australia, in 1982. The Solar Team at Eindhoven University of Technology in the Netherlands demonstrated 'the first solar-powered family car' at a solar meet in Australia in 2013. It seats four, has a trunk, and has a range of about 300 km[8].

The first solar-powered airplane, the Solar Challenger, was built by Paul MacCready (1925–2007) in 1981. It was able to fly across the English Channel[9]. The reference provides more information about solar flight.

Perhaps the first solar building, a passive solar building, in the US was built in 1947[10]. There is a proclaimed fully solar house in the UK (for $1.2 million) that is also connected to the grid. It uses a combination of photovoltaic panels on the roof for electricity and thermal tubes on the roof and underground for heat. It has thermally insulated walls and triple-glazed windows[11]. One anonymous blogger in that article notes that he had his first all-solar home in 2008 by converting an old barn. The Bridgers–Paxton building was the first commercial building heated by solar water heating.

The Wolfe family installed a photovoltaic system on their roof in 2012[12]. It has proved economical.

A different way of converting solar energy into another form has been investigated at MIT. It is a double layer of carbon flakes and carbon foam that floats on water. Capillary action draws the water up where it is turned to steam by the

[4] California Solar Center, (2001); US Department of Energy, (2002).

[5] Coblentz W 1913 *US Patent* 1077219.

[6] William G Cobb demonstrates the first solar-powered car, *History Channel*, online.

[7] Ford to Introduce First Solar Powered Car, *Ecowatch* (2014); *BBC News* (2014); *Computerworld* (2014), all online.

[8] *Clean Technica*, online; Technische Universiteit, Eindhoven, online (2013).

[9] Solar Research and Dryden, Dryden Flight Research Center NASA, online.

[10] *The history of solar*, Office of Energy Efficiency and Renewable Energy, online.

[11] V Woollaston, *Mail Online*, online (2013).

[12] W Wolfe, personal investment.

sunlight that was absorbed by the top black layer. It is not clear how the steam is or will be captured for useful purposes[13].

Solar panels will improve by leaps and bounds, by better quantum efficiency with additional materials, better photon capture by reflection techniques and even whistling modes. The reduction in cost and improvement in efficiency will lead to widespread use and a much greater percentage of our electricity grid. But it will never be enough for all our uses—autos, planes, trains, etc. A society that is fossil-fuel free will need nuclear energy.

[13] Daukantas P 2014 *Opt. Photon. News.*

IOP Publishing

Rays, Waves and Photons
A compendium of foundations and emerging technologies of pure and applied optics
William L Wolfe

Chapter 38

Sources—and there was light

The very first source, of course, was when God said, 'Let there be light[1].' Or, if you prefer, when the Big Bang occurred and the Universe was one big ball of radiating plasma. Humans first saw the light in the Sun, our primary source of light, heat, and life. Then ancient man discovered flame, and he could cook and see by its light. Throughout much of history we used flame in one form or another to see. Kerosene lamps were a standard. Probably ancient man also used indented rocks and shells to hold animal fats to make a controlled fire. Since then there were arc lamps, fluorescent lights, light emitting diodes, and lasers. I have delegated lasers to a separate chapter because they are such an interesting and important light source.

A **lantern** is a form of lighting. Typically it is a frame with transparent panes of some material enclosing the source of light. They were cited in use in Ancient Greece[2], Egypt[3], and Israel[4], as well as China. We are all familiar with the decorative **Chinese lanterns** of different shapes and colors. Some are even said to be illuminated with fireflies[5]. The **Jack O' Lantern**, our carved pumpkin of Halloween, is said to have come to us from the Irish. Stingy Jack tricked the devil several times, one of which got the devil to say that he would not claim Jack's soul. When Jack died, God would not let him into heaven as a result of his misdeeds. So Stingy Jack was banned from heaven and hell, and he roamed the black void with only a coal to light the way. He soon carved a turnip to carry it in. The Irish did not have pumpkins; we do, and they are better than turnips. So this is Jack of the lantern, or Jack O' Lantern[6].

---

[1] *Genesis* 1:3.
[2] Wikipedia on lanterns.
[3] Ibid.
[4] *John* 18:3.
[5] See footnote 2.
[6] History, online.

The **Argand lamp** was a specially designed wick-type device that caused air to pass through the center of the wick and around its sides, thereby increasing and stabilizing the output. It was invented in 1781 by François Pierre Ami Argand (1750–1803)[7].

There is a wide variety of **kerosene lamps** with different designs of the wick and the airflow. Probably the first of these was by al Abu Bakr Mohammad ibn Zakariya al-Razi (865–925) in Baghdad in the ninth century[8]. A later version was developed by Jan Józef Ignacy Łukasiewicz (1822–1882) in 1853.

The **Clamond basket** was invented and patented by Charles Clamond[9]. It consisted of a mantle made of magnesium oxide, much like today's propane camp lanterns (figure 38.1).

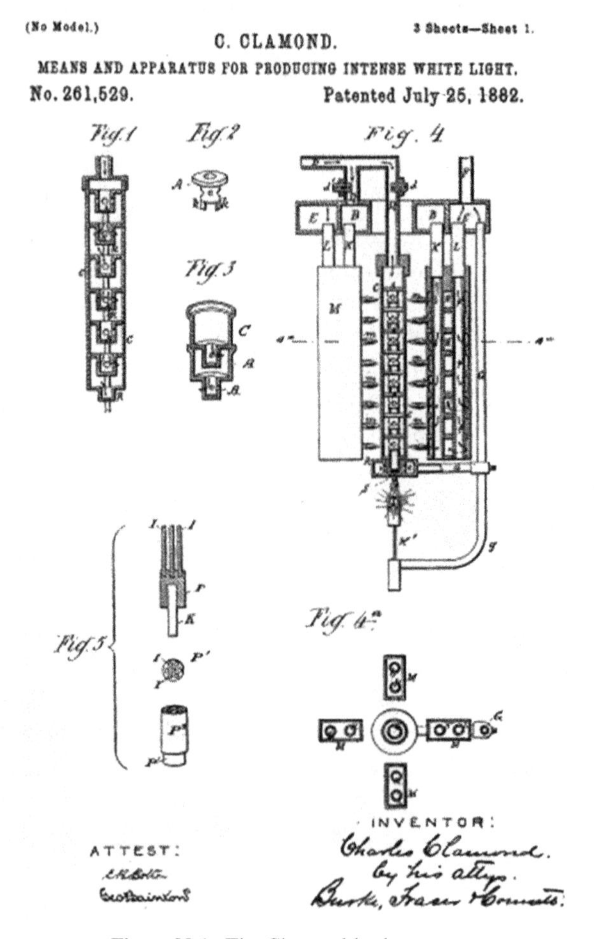

Figure 38.1. The Clamond basket patent.

[7] *Encyclopedia Britannica.*

[8] *Jewish Virtual Library* and *Encyclopedia Britannica*, both online.

[9] Clamond C 1882 *US Patent* 261529.

The **Welsbach mantle** was invented by Carl Auer von Welsbach (1858–1929). It consisted of a mantle made of cotton or other cloth that was impregnated with various oxides. He experimented until he finally chose thorium and cerium dioxides in 1891. That seemed to give the brightest light. We still use these lanterns when we go camping.

The **Davy lamp**, named for Humphry Davy (1778–1829), was a metal mesh-enclosed wick lamp that burned methane that was invented in 1815[10]. Its secret was that the fine mesh prevented the flameouts that might occur due to the methane in the coal mines. It was a miner's lamp. A variation that preceded the Davy lamp was invented by William Reid Clanny (1776–1850) that used a glass enclosure and a water-filled bottom through which air was forced, thus preventing any explosion of methane that escaped[11].

Arc lights or **arc lamps** are light sources that are generated by creating an electric arc between two electrodes by virtue of the potential difference between them. The electrodes can be carbon—thus a carbon arc lamp—or tungsten. They are enclosed in a gas of some sort: neon, argon, krypton, mercury, sodium, or some metal halide. The first arc lamp was constructed by Davy in the early 1800s using charcoal sticks and a bank of batteries[12].

Carbon arcs were used in the United States sometime after the generation of a constant electric voltage was solved[13]. The **Vortek lamp** was an arc lamp with a fused quartz tube that enclosed it and a flow of water around the tube to protect it from the heat of the arc. It was invented by David Camm and Roy Nodwell in 1975[14]. It is notable for its high intensity—over 1.2 million candlepower[15]. Perhaps the most famous use of the carbon arc lamp is as a searchlight due to Elmer Ambrose Sperry (1860–1930) in 1915. It has been used for aircraft defense in the Battle of Britain and now to advertise many county fairs. Sperry was also the inventor of the gyroscope, and I worked for his company for two summers.

Contrary to most common thought, mine included, Thomas Edison (1847–1931) did not invent the light bulb. There were apparently twenty or more who preceded him[16]. The **electric filament lamp** has a long history with many failures. Perhaps the first try was by Davy in 1809, but it was just a glowing platinum bar between two electrodes that did not last very long[17]. It was followed some thirty years later by James Bowman Lindsay (1799–1862)[18].

---

[10] Davy H 1816 *Proc. Trans. R. Soc. Lond.* **106** 1.

[11] Knight D 1992 *Humphry Davy: Science and Power* (Cambridge: Cambridge University Press); *History of Miner's Safety Lamps*, Canadian Mining Museums (2003) available online.

[12] Slingo W and Brooker A 1900 *Electrical Engineering for Electric Light Artisans* (London: Longmans).

[13] *Scientific American* **44**, 14 (1841).

[14] US patent 4021785 (1975).

[15] Approximately 20 000 watts per unit solid angle in the visible, almost all in the searchlight beam.

[16] Friedel R and Israel P 1986 *Edison's Electric Light: Biography of an Invention* (New Brunswick, NJ: Rutgers University Press).

[17] Davis L 2003 *Fleet Fire* (New York: Arcade Publishing).

[18] Fahie J 1902 *A History of Wireless Telegraphy* (New York: Dodd) available online.

Limelight has been a stage lighting. An intense flame of oxygen and hydrogen is directed at a cylinder of calcium oxide—lime. The flame and the material both emit light, and the CaO_2 does not melt until it reaches 2570 °C, so the emission approximates a blackbody at close to that temperature. That would be 2843 K and just about 1 μm. It was discovered in the 1820s by Goldsworthy Gurney (1793–1875)[19]. Although it is no longer used, you can still be in it!

The **Nernst glower**, used mostly in spectrometers, and a forerunner of the incandescent lamp, is a cylindrical rod of zirconium, yttrium, and erbium oxides heated to about 2000 °C. It was developed by Walther Hermann Nernst (1864–1941) in 1897[20] (figure 38.2). It has been largely replaced by the **Globar**[21] that is silicon

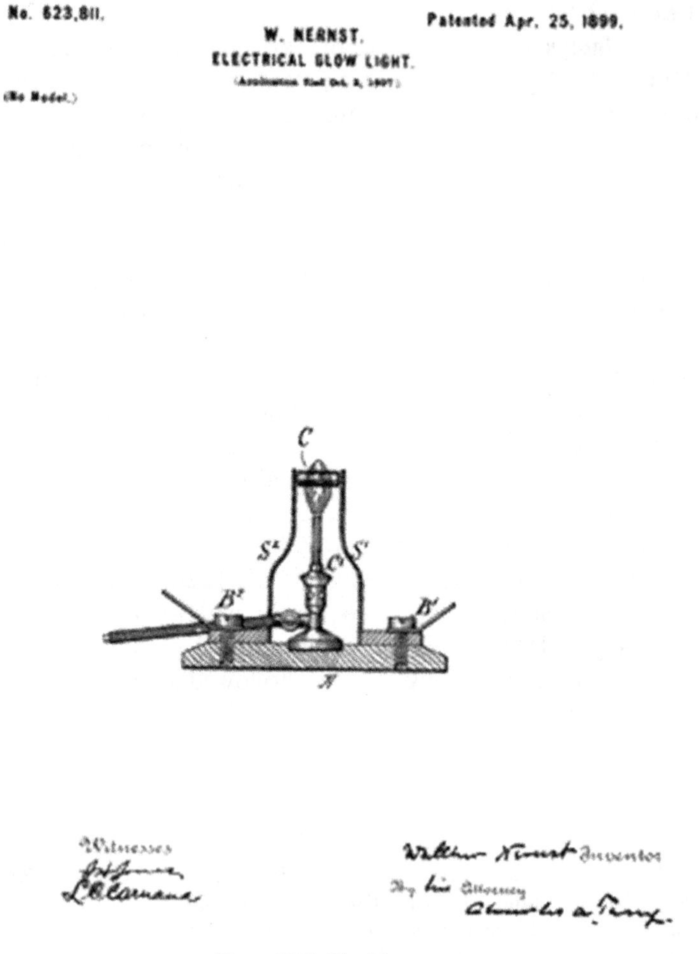

Figure 38.2. The Nernst patent.

[19] James F 1996 *The Correspondence of Michael Faraday* (London: Institution of Electrical Engineers).
[20] Monmouth H 1898 *Science* **11** 689.
[21] US Registration 0200201 (1925).

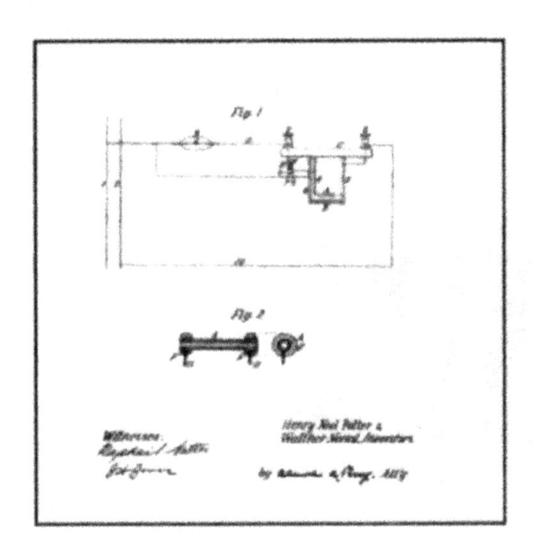

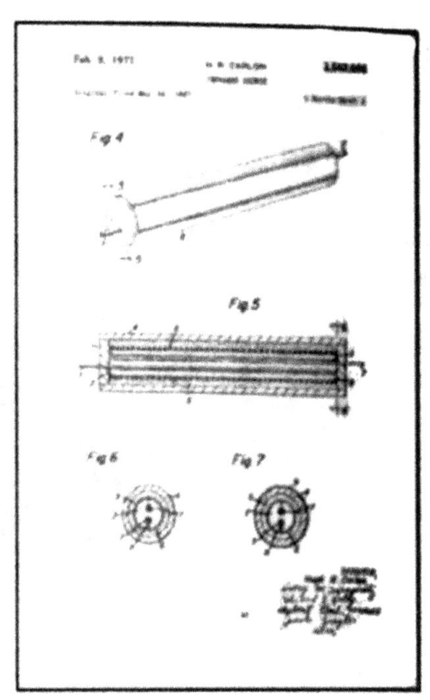

Figure 38.3. The Globar patent.

carbide heated to about 1100 °C (figure 38.3). Both are used because they are a good approximation to a blackbody.

Fluorescent lamps are a form of arc lamp. A high- or low-pressure arc is created within the tube, and the arc provides excitation of the coating on the sides; that is the luminescence or fluorescence. Fluorescence had been observed many years before, notably in fluorite. Alexandre-Edmond Becquerel (1820–1891) in 1857 theorized about coating the walls of Geissler tubes, electric discharge tubes, with luminescent materials to make light bulbs[22]. But it was Peter Cooper Hewitt (1861–1921) who made the first successful fluorescent lamp[23] using mercury vapor. Edmund Germer (1901–1987) developed a high pressure version in 1927[24], and George E Inman patented one that was more economical and produced less heat[25]. There was a flurry of patents and purchases beclouding the issue of prior rights, but these are the early players. By the way, I do not know why it is called fluorescence, meaning to flow, but I do agree with luminescence, full of light. Maybe it is flowing light.

LED's, **light emitting diodes**, had their beginnings in the studies of luminescence, so-called cold light, produced without high temperatures like incandescence (which does). For many years after the discovery, they were merely items of curiosity, partly because the world had the incandescent bulb, and it was good. Partly because it was

[22] Becquerel A 1839 *C. R.* **9** 561; Becquerel A 1840 *C. R.* **11** 702.
[23] Cooper P 1901 *US Patent* 889692.
[24] Germer E 1927 *US Patent* 2182732.
[25] Inman G E 1938 *US Patent* 2259040.

involved in two world wars. In the 1950s, interest was aroused, and more work with the new background in solid state physics and the transistor provided the wherewithal for advancements. At first, these LED's were just indicator lights since they were not very bright, but as more colors were developed and brightness increased, they became viable as flashlights, tail lights, TV and computer displays, point-of-sale scanners, and lighting in general. Lighting accounts for about 25% of our total energy use, and these LED's are far more efficient than incandescents. We will soon see their widespread use, and we should (figure 38.4).

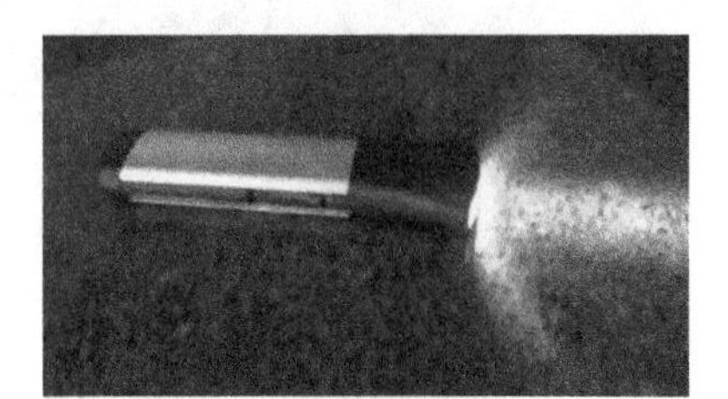

Figure 38.4. LED flashlight

The first versions were invented in 1907 by H J Round (1881–1966), who used a cat's whisker and silicon carbide[26]. A rather unknown scientist in Russia, Oleg Vladimirovich Losev (1903–1942), is said to have done it at the same time but did not publish until later[27]. He measured the characteristics of a diode made of silicon carbide and a metal. It was many years later in 1961 that James R 'Bob' Biard (1931–) and Gary Pittman (1930–2013) developed an infrared LED of gallium arsenide while they were working on solar cells and lasers[28]. The next year in 1962 Nick Holonyak Jr (1928–) produced the first visible LED in the red[29] using gallium arsenide phosphide—GAsP! In 1972, M George Craford produced the first yellow version[30]. Then Shuji Nakamura (1954–) made a blue one in 1979[31]. Ching Wan Tang (1947–) developed the first LED made of organic materials[32] in 1979.

Cherenkov radiation is created when a charged particle, usually an electron, passes through a dielectric medium faster than the phase velocity of light. It has a continuous spectrum with emphasis in the ultraviolet. It is named after Pavel Alekseyevich Cherenkov (1904–1990), who first discovered it[33]. It has limited uses in tagging biomolecules, astrophysics analysis, and nuclear reactors.

[26] Round H 1907 *Electr. World* **49** 309; Shubert E 2003 *Light Emitting Diodes* (Cambridge: Cambridge University Press).

[27] Losev O V 1928 *Phil. Mag.* **6** 1024.

[28] Biard J R and Pittman G 1966 US Patent 3293513.

[29] *Bloomberg Business Week*, May 22 (2005).

[30] Craford M G 1995 *IEEE Spectr.* **32** 52–5.

[31] *Pacific Coast Business Times*, November 12, (2014).

[32] Cherenkov P 1934 *Dokl. Akad. Nauk SSSR* **4** 451.

[33] Cherenkov P A 1979 *US Patent* 4164431.

IOP Publishing

Rays, Waves and Photons
A compendium of foundations and emerging technologies of pure and applied optics
William L Wolfe

Chapter 39

Spectacles—better to see you with

One of my good friends once said to me, 'Since you have so much experience in optics, you can fix my glasses any time.' I replied, 'Only if they are full.' But eyeglasses, spectacles, are the subject of this chapter.

The story of eyeglasses begins in Venice or Pisa. They were double convex lenses that were on a variety of frames—whatever would work. The first important advance was the bifocal by Benjamin Franklin about the time of our Revolutionary War. Then there were trifocals some fifty years later, and then multifocals, or continuously variables, at about 1900. The first spectacles that could correct for astigmatism came in 1825. Safety glasses appeared on the scene in 1880, sunglasses in 1929, and Polaroids in 1932. Photochromic glasses showed up in the 1960s, and contacts at about the same time. The non-glasses, Google glasses, arrived in the twenty-first century. Self-adjustable glasses and bionic glasses for the really blind are still under development.

In chapter 17 on lenses, it was reported that the Nimrud lens was thousands of years old, but that was not a spectacle. It has also been reported that Nero used a gem to better watch the gladiators, lions, and Christians. That would be a spectacle for a spectacle. But it was not. It was just a lens. Seneca (4 BC–65) is reported to have used a globe of water as a magnifier, but that, too, is not an eyeglass; it is a magnifier. Most historians consider that the first spectacles appeared in Pisa or Venice around 1285. They were double convex lenses for each eye and could therefore only correct for farsightedness (hyperopia) and lack of accommodation (presbyopia), and they were put in frames that could be balanced on the nose. Salvino degli Armati (~1258–1317), also known as Alessandro Delia Spina, is credited by some with their invention[1] but disputed by others[2]. It does seem from all

[1] Manni M 1738 *Historical Treatise on Eyeglasses, Invented by Salvino Armati, Florentine Gentleman; Creación y evolución de las gafas—SlideShare* online.
[2] Ilardi V 2007 *Renaissance Vision from Spectacles to Telescopes* (Philadelphia: American Philosophical Society); Ronchi V 1946 *Rivista di oftomaliga* **1** 140; Rosen E 1956 *J. Hist. Med. All. Sci.* **11** 13.

doi:10.1088/978-0-7503-2612-4ch39

these sources that spectacles were born in Pisa and Venice. Good discussions of the origin and its uncertainty are in those first two references.

Temples, the structures that go back over the ears, were invented by Edward Scarlett (1677–1743) in 1730[3]. Before that, people used strings, ribbons, thongs, and whatever worked.

Benjamin Franklin (1706–1790) is well known for his invention of **bifocals** in 1784, and there is some evidence that he used them about 50 years earlier[4]. Bifocals, logically enough, are eyeglasses that have lenses with two different foci. (I prefer foci to focuses—and octopi to octopuses.) The upper part corrects for distance vision, whereas the lower part corrects for close vision, usually reading. As with many inventions, there is a bit of uncertainty. Benjamin West (1738–1820), a friend of Franklin, is mentioned as an inventor, as is Joshua Reynolds (1723–1792), a predecessor of West at the British Royal Academy, but there is no substantiation[5].

Trifocals were introduced by John Isaac Hawkins (1772–1855) in 1827. He also coined the term 'bifocals' and credited Franklin with their invention[6]. Trifocals have three different foci; there is one in the middle for intermediate distances.

Cylindrical lenses that could correct astigmatism were invented by George Biddell Airy (1801–1892) in 1825[7]. He is also the Airy of whom the diffraction pattern is named.

Continuously variable corrective lenses—sometimes called varifocals, Varilux, Progressive, and Multifocal—were invented by Owen Aves in 1907[8]. This was followed by versions developed by H Newbold in 1913, William Stewart Duke-Elder (1898–1978) in 1922, and Irving Rips (1935–2014), the founder of Younger Optics, in 1955 to make you look younger. They were called the Younger Seamless Bifocal, the first commercially viable version. Varilux is the trade name of a modern version of multi-focal eyeglasses. They were designed by Bernard Maitenaz (1926–) in 1959[9]. These lenses have continuous varying focal length from top to bottom, with long-distance correction on the top to close-viewing on the bottom.

Invisilines are bifocals or trifocals that do not have a line across them; they are seamless with a smooth change in magnification, but they are not continuously variable. They were also the creation of Irving Rips in 1955[10].

Modern sunglasses were invented by Sam Foster (1883–1966), of Foster Grant, in 1929[11]—'on the boardwalk in Atlantic City♫.' The answer **to** 'Who's behind those Foster Grants?' is Sam. Chinese judges used smoked glasses to conceal their expressions during hearings back in 1300. The Inuit used protective eye covers

[3] The Foundation of the American Academy of Ophthalmology, online.

[4] The College of Optometrists, online.

[5] Ibid.

[6] Ibid.

[7] Gillespie C 1970 *Dictionary of Scientific Biography* (New York: Scribners).

[8] Aves O 1907 *UK Patent* 15735.

[9] Leroux J 2007 *L'épopée Varilux* (Paris: Perrin); Maitenaz B Wikipedia online.

[10] Vision Monday, December 2013, online.

[11] P Ament, *Sunglass History—The Invention of Sunglasses*, online.

with slits in them around 1000–1600[12]. In 1752, James Ayscough (1720–1759) used tinted glasses that he believed corrected for some eye impairments.

Edwin H Land (1909–1991) invented **Polaroid** in 1929[13] and tried the material in sunglasses in 1936[14]. He founded the Polaroid Company in 1937. Polaroid material was originally a polymer that contained long needles of herapathite that were oriented in a particular direction and thereby polarized or analyzed light. Herapathite was discovered by William Bird Herapath (1820–1868) in 1852[15]. Later versions used PVA, polyvinyl acetate, and other organic chains. Most natural light is polarized, notably the light reflected from a highway or bodies of water. The orientation of polarization in **polarized sunglasses** is such that it cancels much of the reflected glare from those objects. I love them for fishing!

Photochromic lenses, or photochromatic lenses, were invented by Roger Araujo in the 1960s. The photochromatic effect—or chromatism, or chromaticalism—has been observed in animals and a variety of inorganic and organic materials. The incidence of optical radiation, particularly in the ultraviolet, changes the chemical bonds and results in a slightly different substance that then can absorb light. The effect was discovered in the 1880s by Willy Marckwald (1864–1942), who labeled the effect phototropy. Yehuda Hirschman later studied the effect and labeled it photochromism[16].

A cautionary word about sunglasses: always make sure they have ultraviolet protection. The darkening of the glasses results in a dilation of the pupil, making it more susceptible to damaging ultraviolet rays.

Lorgnettes are spectacles on a stick. They have been used mostly as fashion accessories by aristocratic ladies, especially at the opera. Therefore, they are sometimes called opera glasses, but opera glasses are really a form of binocular. They were invented by George Adams (1720–1773) in 1790 (see Binoculars, chapter 3).

Scissors-glasses (binocles-ciseaux) in some cases replaced monocles. One explanation is that they are two monocles connected in a Y-shaped frame that folded. Some were highly ornamented and hung on a fancy ribbon or chain. They were quite fashionable in the 1800s. They are, translated from the French, scissor binoculars.

Pince-nez glasses had no temples or sticks (like lorgnettes) but had two eyepieces that were held in place by squeezing the nose with a spring. Pince nez—squeeze nose. Although they existed before the invention of temples, they did not really gain popularity until they reappeared in 1840[17].

Browline, **cat's eye**, and **bug eye** are three different styles of frames. Their names do a pretty good job of describing their appearance.

---

[12] Photography within the Museum, Canadian Museum of Civilization, online.
[13] Land E 1929 *US Patent* 1918848.
[14] Advertisement in *Life* 7, 71 (1939).
[15] Herapath W 1852 *Philos. Mag.* **3** 151.
[16] Hirshberg Y 1956 *J. Am. Chem. Soc.* **78** 2304.
[17] Wikipedia.

Contact lenses go back a long way. Leonardo da Vinci (1452–1519) was probably the first to air the concept in 1508[18]. René Descartes (1596–1650) suggested placing a lens directly on the eye—just a suggestion. John Herschel (1792–18871) provided the first practical design in 1823, using a mold of the eye. The first one was manufactured in 1887 by F A Muller, and the first plastic one in 1936 by William Feinbloom (1904–1985), actually with a center of glass. Then one was made completely of plastic by Kevin Tuohy (1919–1968) that covered just the cornea in 1948[19]. Soft plastic versions were developed in the 50s, and the first commercially available one was produced by Bausch and Lomb in 1971[20].

Magnetic contact lenses are being investigated by the Navy. A sensor on the head would detect any changes in the magnetic field of the contact to determine the direction of vision. But this is only a gleam in the eye at this point.

A telescopic contact lens has been developed for those with damaged retinas by Eric Tremblay and his team[21]. It magnifies the image and aims it to a healthy part of the retina.

Safety glasses do not correct impaired vision, but protect the eyes from being impaired. They come in at least three varieties: glass in helmets, like welders; cups that fit tightly over the eye; and formats similar to spectacles. They were invented by P Johnson in 1880 as an 'eye protector.' They were incorporated in gas masks for World War 1 by Garrett Morgan (1877–1963). One interesting article from *Good Roads Magazine* in 1897 advertises the Lamb Eye Shield for bicyclists for 25 cents, postpaid! Originally made of layered plastic, most are now polycarbonate.

Google Glass is not a means of improving vision, but instead a means of displaying computer information. The entire system is a small computer with a GPS, a 5 megapixel camera, and other apps. The display is heads up with 640 by 360 pixels and is just outside of and above the eye. They really do look like a pair of glasses. They were developed by Babak Parviz (1973–), Steve Lee, Sebastian Thrun (1967–), and maybe others at Google. A prototype was on hand in 2011. It went commercial soon after and has several improvements since[22].

Self-adjustable glasses have recently made their appearance. They have been promoted for Third World countries and as **Superfocus** glasses for those with more cash. The simple ones use lenses of flexible plastic with a liquid inside. A plunger on the side can be used to apply pressure to the fluid, thereby making changes in the curvature of the lens. This does not correct for astigmatism but does handle nearsightedness and farsightedness, as well as accommodation problems—myopia, hyperopia, and presbyopia. The adjustments can be made without the help of an optometrist, who are few and far between in Third World countries. The so-called Superfocus glasses add a corrective lens[23].

---

[18] Heitz R and Enoch J 1987 *Advances in Diagnostic Visual Optics* (New York: Springer) p 19.

[19] Tuohy K 1948 *US Patent* 2510438.

[20] *The History of Contact Lenses*, online.

[21] Tremblay E 2013 *Opt. Express* **21** 15980; Tremblay E 2013 *Opt. Photon. News* **12**.

[22] Google Glass, Wikipedia.

[23] The Future of Glasses, on the net.

Google has just announced its forthcoming contact lens that monitors glucose. If they had a Google glass, is this a **Glucoseglass?** It monitors the glucose content of tears and can use LED's for warning, but it also sends the information to a receiver elsewhere[24]. I have not been able to determine the method of measuring the glucose, but it is probably a miniature colorimeter.

A similar device has been designed to measure the onset of diabetic autonomic neuropathy. It consists of a pupilometer mounted on eyeglass frames[25]. An infrared LED illuminates the eye, and the reflected light is again reflected by a beam splitter to an infrared camera. Software does the diagnosis based on the imagery.

RK, or **radial keratotomy**, is the process of reshaping the eye by cutting radial incisions in the cornea. It was first accomplished by Tsutomu Sato (1920–1960) in 1936[26].

LASIK, **laser-assisted *in situ* keratomileusis** (reshaping the cornea), is a way of eliminating eyeglasses. This reforming of the cornea can correct for myopia, hyperopia, and astigmatism, but not for presbyopia. The predecessor to this work was carried out by José Ignacio Barraquer Moner (1916–1998) in the 1950s. He cut thin layers of the cornea, about 10 μm thick, to reshape it.

The most critical part of LASIK is the determination of the shape of the cornea and the amount of tissue to remove at each particular place. One way of measuring the eye for these LASIK operations is with **optical coherence tomography**, OCT. This technique uses broadband light to form an interference pattern in a region where the two interfering beams have traveled just about the same distance, within one coherence length (see appendix C). Light from other distances will not interfere. By scanning this spot over a two-dimensional area, a pattern can be generated, and then the depth changed and the scan repeated to get a three-dimensional image of the cornea or other part of the eye. The first of these was presented at a meeting in Germany by Adolf Friedrich Fercher (1940–2017) in 1990[27].

Not a spectacle, but perhaps spectacular, is the new **bionic eye**. A micro camera is mounted on the head, much like a Google Glass. It records images as electrical signals, and they are transmitted to the ocular nerve. This gives limited eyesight to those with maladies like retinitis pigmentosa. The idea of sending signals to the visual cortex dates back to many dreamers, but did not become a reality until the 21st century[28], when it was accomplished by William Harvey Dobelle (1941–2004), although developments were underway since 1968 by Giles Skey Brindley (1926–)[29]. Dobelle's apparatus consisted of about 60 sensors and connections; he figured he could improve to over 500, and today we should be able to do better electrically, if not neurologically. Present systems are still in black and white and provide only outlines of things. But if we can do that, we will surely do better in the future.

---

[24] Mendoza M 2014 *Arizona Daily Star* January 17 (via the Associated Press).

[25] Bembia H 2014 *Opt. Photon. News*.

[26] Sato T 1939 *Acta Soc. Opthtahmol. Jpn.* **43** 544.

[27] Fercher A 1990 *Proc. of the Int. Conf. on Optics in the Life Sciences (Garmisch-Partenkirchen, Germany)*; Fercher A *et al* 2003 *Rep. Progress Phys.* **66** 239–303.

[28] Dobelle W 2000 *ASAIO J.* **46** 3.

[29] Brindley G and Lewin W 1968 *J. Physiol.* **196** 479.

IOP Publishing

Rays, Waves and Photons

A compendium of foundations and emerging technologies of pure and applied optics

William L Wolfe

Chapter 40

Spectra of spectroscopy—ROYGBIV

Spectroscopy, the technique of using the different wavelengths of light to obtain information or perform a task, probably began with Newton. He first showed that 'white light' from the Sun consisted of many colors in about 1700. Then there were solar observations by Fraunhofer 100 years later, and the realization that a spectrum is a unique identifier of a substance by Bunsen and Kirchhoff about 50 years after that. Early spectrometers used prisms; gratings were often better and came later. Then many different arrangements were invented over the years. These included the double pass and double beam. The latter really made a difference for me. Instead of carefully adjusting the slit and running a calibration curve, the system did it for me—after I was a student. In the 1950s, there was a concerted effort to improve efficiency in both resolution and sensitivity. The final result seemed to be the interferometer spectrometer of George Vanasse. There are now many different kinds meant to accomplish different tasks, from the interferometer spectrometer—that has both a multiplex and a throughput advantage—to the single color or multicolor devices for the medical lab analyses, cancer detection, and pollution monitoring.

It is clear that spectroscopy starts with the observation of a spectrum. The very first of these is the rainbow seen by Noah after the rain had stopped[1] (figure 40.1).

Figure 40.1. Rainbow.

[1] *Genesis* 9:13.

doi:10.1088/978-0-7503-2612-4ch40

The Romans were known to observe the spectra formed by prisms[2], but it was many years before optical use was made of this. It seems it was Isaac Newton (1643–1727) who first did it[3] (figure 40.2). Although he describes many experiments involving color and uses a small aperture in a window and a prism, that is not what I would consider a true spectrometer—but it worked. He does describe the fact, apparently for the first time, that **white light consists of many colors**, and these colors can be separated by a prism. Apparently, he never did use it to measure anything. By the way, a spectrometer is a meter that records a spectrum, whereas a spectroscope is one that uses a scope to present a spectrum to the eye. And a spectrograph displays the spectrum on film, or perhaps an oscilloscope.

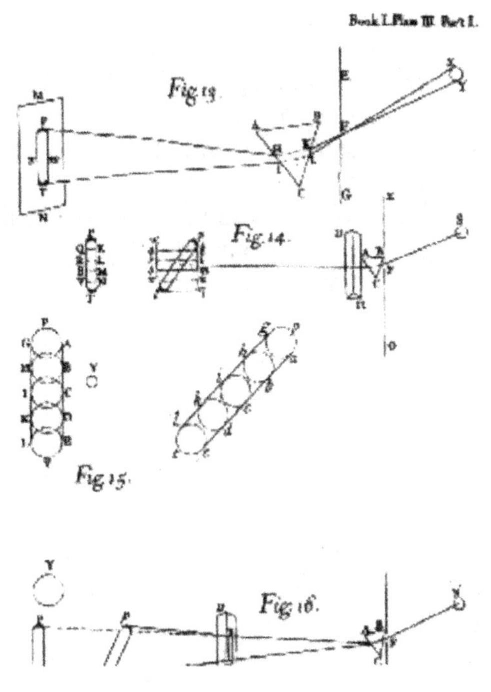

Figure 40.2. Newton's sketch.

Joseph von Fraunhofer (1787–1826) was famous for identifying the **lines in the solar spectrum** in 1814[4], but he did not realize that these lines could be identified with particular elements. Apparently, William Hyde Wollaston (1766–1828) observed these earlier in 1802[5].

[2] Brand J 1995 *Lines of Light: The Sources of Dispersive Spectroscopy, 1800–1930* (London: Gordon and Breach).
[3] Newton I 1704 *Opticks* http://www.gutenberg.org/ebooks/33504.
[4] Kirchhoff G 1859 *Monatbericht der Königlichen Presussiche Akademie der Wissenschaften zu Berlin* **vol. 662**.
[5] Wollaston W 1802 *Philos. Trans. R. Soc.* **92** 365.

Gustav Robert Kirchhoff (1824–1887) and Robert Wilhelm Eberhard Bunsen (1811–1899) were the first to point out that a spectrum was a **unique identifier** of a substance[6]. They established spectroscopy as the science of analyzing materials—flames, gasses, atoms, molecules, but not yet pollutants! Kirchhoff's law in 1859, that absorptivity and emissivity are equal, allowed him to interpret the dark solar lines as certain elements in the photosphere that absorbed some of the light from the interior of the Sun.

The next advance in spectroscopy was replacing the dispersive prism with a **diffraction grating**. For a discussion of the history of diffraction gratings see chapter 6 on diffraction.

Various arrangements were devised by a myriad of investigators. These included using the disperser twice, the double pass, and using a reference beam, the double beam. I use the term disperser to indicate any device that separates the light into its different wavelengths—prism, diffraction grating, and the like. And many different arrangements were devised to reduce aberrations.

These arrangements, called **mounts**, were developed over the years. The **Littrow mount**, after Heinrich von Littrow (1820–1895), using a mirror behind a prism, allowed for a double pass[7], wherein the beam passes through the disperser twice to get increased separation. Light comes from the left, which is the almost universal convention in optical circles, passes through the prism, and is reflected from the mirror back through the prism (figure 40.3).

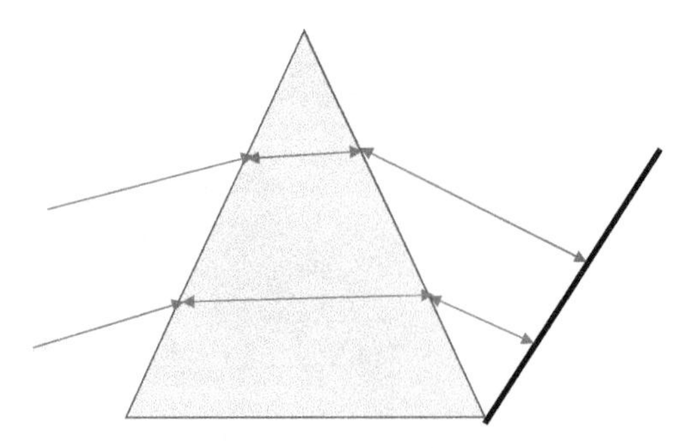

Figure 40.3. Littrow mount.

The **Czerny–Turner mount** uses two mirrors, preferably paraboloidal (figure 40.4). The first collimates the light from the entrance slit and directs it to a grating. The second collects the dispersed light and focuses it to an exit slit. Each is tilted to accomplish this, and the opposite tilts tend to compensate for the resultant off-axis aberrations. It was invented by Marianus Czerny (1896–1985) and Arthur Francis Turner (1906–1996) in 1930[8], who studied in Czerny's lab at the time (figure 40.5).

[6] Kirchhoff G 1860 *Monatsberichte, Akademie der Wissenschaften, Berlin* vol. 662.
[7] Littrow H 1863 *Sitzungberichte der Kaiserlichen Akademie der Wissenschaftern* vol. 47 p 26.
[8] Czerny M and Turner A 1930 *Z. Phys.* **61** 792.

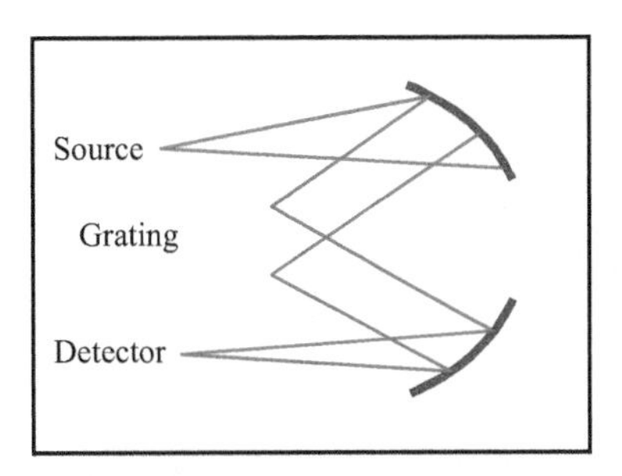

Figure 40.4. Czerny–Turner mount.

Figure 40.5. Francis Turner.

The **Ebert–Fastie mount**, invented by William George Fastie (1916–2000) and Herman Ebert (1861–1913), uses a single mirror rather than two[9]. The mirror must be larger (than those of the C–T arrangement), and it uses one part to collimate and the other to focus. They remain both off-axis and 'tilted' in opposite directions. It has the advantage over the C–T that two mirrors do not have to be adjusted, but it has less freedom and requires a larger mirror. It was originally designed by Ebert in 1889 but reinvented by Fastie in 1952.

The **Rowland circle** is a circle that has a radius that is the same as the radius of a spherical concave grating. It was discovered by Henry Augustus Rowland (1848–1901). An entrance slit on the radius of the circle is focused elsewhere on the circle. One such arrangement is the **Seya–Namioka mount**. Another is the **Paschen–Runge** version.

During the 1950s, considerable attention was paid to the efficiency of various kinds of spectrometers. Two efficiency factors were defined: the **throughput**, or **Jacquinot advantage**[10], and the **multiplex**, or **Fellgett advantage**[11]. The first

[9] Ebert H 1889 *Ann. Phys. Chem.* **38** 489; Fastie W 1952 *J. Opt. Soc. Am.* **42** 641.

[10] Jacquinot P 1954 *J. Opt. Soc. Am.* **44** 761.

[11] Felgett P 1949 Theory of infrared sensitivities and its application to investigations of stellar radiation in the near infra-red *PhD Thesis* University of Cambridge.

indicates how much flux could get through the system, the second as to how many wavelengths of light are being viewed at the same time. Both of them improve the efficiency of the spectrometer system. They were named for Pierre Jacquinot (1910–2002) and Peter Fellgett (1922–2008).

It was George Vanasse (1917–1987) who first introduced the **Fourier transform interferometer spectrometer**, which used a Michelson interferometer or perhaps its first cousin, the Twyman–Green, as a spectrometer[12]. (The Twyman–Green is a Michelson that uses collimated light.) One obtains the interference pattern by scanning the moveable mirror and then performs a Fourier transform on it since the spectrum and the interferogram are Fourier transform pairs.

During this period, many and various instruments were proposed, including the mock interferometer[13], but the Fourier Transform InfraRed, or FTIR, is the one that has endured the test of time.

An excellent online article is entitled *A timeline of atomic spectroscopy*[14], which includes more and different information than is recorded here.

It is now possible to buy an attachment for a smartphone that can record spectra. This can be used to examine vegetables for ripeness or for a variety of medical evaluations. The standard old spectrometer is now miniaturized with tiny photo-formed gratings and nano lenses. Since spectrometers are so useful for so many analytical functions, expect to see them in many miniaturized and cheap forms in the future. Expect to see them in the form of laser devices and interferometric forms as well.

[12] Vanasse G 1967 *Progress in Optics* vol 6 ed E Wolf (Amsterdam: Elsevier) p 259; Sakai H and Vanasse G 1966 *J. Opt. Soc. Am.* **56** 357.

[13] Mertz L 1967 *J. Phys. Colloques* **28**.

[14] Spectroscopy Editors 2006 A timeline of atomic spectroscopy *Spectroscopy* 21 October http://www.spectroscopyonline.com/timeline-atomic-spectroscopy.

IOP Publishing

Rays, Waves and Photons

A compendium of foundations and emerging technologies of pure and applied optics

William L Wolfe

Chapter 41

Speed of light—faster than a speeding bullet

For many years, and by many people, it was believed that light had an infinite speed[1]. That is easily understood because it does travel so fast. In fact, we now know that nothing travels faster. In English units, it is approximately 186 000 miles per second; in metric units, 300 000 km s^{-1}—about one foot per nanosecond. We know these values to much greater precision than stated here, partly because it is one of the fundamental units of nature and has been investigated intensely. This discussion is about the phase velocity of light, the common concept of its speed. The relationship and difference between the phase velocity and the group velocity will be discussed below. They are only slightly different. But how did we measure this incredibly large number in both the vacuum in air and in other media?

41.1 Speed in vacuo

It is rumored that Hero of Alexandria (~10–70) argued that if you close your eyes at night and then open them, the stars are immediately visible. Thus, light from the far away stars gets to you without any delay. But we think Empedocles (490–430 BC) believed differently. He thought we saw by sending beams from our eyes, and the speed was finite[2]. The first known attempts to measure the speed of light were by Galileo (1564–1642) in 1538[3]. He employed people to stand on hilltops and relay the light flashes generated by lamps and shutters. Of course, we now know that what they recorded was the reaction times of the participants. If the two hilltops were about 10 miles apart, a reasonable assumption, the time for light to travel between them is 10/186 000 = 57 μs. Our reaction time is about 0.1 s (100 000 μs). Galileo concluded that light traveled 10 times the speed of sound (which would be about 7700 mph, or about 128 miles per second). He had an additional problem: there were

[1] Descartes R 1637 *La Dioptrique*.

[2] S Schaffer, BBC TV series (2004).

[3] Galileo G 1538 *Two New Sciences* (Leyden: Louis Elsevier).

doi:10.1088/978-0-7503-2612-4ch41

no clocks; there were hourglasses, but not clocks. If there were any other such terrestrial attempts shortly thereafter, they were also doomed to a similar failure.

The **first reasonably accurate measurement** was made by Ole Christensen Rømer (1644–1710) during eight years and culminating in 1675 using Io, a moon of Jupiter[4]. He noticed during those years that the time between the eclipses, which happened every time Io orbited Jupiter, every 1.8 Earth days, was getting shorter. The time was shorter as the Earth in its orbit got closer to Jupiter. He found a difference of 22 min between when the two planets were on the same side of the Sun compared to when they were on opposite sides. Dividing this into the value of the Earth's orbital diameter yielded 200 000 km s^{-1} for the speed of light. The error, of course, was due to errors in the estimate of the time it took and in the value of the diameter of the Earth's orbit. Some say that Rømer did not actually make the calculation, but that Christiaan Huygens (1629–1695) used his data to do so. Huygens did do the calculation[5]; whether or not Rømer did is up to speculation.

In 1728, James Bradley (1693–1762) measured the speed of light by observing the **aberration of light**[6] (figure 41.1). This is not the aberration one gets from lenses, but an apparent displacement of a moving source due to the finite speed of light. It is analogous to the apparent displacement of a jet flying above you. Your eye places it ahead of where the noise seems to come from. He concluded that it was 301 000 km s^{-1}. This phenomenon involves both the motion of the observer and the finite speed of light. A star located at position A will seem to be at angle b instead of a because the telescope moves from right to left, as shown, during the time that light travels from A to D.

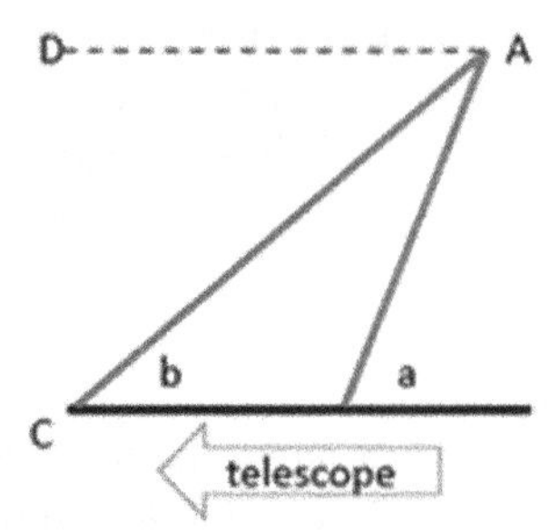

Figure 41.1. Bradley's method.

In 1849, Armand Hippolyte Louis Fizeau (1819–1896) measured the speed of light in 1849 with a **rotating toothed wheel** and a remote mirror[7] (figure 41.2). The wheel's rotation rate was varied until the light emanating from an opening was incident on the same opening after it had moved. A detector behind the wheel measured when the light came through the opening. By measuring the time between

[4] O Rømer, unpublished presentation to Academie Royal des Sciences, December 1676.

[5] Huygens C 1690 Treatise on Light; translated into English by Silvanus P Thompson, Project Gutenberg e-text.

[6] Bradley J 1729 *Philos. Trans.* **35** 637.

[7] Bergstrand E 1956 *Encylcopedia of Physics* vol 24 (Berlin: Springer) p 1.

the light going out and coming back, and knowing the distance to the remote mirror, he could calculate the speed. He concluded it was 313 300 km s^{-1}. Not bad, but not as close as Bradley.

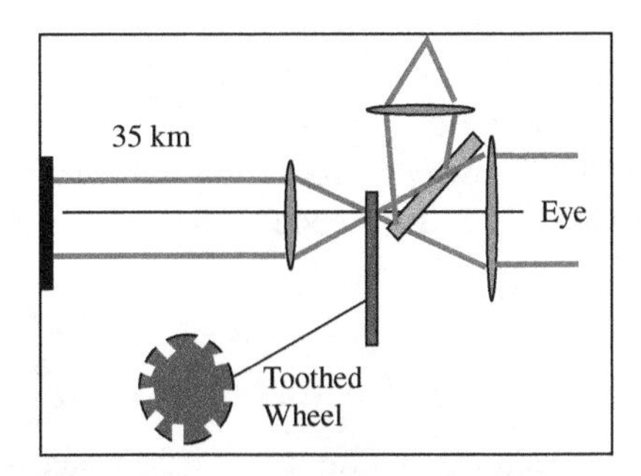

Figure 41.2. Fizeau's instrument.

Jean Bernard Léon Foucault (1819–1868) in 1862 found the speed of light to be 299 796 km s^{-1}. He used a technique similar to that of his countryman, Fizeau, but he used a multi-faceted mirror, a polygonal one[7] (figure 41.3). Now the light came back from a mirror 35 km away and returned to hit a facet at a slightly different angle, where that angle θ is given by $2\omega s/c$, where ω is the angular rotation rate, and s is the separation distance. Then c is determined. The measurement of that angle provided the information as to how much the polygon had rotated and therefore the time lapse. He was pretty close to today's accepted value. The closest so far.

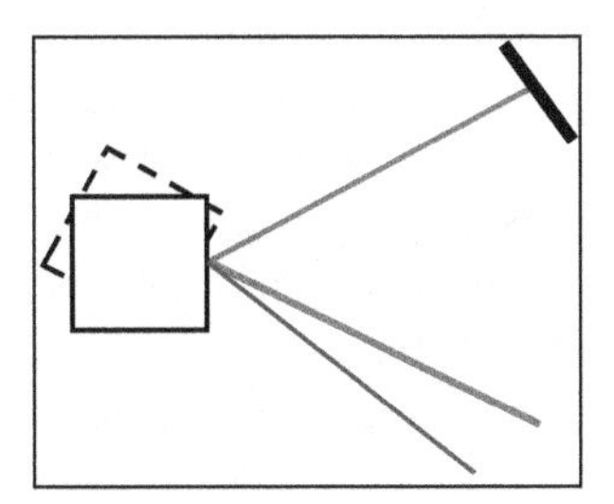

Figure 41.3. Foucault's arrangement

It should be noted that in 1838 Dominique François Jean Arago (1786–1853) reported to the French Academy on a method for determining the speed of light in air, water, and gas, and this was the technique that was used by both Fizeau and Foucault. These are described below.

Albert Abraham Michelson (1852–1931) was the first American to receive a Nobel Prize for physics. He is known for his work on the speed of light and the luminiferous ether, both of which he explored with his interferometer (see Interference, chapter 15). He performed these experiments[8] at what is now Case Western Reserve University. It has been said that when he was a cadet, Michelson was told by the commander of the United States Naval Academy in Annapolis that he should spend more time on tactics and strategy than on his experiments. Thank goodness he didn't.

Modern values of the speed of light are maintained at the US National Bureau of Standards (now the National Institute of Science and Technology) and the British National Physical Laboratory. The corresponding values are 299 792.4574 ± 0.0011 and 299 792.4590 ± 0.0008. The difference is 0.0016, well within the combined uncertainties.

41.2 The designation of c for the speed of light[9]

It was not until 1894 that c became the universally accepted symbol for the speed of light (in vacuo). It had been called V by many, and c was even $\sqrt{2}c$ by Wilhelm Eduard Weber (1804–1891) and Rudolf Hermann Arndt Kohlrausch (1809–1858). But Paul Karl Ludwig Drude (1863–1906) redefined it as it is today. Einstein originally used V but later switched to c. He also pointed out that light travels at a speed c no matter what the speed of its source. There has been attribution to Isaac Asimov that c was for celeritas, and others point out that it was for constant[10].

41.3 Propagation of light

The means by which light is propagated has long been discussed. One argument made use of the analogy to a bunch of swinging balls, like many of us have observed. The ball on the left is allowed to hit the array of the balls in the middle, and the ball on the right moves away almost immediately (due to the rapid transfer of momentum). This gave rise to two conclusions: the propagation was very fast but finite, and it happened by one atom banging into the next. Huygens stated that light was the movement of matter between the source and the receiver[5]. Others made analogy to the propagation of sound, which does use the air molecules. In general, all during the eighteenth century physicists believed there was a very strange substance called the luminiferous ether—with strange properties[11]. It even existed in vacuums because light was seen to propagate in evacuated bell jars. But Michelson and Edward Morley (1838–1923) proved there was no such thing—by trying to prove it was so[12].

[8] Michelson A 1881 *Am. J. Sci.* **22** 120; Michelson A 1887 *Am. J. Sci.* **34** 333; Michelson A 1887 *Philos. Mag.* **24** 449.

[9] Gibbs P 2009 *Why is c the Symbol for the Speed of Light?* (Riverside: University of California) available online.

[10] Asimov I 1976 *Asimov on Physics* (New York: Doubleday).

[11] Whittaker E 1910 *Theories of Aether and Electricity* (London: Longman).

[12] Michelson A and Morley E 1887 *Am. J. Sci.* **34** 333.

41.4 Speed in relatively dense media

The idea that light traveled at a different speed in media denser than air (or vacuum) first was enunciated by Fizeau, who was also able to show in 1851 with the apparatus sketched that light traveled slower in a medium than in air (figure 41.4). There was air in one tube and water in the other. The light comes from the bottom, is split by a beam splitter (the gray line) and directed to each of two tubes. The beams are reflected back by the mirrors on the right, combined by the beam splitter (now a beam combiner), and sent to an interferometer on the left that is not shown. The interfering beams show an interference pattern and the phase shift between them. The system can be calibrated by filling both tubes with air at first.

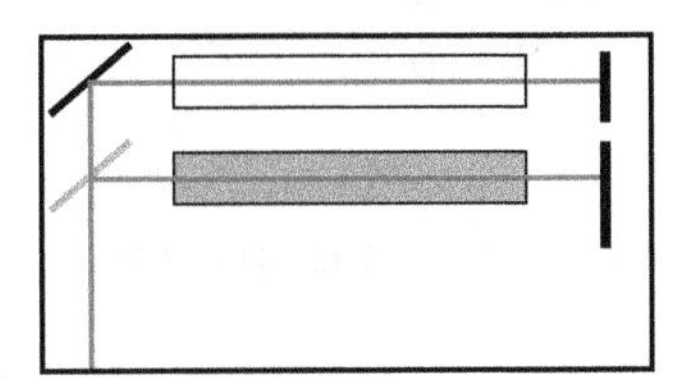

Figure 41.4. Fizeau's apparatus.

Newton also understood this and even formulated a version of Snell's law based on his corpuscular theory of light[13], and Huygens did too[8]. Unfortunately, Newton's formulation required the speed of light to be greater in a medium than in a vacuum. Huygens even got the refractive index of water about right: 3/2 = 1.5. Some think that this was the final nail in the coffin of Newton's corpuscular theory. Young's experiment showed interference; the Arago spot showed wave-generated diffraction, and this shows that light goes slower in denser media.

41.5 Speed in moving media

Fizeau investigated the effect of a moving medium on the speed of light. He used two tubes with water flowing in them, as shown—very similar to his other experimental arrangement (figure 41.5). The water is assumed to flow from left to right in the top tube and right to left in the bottom one, although it could be the opposite. Then the beam of light travels up to the beam splitter, is reflected and passes through the bottom tube against the flow of the water and is then reflected to the top tube where it again moves against the flow of the water. The beam that is transmitted by the beam splitter does just the opposite. The two beams are then combined by the beam splitter (now a beam combiner), and they enter an interferometer. The interferometer then measures the phase shift by the interference thus generated. The drag coefficient was found to be $\delta = (\lambda/n)\ (dn/d\lambda)$, where λ is the wavelength of the light, n is the refractive index and v is the velocity of the medium. So the speed of light is $c/n + \delta v$. This turns out to be a very small change.

---

[13] Newton I 1704 *Opticks*.

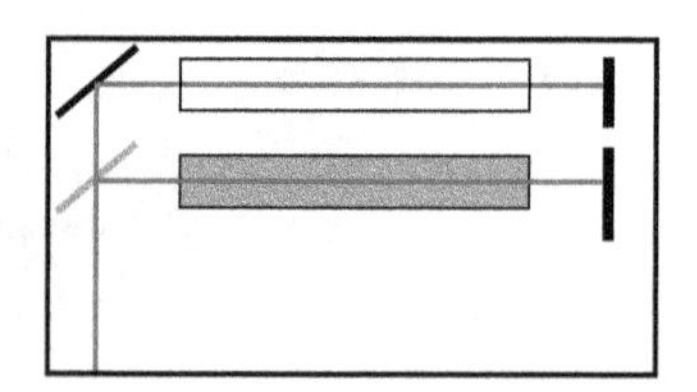

Figure 41.5. Moving media experiment.

41.6 Speed in moving frames of reference

In 1905, Albert Einstein (1879–1955) postulated that the speed of light was independent of the motion of the source, or the motion of its frame of reference, as part of his theory of relativity. (The Doppler effect states that the frequency of light is affected by the motion of the source, but it does not involve the speed of the light.)

41.7 Wavelength, temperature, and pressure variations of the speed of light in air

The speed of light in air depends upon the temperature and the pressure of the air as well as the wavelength of the light, and it is related directly to the value of the refractive index of the air (which is defined as the ratio of the speed in vacuum to that in the air c/v) at the specified temperature and pressure.

The refractivity $(n - 1)$ of air was first estimated by Newton based on astronomical measurements, and the first crude laboratory measurements were made around 1700. The first reasonably accurate measurements were made[14] by Arago and Jean-Baptiste Biot (1774–1862). Many other measurements have been made from the nineteenth century on. A classical report is that of Barrell and Sears[15]. It is not quite right and has been superseded by the Edlén formula[16], developed by Bengt Edlén (1906–1993). This, in turn, was superseded by Edlén himself when he realized a small error[17]. Then others published still better versions[18,19,20], including the effects of water vapor.

The calculation of the speed of light in air is a very difficult prospect, as evidenced by the many investigators attempting to do so. It is a function of the wavelength, especially in the infrared where there are so many absorption bands, causing dispersive effects. It also depends upon the pressure and temperature of the air, as

[14] Biot J and Arago F 1806 *Memories de la Classe des Sciences Mathematiques et Physiques de l'Institut national de France* **7** 301.

[15] Barrell H and Sears J 1951 *J. Opt. Soc. Am.* **41** 295.

[16] Edlén B 1953 Dispersion of standard air *J. Opt. Soc. Am.* **43** 339.

[17] Edlén B 1966 The refractive index of air *Metrologia* **2** 71.

[18] Owens J 1967 Optical refractive index of air: dependence on pressure, temperature and copmposition *Appl. Opt.* **6** 51.

[19] Peck E and Reeder K 1972 Dispersion of air *J. Opt. Soc. Am.* **62** 958; Jones F 1981 The refractivity of air *J. Res. Natl. Bur. Stand.* **86** 27.

[20] Matsumoto H 1982 *Metrologia* **18** 49; Birch K and Downs M 1988 *J. Phys. E:Sci. Instrum.* **21** 694; Beers J and Dixon T 1992 *Metrologia* **29** 315.

these affect its density. Perhaps most complicated of all is the composition. Of course, air consists of oxygen, nitrogen, carbon dioxide, carbon monoxide, methane, other gasses, and water vapor. With this many variables, the problem is almost impossible to solve. The propagation of light in other materials is discussed in the chapter on refraction.

41.8 Constancy of the speed of light

Some have speculated that the speed of light in vacuo has slowed down over time[21]. It is highly unlikely because c is a fundamental atomic constant and many other things would change. (They claim that those have changed, too.) In addition, the claims are relatively weak. Einstein considered it changed and reversed his opinion[22]. These considerations relate to the speed of light in vacuo, not in materials. A fairly modern and very exhaustive paper on this subject includes lists of the measurements considered, their errors, and their differences from the accepted value[23]. The standard deviation of all the measurements is 553.29 m s^{-1} compared to 299 752 468 m s^{-1}, 0.000 018 or 0.000 000 18 percent. All the literature I could find was in obscure journals, and the data went back about 300 years to when measurements were quite uncertain. I am not an expert in this area, but I doubt that there are any variations; I agree with Einstein!

41.9 Slow light[24]

Interest in light that travels very slowly has arisen very recently, although it was recognized many years ago[25] by Hendrik Antoon Lorentz (1853–1928) when he investigated the dispersion relations. Near an absorption line, both the absorption and the refractive index change drastically (see Refraction, chapter 22). The refractive index increases and then decreases dramatically (see appendix B). As a result, the velocity of light in the medium, c/n, greatly decreases as n greatly increases; light slows down. It can even be stopped entirely[26]. The absorption occurs because there is a resonance between the absorbing atoms or molecules and the light. So the phenomenon can be engineered with several different optical configurations, like, for instance, Bragg gratings[27]. It can also be accomplished with artificial resonant structures like Fabry–Pérot cavities and Moire structures. Slow light structures have potential for use as delay lines and storage facilities for optical computers.

[21] Moffat J 1993 *Int. J. Mod. Phys.* **2** 351; Lamoreaux S and Torgenson J 2004 *Phys. Rev.* D **69** 121701; Lamoreaux S and Torgenson J 2004 *Phys. Rev.* D **69** 063506.
[22] Einstein A 1907 *Jahrb. Radioactiv. Electron.* **4** 411; Einstein A 1911 *Ann. Phys.* **35** 898.
[23] *Constancy of the Speed of Light*, Lambert-Dolphin's Library, online.
[24] Khurgin J and Taylor R 2009 *Slow Light: Science and Applications* (London: Taylor and Francis); Khurgin J Slow Light in Various Media: a Tutorial Johns Hopkins University, online.
[25] Lorentz H 1880 *Wied. Ann.* **9** 641.
[26] Philips D, Fleischhauer A, Mair A, Walsworth R and Lukin M 2001 *Phys. Rev.* **86** 783.
[27] Khurgin J 2010 *Adv. Opt. Photon.* **2** 287.

41.10 Phase and group velocity

They are related by $v_g = v - \lambda dv/d\lambda$, where v is the phase velocity, v_g is the group velocity, and λ is the wavelength in the medium. The relationship between the refractive indices is

$$n_g = n/(1 + (\lambda/n)\mathrm{d}n/d\lambda)$$

The phase velocity of a light wave is associated with the speed a monochromatic wave travels, one of a single frequency or wavelength. But there is no such thing as a truly monochromatic wave. One way to look at it is that if it were truly monochromatic, it would have absolutely no line width and therefore no energy. So light has to exist in an *almost* monochromatic form of small wave packets or groups of waves of almost the same frequency or wavelength. The group velocity is the speed of one of these wave packets or groups. It is therefore apparent that it depends upon the variation of the refractive index with wavelength. This situation was probably first discussed by Arnold Sommerfeld (1868–1951) and Léon Brillouin (1889–1969)[28]. You may wonder what all the early investigators were really measuring when they measured the speed of light—the phase or group velocity. It had to be the group velocity and not the phase velocity because there is no such thing as a monochromatic wave. But the two are very close in value. For a typical wave, say in the yellow and a typical change of index with wavelength, $dn/d\lambda$, of about a few parts per million, the two are about 0.000 01 percent different.

I expect that the speed of light in vacuo will remain unchanged and its value will be confirmed with a few more decimals. Slow light in appropriate materials will be used as a buffer or memory in optical computers

[28] Sommerfeld A 1914 *Ann. Phys.* **44** 177; Brillouin L 1914 *Ann. Phys.* **44** 203.

Chapter 42

The stereoscope and stereoscopy—from both sides now

Stereoscopy, solid vision, allows us to see two-dimensional things in what seems to be three dimensions. We see three-dimensional objects naturally by virtue of the fact that we have two separated eyes. Each one sees the object at a slightly different angle, and our mind does the triangulation. We naturally see three dimensions at a limited range, about 10 feet to 100 feet—a range in which the triangulation angles are sufficiently large. There are two distinct methods for displaying three-dimensional images. One is a true three-dimensional representation, like holography, generated by interference effects (see Holography, chapter 11). Stereoscopic techniques do not provide parallax and the ability to see more of the image by moving one's head. The natural perception of real three-dimensional viewing is obtained not only by triangulation but also by the accommodation of the eye and the movement of the head. Many may argue with this limited range of three-dimensional vision. 'Of course we can see in 3D at longer ranges.' It seems that way because of the clues of relative sizes and what is behind what.

In 1838, Charles Wheatstone (1802–1875), of Wheatstone Bridge fame, arranged two views of the same scene at slightly different angles and combined them with mirrors[1]. David Brewster (1781–1868) improved Wheatstone's device in 1849 by using prisms rather than mirrors. Other improvements came from Jules Duboscq (1817–1886) and notably by Oliver Wendell Holmes (1809–1894), who made what is undoubtedly the most popular version in 1861[2].

Three-dimensional movies and television are now with us. They are based on the same principles of triangulation. For most movies, two images of different polarization are presented and viewed with polarized glasses—with each lens polarized differently. Other methods include fast shuttering between the two images, faster

[1] Wheatstone C 1838 *Contributions to the Philosophy of Vision* (London: Royal Society) available online.
[2] Holmes O 1906 *The Stereoscope and Stereoscopic Photographs* (New York: Underwood & Underwood).

doi:10.1088/978-0-7503-2612-4ch42

than the eye response. Different colors can be used. One eye sees reds, blues, and yellows of slightly different wavelengths than the other. The first of these was by Joseph d'Almeida (1822–1880) in 1858, who used different colors, red and green.

Another procedure is tomography—the generation of a three-dimensional image from layers of one-dimensional images. Section and graph = tomography. It was invented[3] by Godfrey Newbold Hounsfield (1919–2004). His patent application was in 1978; it was awarded the Nobel prize in 1982[4].

Three-dimensional printing is, perhaps, the opposite of optical tomography. Instead of creating an image by taking layer-by-layer photographs, an object is formed by laying down successive layers of materials from an image, an object, or even a computer-aided design model. It was first done by Charles W Hall (1939–) of 3D Systems Corporation in 1984[5]. This process is also called layered manufacturing, additive manufacturing, and solid freeform manufacturing.

It is a matter of time and materials, but we will see holographic TV eliminate the need for polarized glasses. Additive manufacture, 3D printing, has now created a boat that is 25 feet long and weighs 5000 pounds. Also, most of the parts for experimental cars. Its time has come and will be used routinely especially for experimental models and 'one-offs'.

[3] Hounsfield G 1973 *Br. J. Radiol.* **46** 1016.

[4] Nobel press release, October 1979; available online at Nobelprize.org.

[5] National Science Foundation 1977 Rapid Prototyping in Europe and Japan; available online.

IOP Publishing

Rays, Waves and Photons
A compendium of foundations and emerging technologies of pure and applied optics
William L Wolfe

Chapter 43

Telescopes—truly astronomical

The history of telescopes begins with Hans Lippershey and Zacharias Janssen—or their children. They discovered that one could arrange two lenses in such a way as to make distant objects seem larger. The next step was by Galileo who improved their design and used it to prove heliocentricity. Several improvements were made, but probably the next major step was the invention of the all-reflecting telescope, generally attributed to Newton but probably due to Zucchi. Newton believed that all materials had the same dispersion so that a refracting achromat was impossible. Over the years, a variety of designs of reflecting (catoptric), refracting (dioptric), and combinations (cata-dioptric) of elements were generated, but probably the next seminal change was the adaptive optics telescope suggested by Horace Babcock in 1953. The mirror shape is changed while it is in use. This allows for the correction of atmospheric turbulence. It was followed about 20 years later by the invention and construction of a multiple mirror telescope by Aden Meinel in 1979. Almost all major terrestrial telescopes use both of these advances.

Hans Lippershey (1570–1619) and Zacharias Janssen (~1580–1638) are said to have invented the first telescope in 1608, although some speculate that it might have been their children[1].

Galileo Galilei (1564–1642) then improved its design by a factor of 10 in 1609 and used it for celestial viewing[2]. The **Galilean telescope** uses a convergent objective lens to form an inverted image and then a divergent eyepiece to form an erect and non-inverted image—since it forms an enlarged virtual image. In the diagram, the object is at infinity; the short inverted arrow is the real image, and the large one on the left is the virtual image seen by the eye that is not shown. The objective on the left was a plano-convex lens and the eyepiece on the right a double concave one, as shown (figure 43.1).

[1] King H 1955 The History of the Telescope, online.
[2] Drake S 1990 *Galileo, Pioneer Scientist* (Toronto: University of Toronto Press).

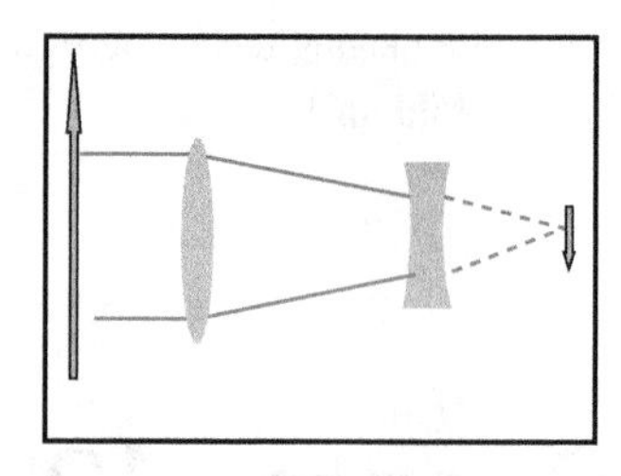

Figure 43.1. Galileo telescope design.

Johannes Kepler (1571–1630) improved Galileo's design in 1611[3] by using a convex lens as an eyepiece. This allows a wider field of view but results in an inverted image. Would you believe that this is called a **Keplerian telescope?** The next improvement was the use of an achromatic lens in 1733 as the objective. See chapter 16 on lenses for the invention of the achromat.

Niccolò Zucchi (1586–1670) probably designed the first **reflecting telescope** in about 1616[4], although Newton is usually given the credit for his invention some 50 years later.

Newton stated that the design of (refracting) telescopes is impeded by the different *refrangibility* (refraction) of the different colors of the rays of light and that conic sections are not right[5]. He did not realize that one could achromatize with two different materials. He proposed two convex–concave lenses with water in between. In 1668, he described what is now called the **Newtonian telescope** that had a spherical primary of glass that is coated with speculum metal on the back—a mirror. He specified that the glass be of equal thickness throughout to keep the images from being colored or indistinct (that is, a concave rear-surfaced mirror). As shown, the secondary was described as a reflecting prism to get the image available to the eye, which is behind a meniscus eye lens (figure 43.2). Today, many use a mirror. He stated that the hole should be no larger than necessary. Although he does not use the term, he recognized this as a field stop for he states it has more advantage at defining things than if it were at the mirror, where it would be an aperture stop. He also recognized the limitations of seeing caused by the turbulence of the atmosphere and suggested viewing from mountain tops.

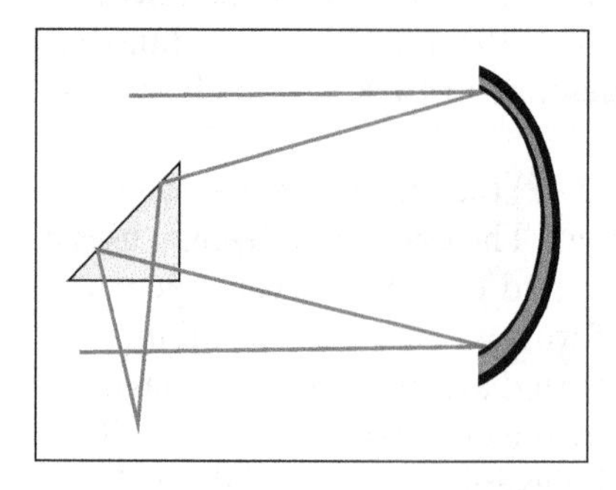

Figure 43.2. Newton telescope schematic.

[3] Kepler J 1611 Dioptrice.

[4] Zucchi N 1652 Optica philosophia experimentalis et ratopne fundamentis constitua.

[5] Newton I 1704 Opticks.

James Gregory (1638–1675) designed in 1668 a reflecting telescope of two mirrors, both of them concave, now called a **Gregorian telescope** (figure 43.3). Gregory's design[6] was not realized until ten years later by Robert Hooke (1635–1703). That would be 1678, after Newton's telescope was published. The Gregorian form consists of a paraboloidal primary and an ellipsoidal secondary. Parallel rays from a distant object are brought to focus by the paraboloid. The first focus of the ellipsoid is coincident with that focus, and all rays that enter that combined focus go to the second focus of the ellipsoid on the far right, the focus of the telescope.

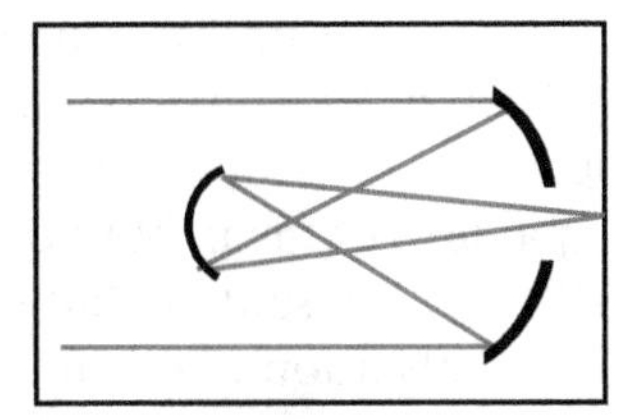

Figure 43.3. Gregorian telescope configuration.

Laurent Cassegrain (1629–1693) invented a similar layout in 1672 that used a convex secondary rather than a concave one[7]. This shortened the overall length of the tube. Both the ellipsoid of Gregory and the hyperboloid of Cassegrain redirect light aimed at the first focal point to the other focal point. Recall that the ellipse has two foci inside, while the hyperbola of two sheets has the foci located on the two sides (figure 43.4).

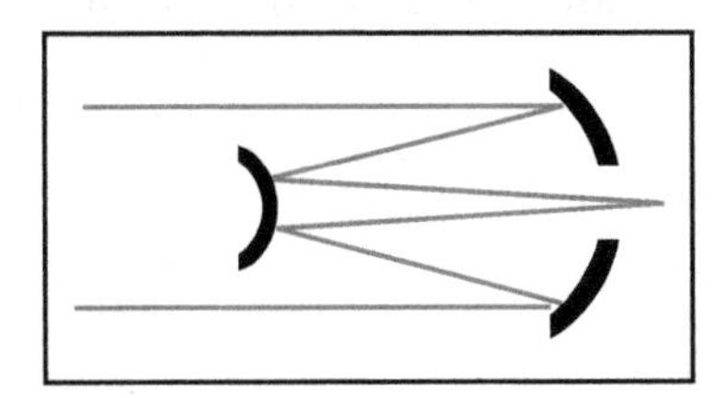

Figure 43.4. Cassegrain telescope configuration.

An interesting variation on these two-mirror designs is that of **Karl Schwarzschild** (1873–1916). It uses a Gregorian configuration but employs a double reflection at the primary. That is, the light reflects from the primary to the concave secondary, back to the primary, and back through a hole in the secondary (figure 43.5).

[6] Gregory J 1663 *Optica Promota* (London: J Hayes).

[7] Cassegrain L 1672 Recneil des mémoires et conférences concernant les arts et les sciences.

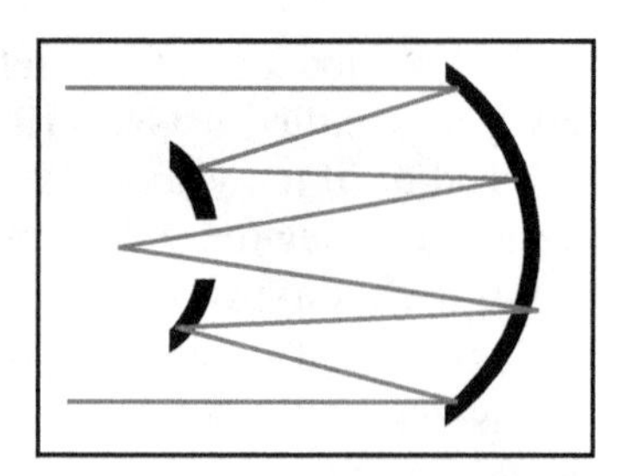

Figure 43.5. Schwarzschild variation.

Bonaventura Cavalieri (1598–1647) is said by some to have pre-empted Newton and Gregory in postulating a reflecting telescope, but his publication[8] appears to be entirely about conics and not about telescopes. As far as I can tell, there is no discussion of telescopes and no pictures of them, but I do not read Latin all that well.

Marin Mersenne (1588–1648) is also said to have published about reflecting telescopes before Newton in his publication about music[9], but I could only find a brief mention of mirrors. The modern version is an afocal version of a Cassegrain.

The **Schmidt telescope** design, created by Bernhard Woldemar Schmidt (1879–1935), was probably the first catadioptric one, using both reflecting and refracting elements[10]. It used a spherical mirror with a stop at the center of curvature. Thus, all rays are on axis; the only remaining aberration was spherical (see Optical Design, chapter 24, for aberrations.) He corrected this with a special curved refracting plate designed to properly distort the wavefront (anti-spherical?) to account for the spherical aberration, at least on axis; it is now called a Schmidt plate. It was realized in 1930. The curvature on the plate is not shown since it is so small (and hard to draw on a computer) (figure 43.6).

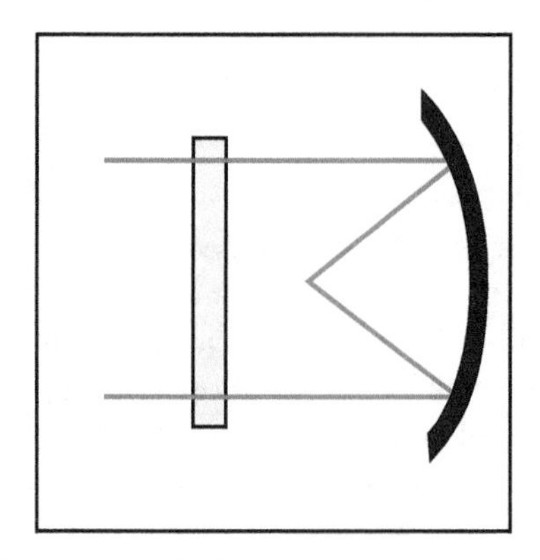

Figure 43.6. Schmidt telescope representation.

[8] Cavalieri B 1632 Lo specchio ustorio overo trattato delle settione coniche.

[9] Mersenne M 1634 *Harmoniques* (Paris: Jacques Villery).

[10] Schmidt B 1931 *Central-Zeitung für Optik und Mechanic* **52.2** 25.

A simple version of a **folded Schmidt** is shown in figure 43.7. An all-reflective version uses a diagonal mirror with approximately the shape of the refractive corrector in place of it. This correcting mirror can be made in either of two ways. One is rotationally symmetric; the other is not. The latter provides better correction, but it is much more difficult to figure. It is, of course, easier with the modern computer-controlled methods of today. One system uses an adaptive mirror to do this[11]. All Schmidts have good wide-field coverage. They do not become aberrated until the obliquity of the corrector in effect changes its shape. I have not been able to find an earlier reference to an all-reflective Schmidt, but I knew about it at least 40 years ago.

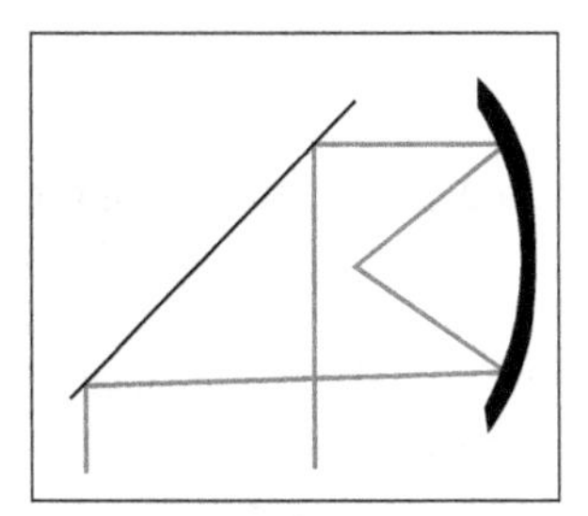

Figure 43.7. All-reflective Schmidt arrangement.

The **Schmidt–Cassegrain** form, designed by James Gilbert Baker (1914–2005) in 1940, places the secondary on the back of the corrector plate. Since it does also have a Cassegrain form, it does not have such a wide field (figure 43.8).

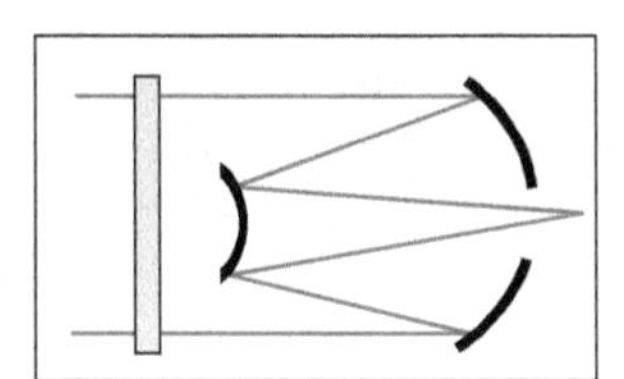

Figure 43.8. Schmidt–Cassegrain.

The so-called **Bouwers–Maksutov telescope** was invented by both men at essentially the same time—February and October of 1941, respectively—but at different places and during World War II[12], so the ideas were independent. Albert A Bouwers (1893–1972) lived in Holland and Dmitri Dmitrievich Maksutov (1896–1964) in Russia. It is a variation of the Schmidt design. The primary is a sphere with a stop at the center of curvature (to eliminate all aberrations except spherical). Then a concentric spherical corrector plate is used, rather than the more difficult one of Schmidt. The curvatures of its two surfaces are both centered on the center of curvature of the primary (figure 43.9). The corrector can be placed before or after the stop—or in both places—and sometimes with a Schmidt plate to boot. The disadvantage of the Bouwers–Maksutov is the deep curves of the corrector plates.

[11] Cui X *et al* 2000 *Proc. SPIE* **4004** 347.

[12] Bouwers A 1950 *Achievements in Optics* (Amsterdam: Elsevier).

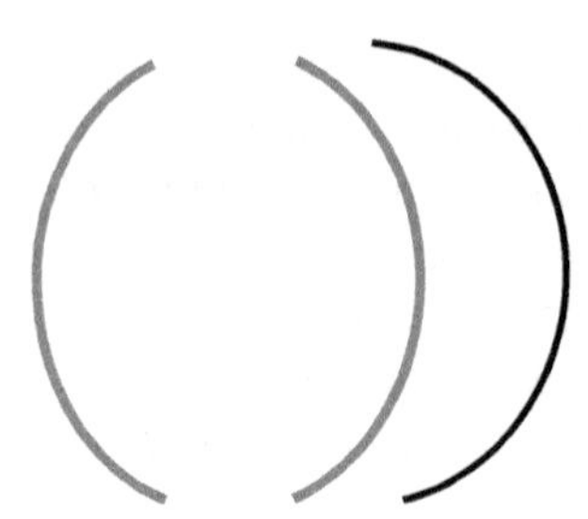

Figure 43.9. Bouwers–Maksutov.

Modern designers have made all sorts of variations on these basic designs, and a host of other names for telescope arrangements have arisen. One can replace the hyperboloidal secondary of a Cassegrain with an ellipsoid and the secondary with a sphere to get a **Dall–Kirkham**. Horace Edward Stafford Dall (1901–1986) designed it in 1928 and published with Allan Kirkham two years later[13]. One can leave out the correctors of the Schmidt or Bouwers and get an uncorrected version. Or one can add a Schmidt plate to a Cassegrain and get, yes, a Schmidt–Cassegrain. The variation most used in today's telescopes is the Ritchey–Chrétien version of the Cassegrain. Originally it specified two hyperboloids, but is now often two higher-order aspheres. The originators were George Willis Ritchey (1864–1945) and Henri Chretien (1879–1956).

In 1789, William Herschel (1738–1822), the same Herschel who discovered infrared and the planet Uranus, used the simple maneuver of placing a paraboloid off axis to avoid the folding mirror of Newton[14]. His system thereby suffered from off-axis aberrations but was unobscured and had only one mirror. It may be considered the first of the off-axis telescopes. It is called a **Herschelian telescope**, of course (figure 43.10).

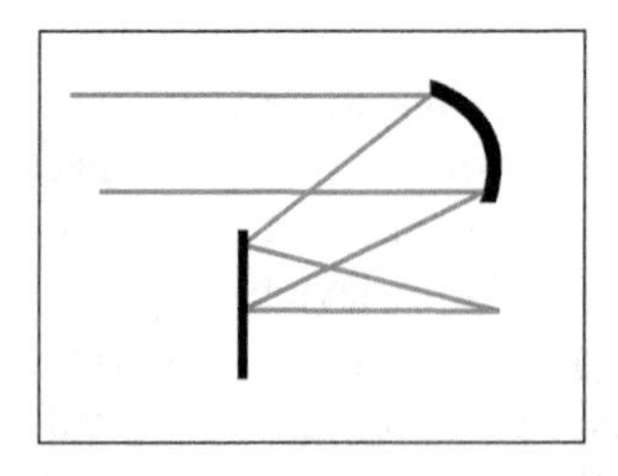

Figure 43.10. Herschelian off-axis telescope.

In 1808, Johann Elert Bode (1747–1826) proposed an off-axis system that incorporated a Cassegrain form—a tilted Cassegrain[15].

[13] Dall H and Kirkham A 1930 *Scientific American.*

[14] Multimedia Catalog and Wikipedia, online.

[15] Bode J 1881 *Astronomische Jahrbuch.*

The next off-axis telescope seems to be the **Schiefspiegler**[16] of Anton Kutter (1903–1985). It is basically an off-axis Cassegrain with the secondary just outside the field of the primary. The two mirrors are tilted in opposite directions to obtain anastigmatic performance (see Optical Design, chapter 24). It can also involve a lens for final correction or a paraboloidal tertiary. Johann Zahn (1641–1797) is sometimes credited with the first of these designs in 1685, as is Niccolo Zucchi (1585–1670) in 1685 with limited success[17]. Of course, as with so many telescope designs, there are numerous variations on the theme, some with three and four mirrors. Perhaps the most notable of these is the **Yolo**, developed by Arthur S Leonard (–2001) in about 1965[18]. It uses a warped concave secondary in contrast to the Kutter design and is named for a county in California. I could not determine why he named it thus; maybe he just liked where he lived.

The Cold War fostered a family of off-axis telescopes. Scatter in the system was a problem, and the scatter from either the secondary or the corrector plate exacerbated this—and the diffraction pattern is simpler, a Jinc function (see appendix A). Most were three or more mirrors. Many were called **three-mirror anastigmats**, or TMA's. Three examples are cited here[19]. The Korsch is of Cassegrain form with a third mirror that sends the light to the focus via a folding flat. It was proposed for this application as an eccentric pupil version.

Another variation of the **unobscured telescope** is what may be called the off-aperture or **eccentric pupil** version. (This has nothing to do with my former students.) It uses a portion of the full primary mirror. It therefore remains on axis optically and does not suffer off-axis aberrations, but it is a harder mirror to manufacture. Some have simply made the larger mirror and used only part of it; others have made the larger mirror and cut it down to size (figure 43.11).

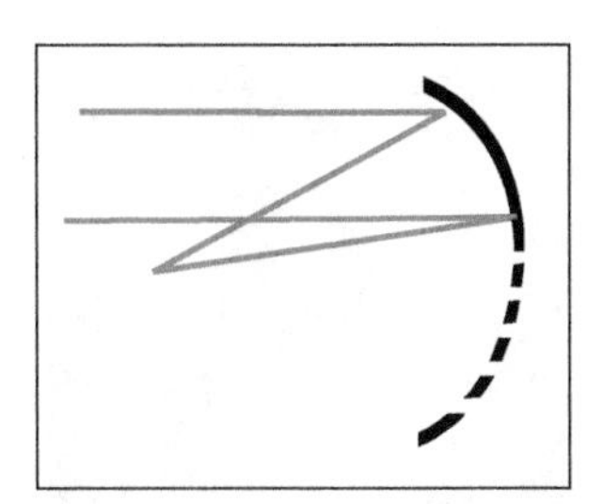

Figure 43.11. Eccentric pupil schematic.

In 1953, Horace Welcome Babcock (1912–2003) first introduced the idea of an **adaptive mirror** for telescopes[20]. Since then, it has become part of almost every large telescope. It consists of a relatively thin blank that can be deformed by actuators just

[16] Kutter A 1953 *Der Schiefspiegler* (Biberach, Germany: Weichardt).

[17] *Some Schiefspiegler History* online via Schiefspiegler telescopes—Helmut Frommert.

[18] *The Yolo Reflector*, a booklet by Arthur S Leonard—The ATM site, online.

[19] Cook L 1989 *US Patent* 4834517; Shafer D R 1978 *Appl. Opt.* **17** 1; Korsch D 1975 *Opt. Eng.* **14** 533; Korsch D 1977 *Appl. Opt.* **16** 2074.

[20] Babcock H 1953 *Publ. Astron. Soc. Pac.* **65** 239.

behind it. The slight deformations can be used to correct for other mirror errors and for atmospheric turbulence in real time[21].

The first telescope to use more than one primary-secondary mirror system and combine the outputs from several of them was the **multiple mirror telescope**[22] constructed at what was then the Optical Sciences Center of the University of Arizona and is now the James C. Wyant College of Optical Sciences. It was conceived by Aden Meinel (1922–2011) in 1979. It was constructed of six surplus 72 inch (1.85 m) mirrors and mounted on Mt Hopkins. It is the model for all modern telescopes.

Wikipedia lists very many **astronomical observatories**—on the ground, in the air, and in space. It is only possible to deal with a few, the main ones that saw advances in the use of telescopes.

The **Yerkes Observatory**[23], run by the University of Chicago, was established in 1897 by George Ellery Hale (1868–1938) with a grant by Charles Tyson Yerkes (1837–1905), and is known for having the largest refracting telescope still in use—102 cm (40 inches) in diameter. They also claim to be the first integrated observatory with more than just a telescope.

The **W. M. Keck Observatory**[24] is on the top of Mauna Kea in Hawaii. It is a multiple mirror version with Zerodur, a very low expansion material (see Mirrors, chapter 22). Each segment is adjusted twice a second with an uncertainty of 4 nm to obtain optimum optical performance. The first telescope was finished in 1993, the second in 1996. The observatory is run jointly by Caltech and the University of California. It uses adaptive secondaries.

The **Giant Magellan Telescope**[25] will reside in the Las Campanas Observatory in Chile at an altitude a little over 2500 m (about 7500 feet), where there is extremely good seeing. It will consist of seven 8.4 m (about 25 foot) diameter mirrors made at

Figure 43.12. Giant Magellan telescope model.

[21] Beckers J M 1993 *Annu. Rev. Astron. Astrophys.* **31** 13.
[22] Personal observation.
[23] Several net sites with that name.
[24] See the net for details.
[25] Giant Magellan Telescope, Wikipedia.

the University of Arizona. Site preparation began in 2012, and completion is planned for 2020. Several universities and institutes in the US, Australia, and Korea are working together on this project. The second of the mirrors was cast on January 14, 2013, and should then take about fifteen weeks to form into a paraboloid and cool to then be figured and polished[26].

The **European Extremely Large Telescope** is also destined for a place in the Andes[27]. Its design consists of 798 hexagonal adaptive mirrors, each 1.45 m (about 4.5 feet) in diameter, providing a full diameter of almost 40 m (120 feet). It is supported by the ESO, the European Southern Observatory, which is a collection of about 15 nations (some not in Europe), who presumably will put up the cash. Approval was attained in 2012; the projected first light has not been specified.

The **Hubble Telescope** was both a dismal failure and a spectacular success. It was a failure because the first version had the wrong figure on the primary. It was a success because after it was repaired, it provided magnificent pictures of the heavens for astronomers to interpret. The project was started in 1970 at the PerkinElmer Corporation, sponsored by NASA and ESA, the European Space Agency. It is a Ritchey–Chrétien design with a diameter of 2.4 m (7.2 feet) and is in a low Earth orbit of 559 km. It was launched in 1990. The primary was ground to the wrong figure because of a misplacement of the null lens during testing—by about 1.3 mm (0.05 inch). This resulted in an error in the conic constant by about one percent. Bob Fischer has described this in detail[28]. The Hubble was corrected with additional optics in a 'rescue mission' in 1993. It has enabled astronomers to get a better handle on the rate of expansion of the Universe, its age, more information about black holes, and discovered another moon of Pluto.

The **James Webb Space Telescope** is in the latter stages of development and testing, funded by many organizations, and due to be launched around 2021. It will be a three-mirror anastigmat with a collection diameter of 6.5 m (about 20 feet)[29]. It is meant mainly for infrared observation and therefore has gold-coated beryllium mirror segments and is located at a Lagrange point to protect it from the heat of the Sun.

NASA has other space observatories in its plans. Those that investigate exoplanets, stars, galaxies and black holes. The LUVOIR (Large UV Optical, InfRard surveyor) is essentially a Hubble on steroids, with a 50 foot diameter. The HabEx (Habitable Exoplanet observatory) will have a telescope diameter of 10–20 feet, but its salient feature is the star shield, a large petal-like structure in front of the telescope to shield the star's light. NASA also proposes the Lynx and x-ray telescope and the Origins Space Telescope, an infrared, cooled system that it meant to peer through interstellar dust at its long wavelength.

---

[26] *Arizona Daily Wildcat*, March 25, 2013.

[27] *Astronomy Magazine*, **4**, 28 (2010).

[28] Fischer R and Talic-Galeb B 2000 *Optical System Design* (New York: McGraw-Hill).

[29] NASA, James Webb Telescope, on the net.

IOP Publishing

Rays, Waves and Photons
A compendium of foundations and emerging technologies of pure and applied optics
William L Wolfe

Chapter 44

Ultraviolet—aglow in the dark

Ultraviolet light, black light (if that is not an oxymoron), is generally considered to be that part of the spectrum between about 400 nm and 100 nm, below which it is considered x-rays. The UV spectrum is divided in a number of ways. One is ionizing and non-ionizing. Another is the vacuum UV versus the rest, and solar blind versus the rest. The vacuum UV has wavelengths shorter than 200 nm because in that region it is strongly absorbed by air; operations must be carried out in a vacuum. The solar blind region from 215 to 315 nm is defined by the fact that the ozone layer in the upper atmosphere absorbs the UV in this region. So you cannot see the Sun in this part of the spectrum. The UV spectrum is more formally divided into the UVA, UVB, UVC, and NUV, which are 315–400, 280–315, 315–400, and 290–400 nm.

This so-called actinic radiation, that which causes a chemical reaction, has a variety of uses and dangers. It is a form of germicide and bactericide; causes fluorescence, sunburn, and cancer; but can also treat psoriasis.

The ultraviolet part of the spectrum with wavelengths just short of the visible violet was discovered in 1801 by Johann Wilhelm Ritter (1776–1810) when he observed the darkening of some silver chloride crystals that were located in the part of the solar spectrum just outside the violet that he created with a prism. Contrary to the discovery by Herschel of the infrared just a year earlier, Ritter was looking for it. He speculated that, since there was heating radiation beyond the red, there might be cooling radiation beyond the violet. That was wrong; it is also energy that also causes heating, but is better known and used for other effects.

The **vacuum UV** was discovered by Victor Schumann (1841–1913) in 1893[1].

The use of ultraviolet radiation as a **bactericide** was first discovered in 1877 by A Downes and Thomas Porter Blunt (1842–1929), who observed jars of sugar solution, some in the Sun and some in the shade on the window sill[2]. Those on the sunny side

[1] Schumann V 1914 *Astrophys. J.* **38** 1.

[2] Downs A and Blunt T 1877 *Nature* **16** 218; Downs A and Blunt T 1877 *Proc. R. Soc. Lond.* **26** 488.

doi:10.1088/978-0-7503-2612-4ch44

of the sill stayed clear, but bacteria grew in the shaded jars. John Tyndall (1820–1893) substantiated these results[3]. Marshall Ward in 1892 showed that it was really ultraviolet light that performed the function. Since then, considerable studies have been carried out to determine the relative effectiveness of different wave-lengths, intensities, and duration, among other things[4].

The incidence of UVB radiation on the skin produces vitamin D. It also produces a **tan**, **sunburn**, and **cancer**. The tan is generated when UV activates melanin and causes it to surface. Sunburn is the dilation of capillaries and resultant reddening. Cancer is probably related to changes in the DNA molecules caused by the ionizing action of the UV[5].

Ultraviolet imagery has been used to detect **art forgeries** by revealing fluorescence, or 'paint overs[6].'

One application I had never considered was that UV radiation can be influential in the **choice of a mate!** 'Although UV rays are substantially absorbed by the cornea and lens, recent evidence indicates that they can effect mate choice, communication and foraging for food and circadian rhythms[7].'

UV radiation has been considered for military applications of at least two types: **missile homing** and **camouflage detection**. The former is based on the fact that there is little or no background in the solar blind region of the spectrum, so any signal is easily detected. Jet and rocket engine exhausts have an ultraviolet component, and this should be detectable. It turns out that such signals are very weak, and the detection is questionable[8]. Infrared techniques are far superior (see Infrared, chapter 12). Visible camouflage techniques are different from those in the UV (and the IR). One company has a material that provides UV camouflage[9].

There is some interest in the use of UV **radios**. They have the high frequencies that allow large bandwidths, but UV light is scattered greatly in the atmosphere and would thereby limit the transmission range. This could be an advantage for secure short-range communication[10].

Ultraviolet radiation has been found an aid in the **preservation of food**. It can protect against salmonella, lengthen the shelf life of strawberries, pasteurize juices, and more[11].

---

[3] Tyndall J 1881 *Proc. R. Soc. Lond.* **2** 478.

[4] Reed N 2010 *Public Health Rep.* **125** 15.

[5] *Encyclopedia Britannica* online under 'ultraviolet'.

[6] *Encyclopedia Britannica* online under 'forgery'.

[7] Hockberger P 2002 *Photochem. Photobiol.* **76** 561.

[8] Private communication with L Biberman.

[9] UVR Defense Tech, on the net.

[10] Adee S 2009 *IEEE Spectrum*, online.

[11] Koutchma T 2008 *IUVA News* **10** 25.

IOP Publishing

Rays, Waves and Photons
A compendium of foundations and emerging technologies of pure and applied optics
William L Wolfe

Chapter 45

One man's glimpse of the future

What will our optics future look like? One can only imagine. But I will look through my glass darkly.

Homes and other buildings will all be topped with solar panels and they will have both DC to AC converters and direct DC for all the LED's in the house. The paths into our homes will be walk-on solar panels with embedded LED's and sensors that interactively show the way. The front door will have a panel that recognizes friends, unlocks the door and announces their arrival. It also keeps the door locked against strangers and obtains their identity and purpose. TV's and movies will be holographic. The pictures on the wall will be generated by computer so that one week we can admire the Mona Lisa and the next a favorite trout stream. Sensors will sense the wall and air temperature of each room and maintain the temperature separately for those that are occupied and unoccupied. There can be a vision of the night sky on the ceiling above our beds as we doze off. We will have responsive night lighting to the bathroom and refrigerator for relief and raids. Shelves and refrigerators may be photographically monitored for content and warnings issued when the stock gets low or out of date. We already have robotic vacuum cleaners, floor scrubbers and snowplows, but we may even see self-washing windows that sense when they are dirty by scatter measurements and apply water and squeegees. Hydrophobic coatings will help, but not be enough.

Traffic will be controlled by lights that respond to the number of cars on both cross streets. Of course, they will be LED's for efficiency and long life. Most, if not all, of our cars will be automated. We can sit and read. They will have infrared night driving assists. We will not own them, but they will be on call. We can order a sporty electric convertible to go to work or a seven-seater for a vacation trip. **It** will drive **us**. No garages. No emission checks. No registration. We may even have automated choppers commuting to work that are controlled by lidars and cameras. And they may be able to take us to nearby trout streams! But we will do more work at home via the internet that is powered by laser diodes and fiber optics.

We will visit our primary care physician via smartphone. We will be able to perform most laboratory tests with smartphone attachments so we can monitor our blood on a weekly basis at home and without cost for blood pressure and contents like oxygen, bilirubin and % CO_2. Large pills with cameras and WiFi may replace **all** endoscopes. We may get special operations from special surgeons via telemedicine, and their hands will be steadier. There will be tiny robots that can travel up and down our veins and arteries. Cleaning them out or repairing them. There will be better methods of detecting cancer based on thermography and fluorescent spectroscopy.

There may be strips of solar panels embedded with LED's and sensors along highways that will both generate electricity and provide lighting and directions. Long-haul trucks will have at most one driver for four or five of them. They will travel about 20 hours a day at about 50 mph to conserve fuel (in the slow lane). Airplanes will be able to land in just about any kind of weather due to enhanced and virtual landing systems. Ultraviolet lidar will be used for clear-air turbulence detection.

The military will use spy satellites with ground resolution twice to three times as good as now, maybe using synthetic apertures. They will add many low altitude, lower resolution CubeSats for more coverage. The family of Sidewinder homing devices will get more colors, maybe **many** more colors to defeat countermeasures. ABM's will have multiple interceptors that are directed by AI and use miniature IR homing sensors to ward off MIRV's. Drones will be all over the battlefield doing recon with visible and infrared cameras. Laser designators will be multi-colored and have more sophisticated pulse codes. Laser weapons will come into limited use, especially for softer targets. I don't see robotic infantrymen, as in Star Wars, but probably remotely controlled tanks.

Manufacturing, especially of automobiles, will increase its use of robots that will become more versatile. McDonalds and its competitors will drastically reduce the number of employees. You will punch in your order or speak it aloud. A robot will even put the cherry on the top of your milk shake. They may even revive the old skaters and deliver your order to your car—with a robot skater, of course. Additive manufacture will improve, increase and be more diverse.

Crops will be monitored by satellite sensors and by drones to sense moisture, ripeness and need for fertilizer. Applications of water and fertilizer will also be done that way for more efficiency, just where it is needed. Grocery shopping could become on-demand. We prepare menus or shopping lists, perhaps with the aid of virtual reality store shelves; we order it; the food is gathered near the farms and sent to us when we need it. Some stores disappear, less food will be wasted, and it should be cheaper. Way off in the distance are opaque greenhouses with LED lights tuned to growing frequencies. The 'days' are long; the 'nights' short. There will be four or five crops a year with Alaskan size produce.

Disabled people will be able to operate computers by eye and prosthetic eyes will be available that will get at least 20/200 vision. Bridges and buildings will have built-in fiber optic sensors that can warn when repairs are needed.

Bibliography—I read you

There are many modern texts that are appropriate for the different subjects contained in this book. I have divided this bibliography into the same sets of subject matter as the chapters. I have used some shortcuts in designating publishers. The traditional way would be, for example, McGraw-Hill Book Company, New York, Toronto, London, 2000. But all these publishers have expanded their bases, and the shorter designations should be sufficient for the reader to find the publication. Although most are obvious, a few should be explained. Oxford stands for Oxford University Press, Plenum for Plenum Press. Some of them, usually if old enough, can be read online, e.g. Newton's *Opticks.* A few are out of print but can be obtained in used form from one of the online vendors. The net is a valuable source of much information, even if it has not been critically reviewed. I have found it to be mostly accurate in the areas in which I could make a test or comparison. It is certainly a valuable starting place. Two books that cover many of these topics are M Born and E Wolf's *Principles of Optics* Pergamon with about seven different printings and updates and M Bass, E Van Stryland, D Williams, and W Wolfe's *Handbook of Optics* from McGraw Hill (1995). The book by Born and Wolf is a thorough theoretical treatment of the basic principles, while the *Handbook of Optics* is in four volumes and covers many topics by many different authors.

doi:10.1088/978-0-7503-2612-4ch46

Acousto-optics

An excellent treatment of some of the history and the performance of acousto-optics is A Korpel's *Acousto-Optics*, Dekker (1988).

Cameras and photography

A number of books by Helmut Gernsheim trace the origins of photography and therefore of cameras from the camera obscura to various dates. The list is online at his address.

Communications

A good reference for this field is by G P Agrawal, a pioneer in it: *Lightwave Technology: Components and Devices*, Wiley (2004).

Diffraction

A good book on diffraction gratings, the history, theory, and manufacture is by M C Hutley, *Diffraction Gratings*, Academic Press (1982). A good history of diffraction is by C F Meyer, *The Diffraction of Light, X-rays and Material Particles*, University of

Chicago Press (1934). An excellent treatment of both the history and the use of gratings is *Diffraction Grating Handbook*, Christopher Palmer and Erwin Loewen, Newport (2005).

Displays

A detailed treatment of three-dimensional displays is by J Cheng, *Advances in Optics and Photonics*, 3, 128 (2011).

Eye tracking

Clarence E Rash, ed., *Helmet Mounted Displays*, US Army Aeromedical Research Laboratory, online. Wikipedia, with that title, has a nice review of more than HMD's, with pictures and references.

Fiber optics

Jeff Hecht, *The City of Light: The Story of Fiber Optics*, Oxford (1999) is a thorough treatment by a prolific author of scientific articles. G Agarwal, *Fiber Optics Communication Systems*, Wiley (2010), is a technical treatment of the subject. *Fiber Optic Tutorial* online is a semi-technical discussion of the subject.

Infrared

One of the earliest books on infrared technology was written by Smith and co. of the Royal Radar Establishment in Great Britain: R A Smith, F E Jones and R P Chasmar, *The Detection and Measurement of Infrared Radiation*, Oxford (1957). The next one chronologically was generated from a series of lectures that Dick Hudson gave at the Hughes Corporation: R D Hudson, *Infrared System Engineering*, Wiley. The next three all appeared almost simultaneously. They were written by some of the workers at the Infrared Labs at the Willow Run Research Center of the University of Michigan, those at Honeywell in Minneapolis, MN, and at Aerojet in Azusa, CA. They are, respectively, M R Holter, S Nudelman, G H Suits, W L Wolfe and G I Zissis, *Fundamentals of Infrared Technology*, Pergamon Press (1962); P Kruse, G McGlauchlan and R B McQuistan, *Elements of Infrared Technology*, Wiley (1962); and J A Jamieson, R H McFee, G N Plass, R H Grube and R G Richards, *Infrared Physics and Engineering*, McGraw Hill (1963). Each includes all the elements of the technology, sources, backgrounds, atmospheric transmission, optical materials, detectors, and system design. Some are stronger on some subjects than others.

The Office of Naval Research sponsored the production and publishing of two references on infrared techniques that are dear to my heart (-: W L Wolfe, *Handbook of Military Infrared Technology*, U S Government Printing Office (1965) and W L Wolfe and G J Zissis, *The Infrared Handbook*, same publisher, (1978).

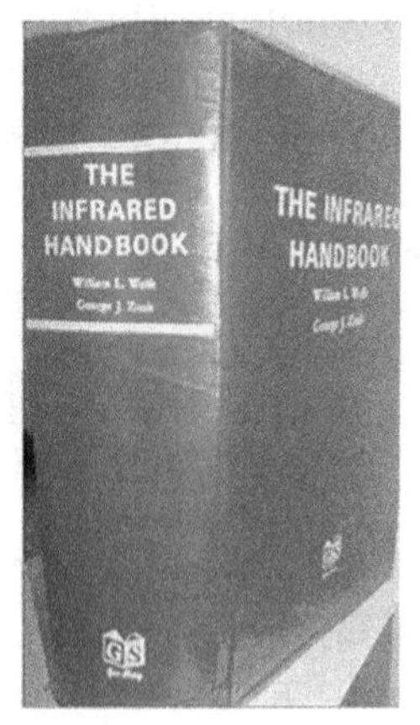

That was followed by an eight-volume series: J Ascetta and D Shumaker, *The Infrared and Optics Handbook*, same publisher. Each volume is a collection of chapters written by experts in the field. Each covered all the appropriate topics in infrared technology.

Still, the best reference on thermal imaging systems is J M Lloyd's *Thermal Imaging Systems*, Plenum (1975). It emphasizes the use of linear systems techniques, including spatial spectral analysis. A treatise that is mostly about detectors is that by E L Dereniak and G D Boreman, *Infrared Detectors and Systems*, Wiley (1996). Two books that arose from SPIE tutorials are both by me: W L Wolfe, *Introduction to Infrared System Design*, and *Infrared Design Examples*, both by SPIE Press, dated 1996 and 1999, respectively.

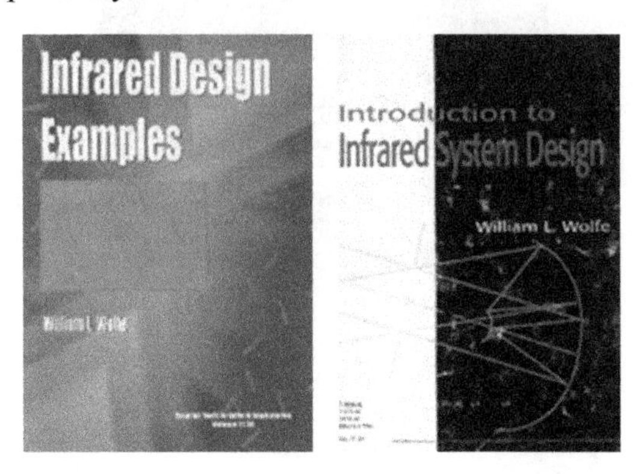

Two nice reviews of lead salt detectors as of 1960 by Trevor Moss are Proc IRE **43**, 1869 (1955) and *Advances in Spectroscopy*, Interscience (1959).

Anton Rogalski has written a very nice review of the development of infrared detectors and arrays from the discovery of the infrared spectrum to the date of his publication: A Rogalski, History of infrared detectors, *Opto-Electronics Review* 20, 279 (2012); available online.

A fairly comprehensive review of all the many versions of the Sidewinder missile is given in *The Sidewinder Story* on the net. Herb Kaplan has written a nice treatment of many infrared applications: H Kaplan, *Practical Applications of Infrared Sensing and Imaging Equipment*, SPIE Press (2007).

A host of books on infrared techniques of all sorts can be found on the internet. I have not had the opportunity to read them all, but some are quite good and some not so much.

Interference

One very nice book on the subject of filters is H Angus Macloud's *Thin-Film Optical Filters* 4th edn, CRC Press (2010). Another is Phil Baumeister's *Optical Coating Technology*, SPIE Press (2004). Phil was a professor for years at The Institute of Optics, and Angus was a professor at the Wyant College of Optical Sciences.

Lenses

The very best book on photographic lenses is by Rudolph Kingslake, *History of the Photographic Lens*, Academic (1989). It takes the same tack I have; it treats the history of each type of lens individually. It is nice to know that I inadvertently followed the path of an icon. A nice review of gradient index optics is by Duncan Moore, *Applied Optics* **19**, 1035 (1980). Another is the book by E W Marchand, *Gradient Index Optics*, Academic Press (1978). A description of both the lens and its performance is given by Warren Smith, Chapter 14 in W L Wolfe, ed., *Optical Engineer's Desk Reference*, OSA/SPIE (2010).

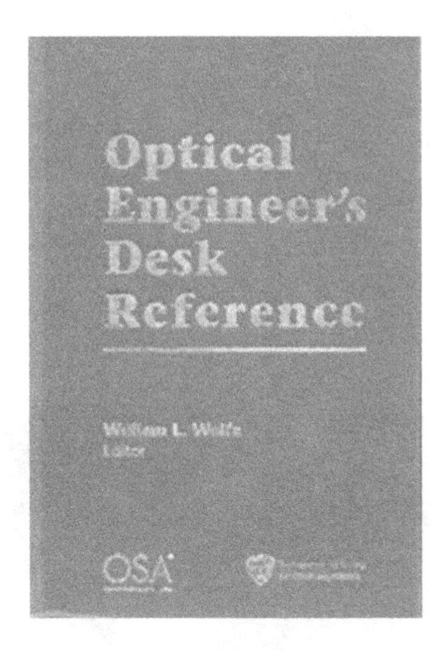

Light

A Zajonc, *Catching the Light: The Entwined History of Light and Mind*, Oxford (1993). C Roychoudhuri and R Roy, *The Nature of Light: What is a Photon?*, OPN Trends, The Optical Society of America (2003). Wikipedia has a decent treatment.

Lithography

Two very nice reviews are John Bruning's *Optical Lithography—30 Years and Three Orders of Magnitude*, *Proc. SPIE.* **3050** (1997); and Optical lithography—40 years and holding, *Proc. SPIE.* **6520** (2007).

Mirrors

The article by Jay Enoch is the best I have found on the early history of mirrors: J M Enoch, *Optometry and Vision Science*, **83**, 775 (2006).

Nonlinear optics

A good reference is R W Boyd's *Nonlinear Optics*, Elsevier (2008).

Optical design

Books on optical design did not really blossom until late in the 1950s, although there are two notable exceptions. These are those by A E Conrady and R Kingslake.

Perhaps the most widely known book on optical design considerations, although it is a bit broader than that, is by Warren Smith, *Modern Optical Engineering*, McGraw-Hill, (1990) (2nd edn). Other good texts on this subject are by the Bobs, Shannon and Fischer: R E Fischer and B Tadic-Galeb, *Optical System Design*, McGraw-Hill (2000); and R R Shannon, *The Art and Science of Optical Design*, Oxford (1997).

Max Riedl has published a nice text on infrared optical design: M J Riedl, *Optical Design Fundamentals for Infrared Systems*, SPIE Optical Engineering Press (1995). An article on the net that is very comprehensive about this subject is Wikipedia on Optical Aberration.

Optics olio

Wikipedia gives a nice review of the history of our knowledge of climate change, that is, global warming: History of Climate Change. Wikipedia similarly has good reviews of Additive manufacturing and 3D printing.

Polarization

A comprehensive article on polarizers is by J Bennett, vol II, chapter 3, Polarizers, in M Bass, ed. *The Handbook of Optics*, Optical Society of America, McGraw Hill (1995). A comprehensive book on the subject is R Chipman, W Lam and G Young, *Polarized Light and Optical Systems*, CRC Press (2019).

Prisms

A detailed presentation of prismatic polarizers is in the same article by Jean Bennet. I have an article in *The Handbook on Optics* on non-dispersive prisms.

Radiometry and photometry

A nice article on the history of photometry in astronomy is in the *Journal of the British Astronomical Association*, **117**, 4 (2007). A classical treatment of photometry is *Photometry* by J W T Walsh, Dover (1958); it includes a good history as well. A modern treatment of radiometry based on course notes by Jim was edited and

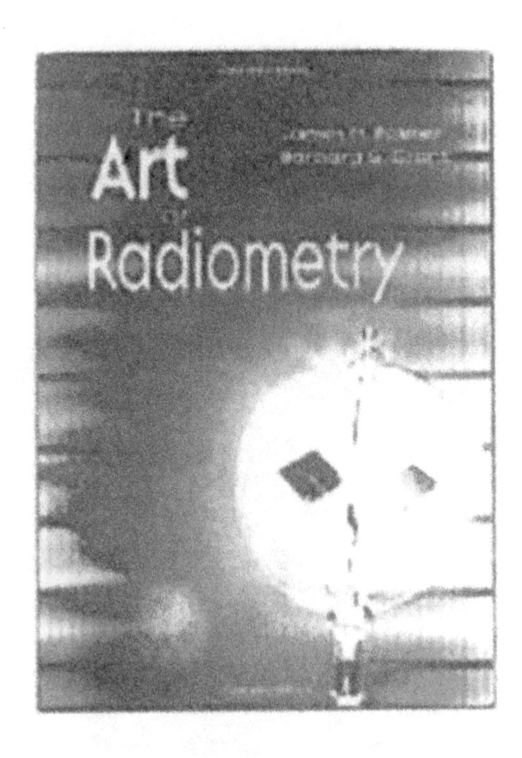
The
Art
Radiometry

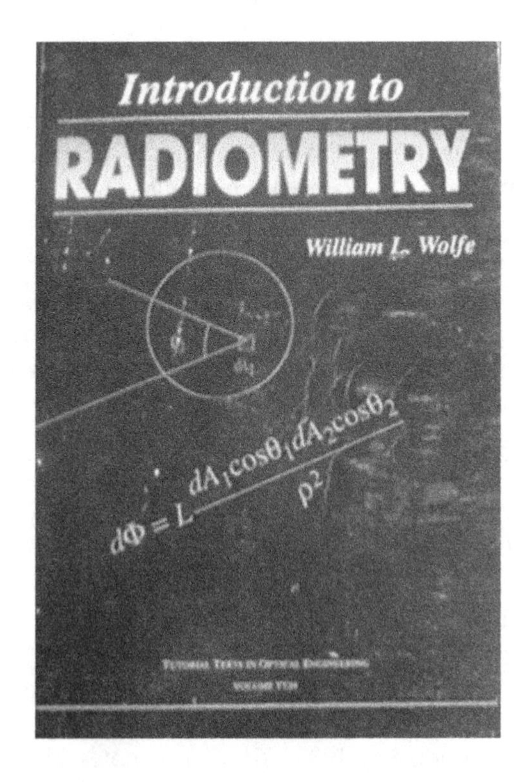
Introduction to
RADIOMETRY
William L. Wolfe
$d\Phi = L\frac{dA_1\cos\theta_1 dA_2\cos\theta_2}{\rho^2}$

expanded by Barbara Grant: J M Palmer and B G Grant, *The Art of Radiometry*, SPIE Press (2009). I contributed, too: W L Wolfe, *Introduction to Radiometry*, SPIE Press, (1998).

Refraction

An excellent bibliography of methods of measurement of refractive index is in Ben Piatt's dissertation: B Platt, *Instruments for Measuring Properties of Infrared Transmitting Optical Materials*, The University of Arizona, (1976).

Relativity

Relativity began with Galileo. So, his books are a source. An English version is available online as one of the references in Wikipedia, *Galilean Transformation.* It really matured with Einstein's special and general versions. There are many excellent books on these, some by Einstein himself: A Einstein, *The Meaning of Relativity*, Princeton University Press (1922, 1956); *Relativity, the Special and General Theory*, Pi Press (2005). There are simplified versions like Bertrand Russell's *The ABC of Relativity*, Allen and Unwin (1969); and Stephen Hawking's *A Briefer History of Time*, Bantam (2005). And there are many texts invoking tensor analysis that will not be cited here.

Remote sensing

A very nice review of recon imagery is *The Journal of Photography and Motion Pictures of George Eastman House*. Number 65 contains a fascinating description of early balloon remote sensing adventures, also available online. An extensive treatment of both the techniques of detecting fires remotely and of assessing their results is the paper by Leigh B Lentile and others, *International Journal of Wildland Fire* **15**, 319 (2006). A modern source is the open access journal *Remote Sensing*, available online. A comprehensive (almost 2000 pages), and expensive (almost $300), overview of remote sensing is H Kramer's *Observation of the Earth and its Environment: Survey of Missions and Sensors*, Springer (2002). Three publications that describe the state of the art as of their dates are *Proc. IEEE* **57** (1969); **73**, (1985); and **82** (1994), in special issues of that journal.

Scattering

Probably the best modern books on scattering are Van de Hulst's *Light Scattering from Small Particles*, Dover (1957); C Bohren and D Huffman's *Absorption and Scattering of Light by Small Particles*, Wiley (2007), and on particulate scattering; and A Maradudin's *Light Scattering and Nanoscale Surface Roughness*, Springer (2007), on scattering from surfaces.

Scopes

A nice, detailed treatment of endoscopes is J Emonson's History of the instruments for gastrointestinal endoscopy, *Gastrointestinal Endoscopy*, **37**, 42 (1991).

Sources

A nice history of the first LED on the net with very good references is Nikoly Zheludev's *The life and times of the LED—a 100-year history.* A remarkable history of the laser is Mario Bartolotti's *The History of the Laser*, Bollati Boringhieri Editore Torino (1999), remarkable for the side stories in it. Jeff Hecht has written a nice article on X-ray lasers in *Optics and Photonics News*, May 2005.

Spectacles

A comprehensive discussion of safety glasses is online at The Optical Vision Site. A nice review of modern optical coherence tomography is Quantifying the cornea's optical performance, *Optics and Photonics*, 34, April 2014, Optical Society of America.

Spectroscopy

Probably the two best books on spectroscopy in general are: R A Sawyer, *Experimental Spectroscopy*, Dover (1944, 1951, 1963); and G Harrison, R Lord and J Loofbourrow, *Practical Spectroscopy*, Prentice Hall (1948) and online.

Speed of light

A comprehensive work on slow light is P W Milonni's *Fast Light, Slow Light and Left-Handed Light*, Institute of Physics, (2005). Another is J B Khurgin's *Advances in Optics and Photonics*, **2**, 287 (2010).

Stereoscopes

W Darah, *The World of Stereographs*, Gettysburg, PA, Darah (1977).

Telescopes

Henry King, *History of the Telescope*, London, Charles Griffin (1955); A Kutter, *Der Schiefspiegler;* Louis Bell, *The Telescope*, McGraw Hill (1922); John Herschel, *The Telescope*, Edinburgh (1861).

Ultraviolet

Perhaps the best book on the technology of this part of the spectrum is by Alex Green, *The Middle Ultraviolet: Its Science and Technology*, Wiley (1966). An excellent article on its history as related to biology is P Hockberger's A history of photobiology for humans, animals and microorganisms, *Photochemistry and Photobiology* **76**, 561 (2002).

Miscellaneous

A nice treatment of Xerography is R Schaffert's *Electrophotography*, Focal Press (1975). Another, with a very long title, is D Owen's *Copies in Seconds: How a Lone Inventor and an Unknown Company Created the Biggest Communication Breakthrough Since Gutenberg: Chester Carlson and the Birth of the Xerox Machine*, Simon & Schuster (2004).

Appendix A

Diffraction

The Huygens idea of diffraction is that whenever there is an obstruction, either an aperture or obstacle, the wave front becomes a source of subsidiary waves, often called daughter waves. This is portrayed in figure A1. Each of those circles add up to make what is mostly a plane wave. He called them daughter waves. The edges are not plane, and that is what causes diffraction patterns. Of course, they are spheres in three dimensions.

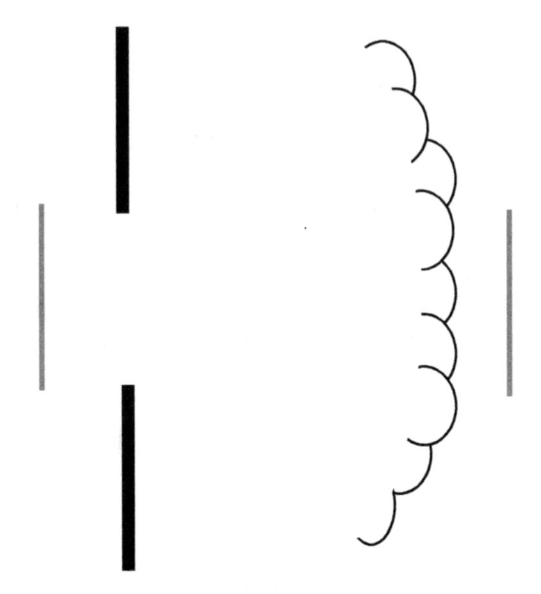

Figure A1.

The **Fresnel–Kirchhoff** integral with which diffraction problems are calculated is

$$iA/2\lambda \int e^{ik(r+s)}/rs[\cos(n,\, r) - \cos(n,\, s)ds]$$

doi:10.1088/978-0-7503-2612-4ch47

where A is the area of the diffracting object, λ is the wavelength, r and s are the respective distances, and the angles (figure A2). Diffraction geometry of the cosines is between the source and its normal and the receiver and its normal to the surface. The term $e^{ik(r+s)}/rs$ is the expression for a spherical wave, and the bracketed term defines its direction. So the integral is the total of all spherical waves coming from an area A in different directions. It is the integral formulation of Huygens principle. It is a difficult integral to evaluate and applies to diffraction in the near neighborhood of the diffracting object, called **Fresnel diffraction**.

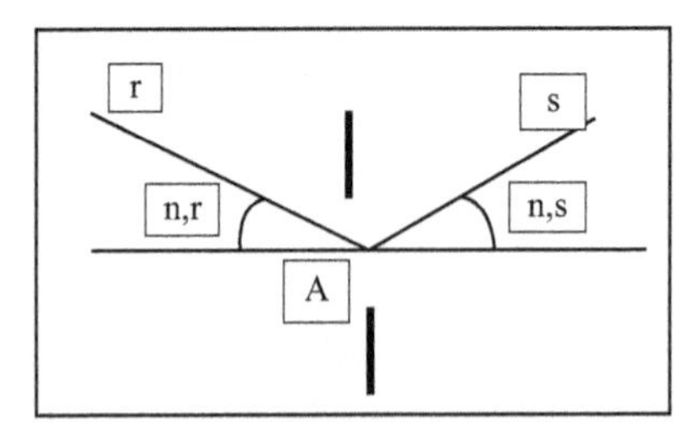

Figure A2.

Fraunhofer diffraction occurs far from the diffracting object; in that case, the integral can be simplified. The factor in the square brackets becomes a constant, and r and s do not change appreciably as the point of integration on the object is changed. The Fraunhofer integral is then

$$iA\cos\delta/(2\lambda rs)\int e^{ik(r+s)}dS$$

where δ is the constant angle with respect to the normal. This is a Fourier transform type of integral and leads to much of Fourier optics. The Fraunhofer diffraction pattern of objects is the Fourier transform of them.

When the Fraunhofer integral is applied to a rectangular aperture, like a slit, the result is a sinc function. In one dimension, the integral becomes (when the integration is from $-a$ to a)

$$\int e^{ikx\sin\theta}dX = \int e^{2\pi x\sin\theta/\lambda}dx = (^{i/kx\sin\theta})[e^{ik\sin a}.e^{ik\sin a}] = 2\sin(ka\sin\theta)/ka\sin\theta$$
$$= \mathrm{sinc}(ka\sin\theta)$$
$$= \mathrm{sinc}((2\pi/\lambda)a\sin\theta)$$

When it is applied to a circular aperture, it becomes a jinc function and is also called the Airy function. In this case, the integral must be put in polar coordinates and the solution is in terms of a Bessel function of the first kind, $J_1(x)$. Thus, $J_1(x)/x = \mathrm{jinc}(x)$. These two functions are shown below. The first zero of the sinc is when $2\pi/\lambda = 2\lambda a\sin\theta$, or when $a\sin\theta = \lambda$. Since the first zero of the jinc function is at 3.83, it occurs when $d\sin\theta = 1.22/\lambda$. For small angles, the sine can be replaced by the angle and the two results are $\theta = \lambda/a$ and $\theta = 1.22\lambda/d$.

A.1 The sinc function

This mathematical function arises when there is diffraction from a square or rectangular aperture. The distribution of the light is then a sinc function, that is, the sine of x divided by x, $\sin(x)/x$, where x can be any variable. In the case of diffraction it includes the wavelength and the dimensions of the aperture.

A.2 The jinc function

This is the equivalent of the sinc function for circular apertures. But since it is a circular geometry, a Bessel function of the first kind arises instead of the sine function. That Bessel function of the first kind is usually indicated by $J_1(x)$.

A.3 The sinc and jinc functions

These are both shown in figure A3. The sinc function is shown as the solid line, the jinc as the dashed line. The jinc is for the diameter of a circular aperture that equals a side of a rectangular aperture. The first zero of the jinc is a little larger than for the sinc. In fact, it is 1.22 times as large, and that is from whence that funny number comes into play for many calculations. It is 3.83 instead of π or 3.14. The jinc is also a little smoother than the sinc.

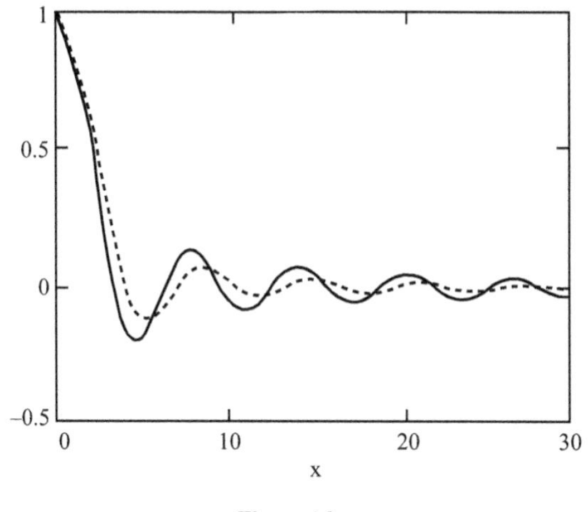

Figure A3.

Although there is a gradual change from Fresnel to Fraunhofer diffraction, a **Fresnel number** is used to separate these two regions. It is $F = a^2/\lambda L$, where a is a characteristic length of the diffracting element (a diameter of a circle or side of a rectangle), λ is the wavelength, and L is the distance from the diffractor. When F is less than one, the diffraction is said to be Fraunhofer or far field and vice versa. Basically, when the distance is large compared to the aperture size measured in wavelengths, it is far field.

Obviously, all these effects are wavelength dependent, so that heterochromatic light will blur these results. The rings will be colored and broadened.

A **diffraction grating** is a collection of closely spaced parallel lines. The resulting pattern generated by shining polychromatic light on one is given by the expression

$$d(\sin\theta_i + \sin\theta_d) = m\lambda$$

where d is the grating spacing, the θ's are the angles of incidence and diffraction, λ is the wavelength, and m is an integer. So for a given angle of incidence and line spacing, there will be a series of spectra occurring at various values for m—that is, $m = 1, 2, 3$, etc. These are the diffraction orders.

A.4 The Arago spot

It is explained in chapter 6 on diffraction that the Arago spot was one of the key experiments, and surprises, that proved the wave theory of light. Recall that it is the calculation of the light distribution behind a circular obstruction. The calculations go like this: start with the diffraction integral and specialize to the geometry involved. The distribution of light $U(q)$ is given by

$$U(q) = (Ae^{jkr0}/r_0)\int[(e^{jkr1}/r_1)(1 + \cos\alpha)/2\lambda]ds$$

where the geometry is defined in the figure and $k = 2\pi/\lambda$, A is the amplitude of the incoming wave, S is the surface of the obstruction, O is the angle of inclination, and j is the imaginary (figure A4).

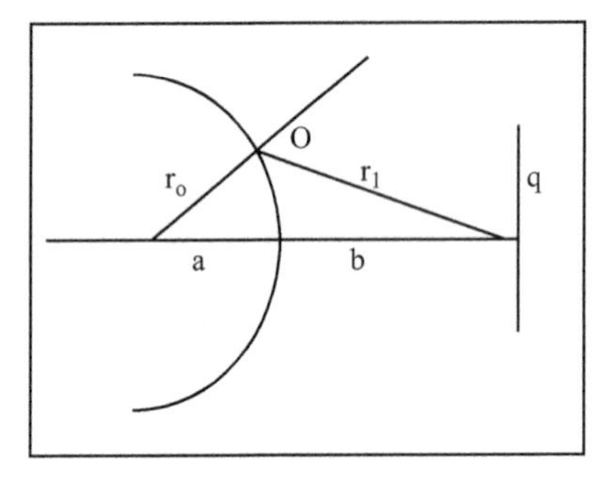

Figure A4.

Then invoke the geometric factors that require that the disk be large with respect to wavelength and is small compared to the distances. This means that the obliquity factor disappears. Then, with polar coordinates, one obtains the following expression:

$$U(q) = (A\exp[jk(a + b)]/abk\int(\exp[jk/2(1/a + 1/b)r^2]rdr$$

This can be evaluated on axis as

$$U(q) = [A \exp(jka)/a][b^2/b^2 + a^2]\exp(jk)\sqrt{(a^2 + b^2)}$$

The intensity is the absolute square of U, and the term in the first brackets is the source intensity:

$$I(q) = [A^2/a^2][b^2/(a^2 + b^2)] = I_0 b^2/(a^2 + b^2)$$

This shows that as b gets large, the on-axis intensity behind the disc approaches the intensity of the source. Figure A5 is for a value of 10. So when the distance behind the disc is equal to the source distance in front of the disc, the intensity becomes equal to that of the source.

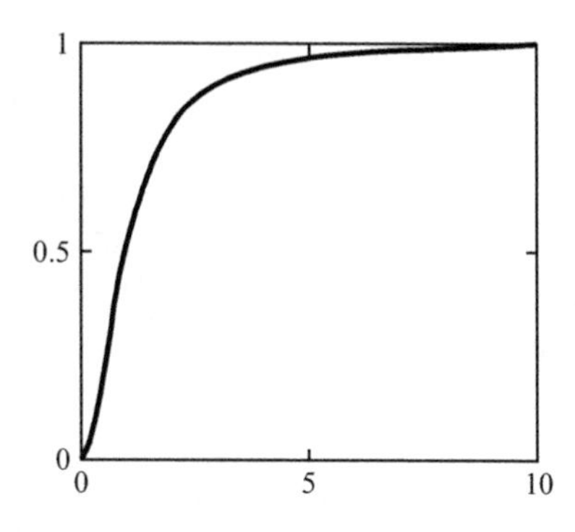

Figure A5.

IOP Publishing

Rays, Waves and Photons

A compendium of foundations and emerging technologies of pure and applied optics

William L Wolfe

Appendix B

Refraction

The refractive index is a measure of the speed of light in various materials. It is the ratio of the speed of light in a vacuum to that of the material it is in. The speed of light in vacuo is designated c, and usually the speed in a material is designated v. Thus, the refractive index, which is almost universally designated n, is given by $n = c/v$. This definition is for the phase velocity of the light and not the group velocity. The group velocity is that for the speed of a wave packet. These are discussed below. The relationship is

$$v_g = v_p[1-(\lambda/n)dn/d\lambda].$$

The two are equal as long as there is no change in index with wavelength (which there almost always is, but has a small value of about 0.000 01 per μm). For yellow light at about 0.5 μm and air, with n about equal to 1, the group velocity is about 0.999 995 times the phase velocity. The phase velocity may exceed c because it conveys no information.

B.1 Refraction from a rare to a dense material

This is the usual consideration of refraction. It makes us fishermen think that a fish is not where he really is. The light is bent toward the perpendicular as it goes into the denser material (figure B1). Snell's law is

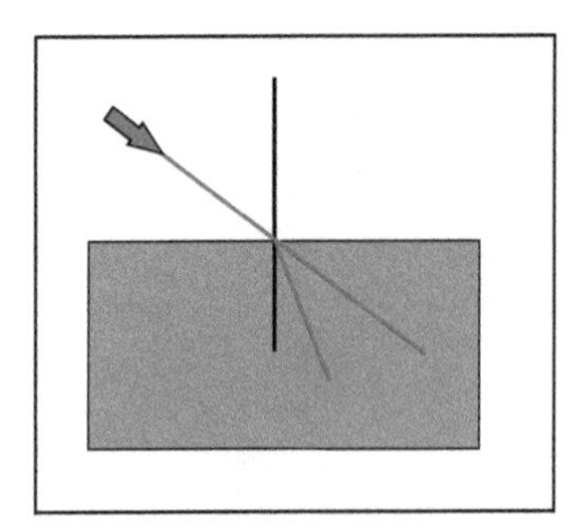

Figure B1. Rare to dense refraction.

doi:10.1088/978-0-7503-2612-4ch48

$$n_1 \sin \theta_1 = n_2 \sin \theta_2$$

Since $n_2 > n_1$, then $\theta_1 > \theta_2$. Cast a little more in front of him.

B.2 Refraction from a dense to a rare material

Snell's law still holds, but the ray refracts away from the normal as it exits the denser material. This is shown schematically in figure B2(a). As the incidence angle gets larger, further from the normal, the exiting ray gets closer to the surface. Finally, when the incidence angle is large enough, the exiting ray goes along the surface and does not exit. That is the angle of total internal reflection, the critical angle.

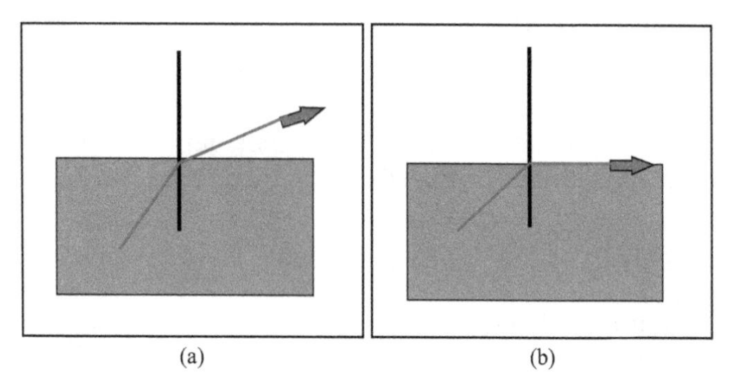

Figure B2. (a) Dense to rare refraction, (b) total internal reflection.

B.3 Total internal reflection

Snell's law is still valid but becomes simpler when the sine is one (figure B2(b)).

$$n_1 \sin(\theta_1) = n_2 \sin(\theta_2); \quad \sin(\theta_1) = (n_2/n_1)\sin(\theta_2); \; \theta_1 = \arcsin(n_2/n_1)$$

where n_1 is the lower index and n_2 is the higher index. Whenever the right-hand side exceeds or equals one, there is no longer refraction. For water where $n = 1.33$ and air with $n = 1$, the angle of total internal reflection is 49°: $\theta_2 = \arcsin[(n_2/n_1) \sin (\theta_1)] = \arcsin [1.33 \sin (49)] = 90°$. For glasses with an index of about 1.5, the angle of total internal reflection is about 42°.

B.4 The polarization angle, Brewster's angle

This is the angle at which there is complete polarization of light upon reflection. This occurs when the refracted and reflected rays are at right angles (figure B3). The argument involves the vibration of the charges in the material. The result is that Brewster's angle is given by

$$\tan(\theta) = n_2/n_1; \qquad \theta = \arctan(n_2/n_1)$$

where the subscripts have the same meanings. For an air–water interface, the polarizing angle is 53°. For glasses it is about 56°.

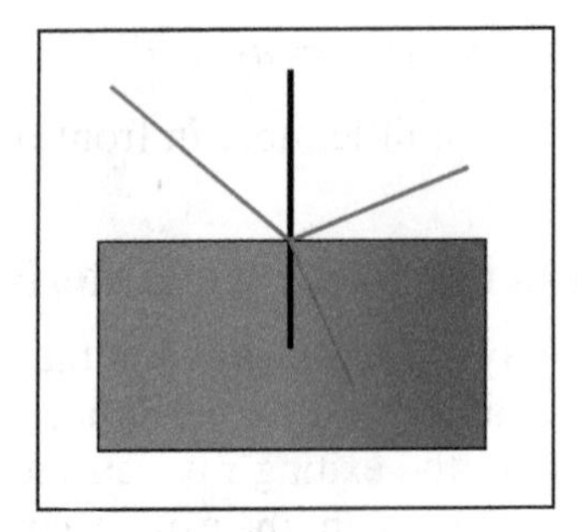

Figure B3. Polarization angle.

B.5 Complex index of refraction

As noted in chapter 31 on refraction, the refractive index can be written as a complex variable, $n - ik$. Then n represents the real part, what we commonly refer to as the refractive index, and k is the extinction coefficient, representing absorption. The solution to the wave equation for the electromagnetic field can be written in the x dimension as

$$E = E_0 e^{j(\omega t - kx)}$$

where the time variation is described by ω and the spatial variation by kx, where k is $2\pi/\lambda$. This would be for free space. However, for the wave in a material, the wavelength in the material must be considered. The frequency of the wave remains unchanged in a material, so the wavelength must change since the speed changes. This leads to the expression

$$E = E_0 e^{j(\omega t - nkx)}$$

Light is absorbed in a material according the Lambert–Beer law and is exponential in nature.

So we add that into the term:

$$E = E_0 e^{j(\omega t - (n - jk)x)} = E = E_0 \; e^{j(\omega t - nx) - kx}$$

Unfortunately, there are several ways to define this complex index of refraction. One is $n - jk$. Another is $n(1-jk)$.

Be careful when reading!

B.6 Refractive index measured by minimum deviation

This is the method that is taught and has been used for many years by various investigators. The expression for the refractive index n when a prism is at minimum deviation is

$$n = [\sin((\alpha + \delta)/2)]/ \sin(\alpha/2)$$

where α is the prism angle, which is measured separately, and δ is the measured angle of minimum deviation. In practice, one selects a wavelength and observes its direction after refraction by the prism. Then the prism is rotated until the output beam moves in one direction and then reverses direction. That position of reversal is the minimum deviation. Then, using the measured prism angle and the measured angle of minimum deviation, one solves for the refractive index. This method is capable of an uncertainty of a few digits in the fifth decimal place (figure B4).

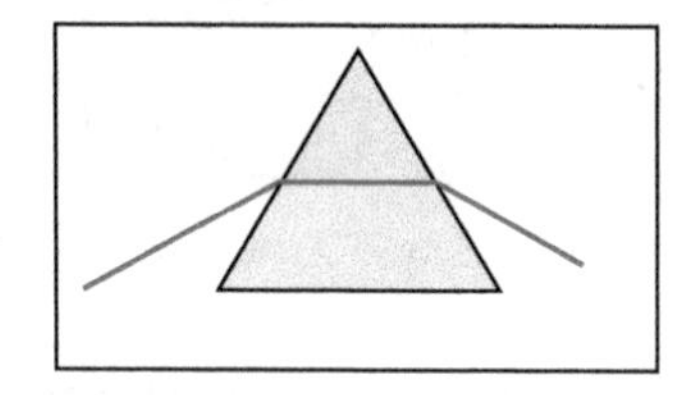

Figure B4. Minimum deviation geometry.

B.7 Refractive index measured by normal incidence

If the incident ray is normal to the front face, there is only one refraction—at the back face. One needs to autocollimate off the front surface (make sure the ray is perpendicular to the face) and measure the angle of refraction from the back of the prism. This method is slightly less accurate than that of minimum deviation (figure B5).

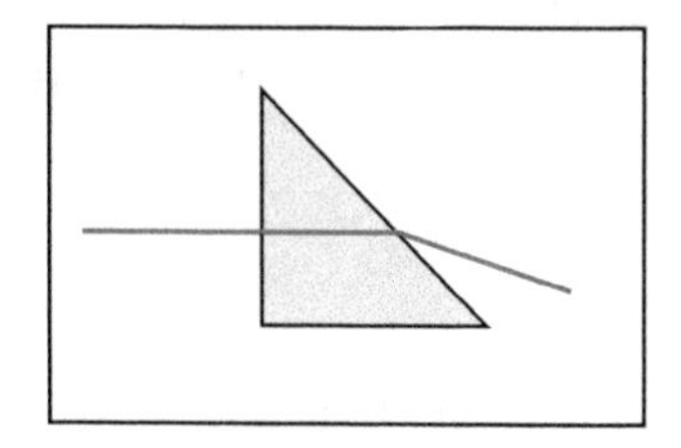

Figure B5. Normal incidence geometry.

One example of this is the apparatus designed by Ben Platt[1] and shown in figure B6. The blue box is a spectrometer for selecting a wavelength. The taller silver cylinder is a dewar that contains the prism (to keep it cold). The smaller silver cylinder contains a detector that senses the exiting ray. It is mounted on a turntable that provides the angle. The units underneath are electronics. This apparatus was used to measure refractive indices of materials with wavelengths from 0.5 to 25 μm and temperatures from 20 K to 500 K, with an uncertainty of

[1] Platt B 1976 Instruments for measuring properties of infrared transmitting materials *Dissertation* University of Arizona.

about one in the fourth decimal place. That temperature range is the reason the prism is in a dewar, essentially a Thermos bottle.

Figure B6. University of Arizona refractometer.

B.8 The Kramers–Kronig relations and the measurement of refractive index

The real and imaginary parts of an analytic function are related by the Hilbert transforms. These might be, for instance, the absorption coefficient and the refractive index, the two parts of the complex refractive index. It can be written as

$$H(t) = \int h(\tau)/(t - \tau)d\tau/\pi$$

where the integration is from $-\infty$ to ∞ and the principal value is taken. In practice, one does not have to measure that far (of course) but over a reasonable range to get an accurate answer. H might be the index and h the reflectance, and t is usually the wavelength. This gives about two or three decimal uncertainty in the index. One example of this is the measurements made by Mike Lang[2]. The measurements were of the refractive index of fused silica and quartz from 0.3 to 25 μm. The apparatus consisted of a spectrometer, a stand for the sample, and a detector. It was a simple measurement of reflectance.

B.9 Dispersion

Dispersion is the change in refractive index with respect to wavelength (or frequency). The theory is based on the absorption properties of the material, like a gas. The electrons move as a simple harmonic oscillator with damping. The form of the expression for the refractive index then becomes

$$\sum N/(\omega_i^2 - \omega^2) - d)$$

where the summation is over all the resonances, N is their number, ω_i is the resonant frequency, ω is the variable frequency, and d is the damping factor. The refractive index decreases and then increases drastically near a frequency of resonance, where the characteristic frequency of an electron bond is the same as the input light frequency. The refractive index becomes sharply positive, reverses to become sharply

[2] Lang M Private communication.

negative, and then levels off to a gradual decrease. This region of drastic change is called the region of anomalous dispersion. It is shown in figure B7).

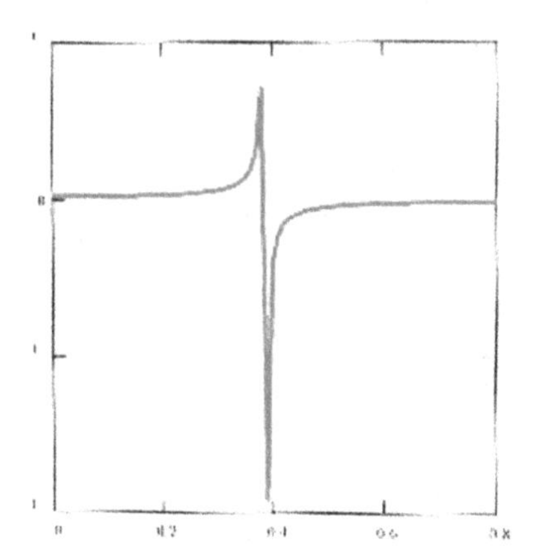

Figure B7. Refractive index at a resonance.

That is a simplified theory of dispersion. Basically, the mechanical resonance of charges in materials. But there are also the dispersion formulas for fitting refractive indices. These are of great value to lens designers. It appears that the formulas of Cauchy and Herzberger are just curve fitting in the region outside of anomalous dispersion. But they do work. I have tested them many times on a variety of infrared materials—by interpolating and checking against measured values—and plotting on giant graphs!

B.10 Waves and wave packets

This is all about monochromatic and quasi-monochromatic light, waves, wave packets, phase, and group velocity. First, a truly monochromatic wave is at a single frequency and can be represented by a sine or cosine function. But such a wave can have no energy, as noted in the main text, and it must be viewed for some finite length of time. If a sine or cosine wave is viewed for a finite time, then it can be expressed as that wave integrated from zero to time T; that is, the wave is

$$W(t) = \int \cos(\omega t)\, dt$$

where ω is the angular frequency $= 2\pi f$, and the integral goes from 0 to T. This results in a sinc function, that is

$$W(T) = \sin(\omega T)/\omega T$$

The sinc function is shown in figure B8. It has a range of frequencies in the first lobe. The first zero of the sinc is at π, where $\omega T = \pi$, or $2\pi fT = \pi$. For the frequency spread to go to zero, T must go to infinity. We never have enough time to observe a truly monochromatic wave!

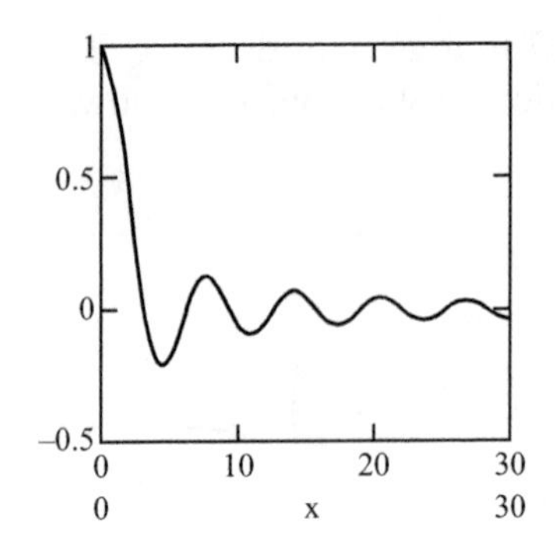

Figure B8.

So a real wave must consist of wave packets, waves of more than one frequency. I have also calculated an example of one of these. Figure B9 shows a wave packet that consists of the sum of a number of waves with frequencies that are almost the same.

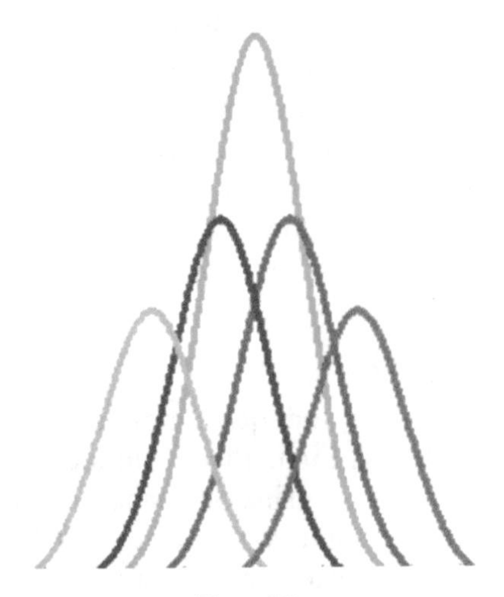

Figure B9.

IOP Publishing

Rays, Waves and Photons

A compendium of foundations and emerging technologies of pure and applied optics
William L Wolfe

Appendix C

Interference and coherence

C.1 Basic interference

Two beams of monochromatic light of equal amplitude interfere when they are superimposed and form an intensity pattern that is based on their phase relationships. If they are perfectly in phase, the pattern is just twice the amplitude of either. In figure C1(a), the two in-phase waves are shown as solid lines, and their combination as a dashed line. If they are exactly out of phase, by 90° or $\pi/2$ radians, they cancel, as shown in figure C1(b). Otherwise, they form another sine wave, one example of which is in figure C1(c). If they are of a different frequency, they generate another waveform. A bunch of these will form a mess! So one way to have incoherent light is for it to be heterochromatic of many different colors and wave lengths. Another is for it to be of many different phases. Since there is no physical realization of a truly monochromatic beam, and any real source emits beams of varying phases, we must deal with partially monochromatic or quasi-chromatic light and light beams that are almost in phase.

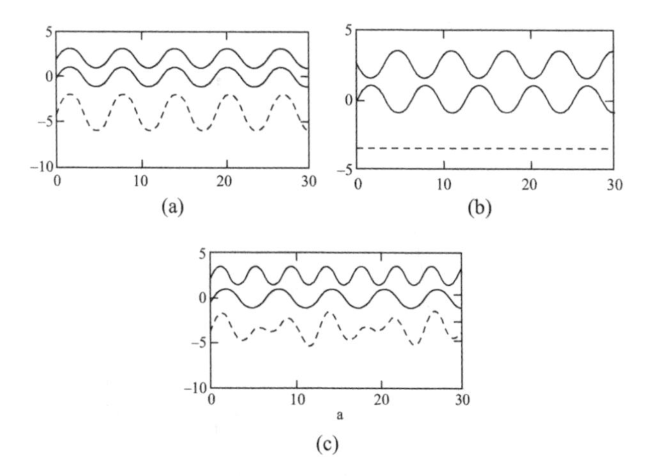

Figure C1. (a) Constructive interference, (b) destructive interference, (c) general interference.

doi:10.1088/978-0-7503-2612-4ch49

A truly monochromatic beam had to have started a long time ago, and we do not have time to wait to see it. Consider a sinusoidal, monochromatic beam lasting for a time T. It is represented by the integral of the sine from $-T$ to $+T$.

$$\int \sin(\omega t)dt = \sin(\omega T)/\omega T = \text{sinc}(\omega T)$$

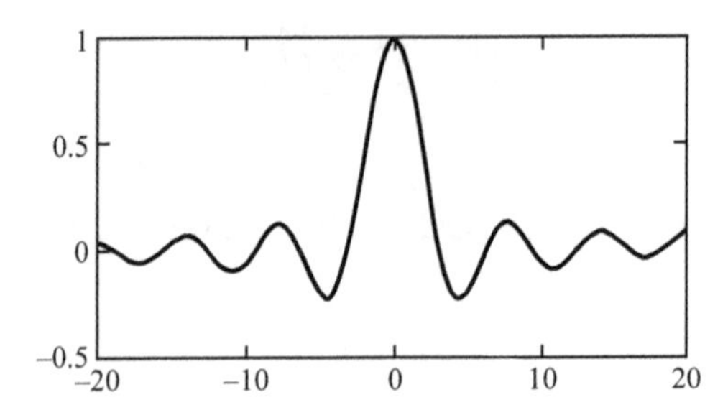

Figure C2. Representative spectral line.

This pattern is shown in figure C2; the spectral width between the first zeroes is proportional to $1/T$. The only way for the wave to be truly monochromatic, to get the width of that central peak to be zero, is for T to be infinite, from $-\infty$ to $+\infty$. That is from way before I was born and until way after I die! And my grandchildren! And you!

C.2 Coherence and partial coherence

These concepts lead to the ideas of the degree of coherence of a beam and what happens with beams that are only partially coherent—like all of them. Figure C3 shows the interference pattern for eleven different waves with frequencies covering an octave, from v to $2v$. It can be seen that there is a bright spot, a maximum, in the middle where they are all in phase, and then a decrease as they get out of phase away

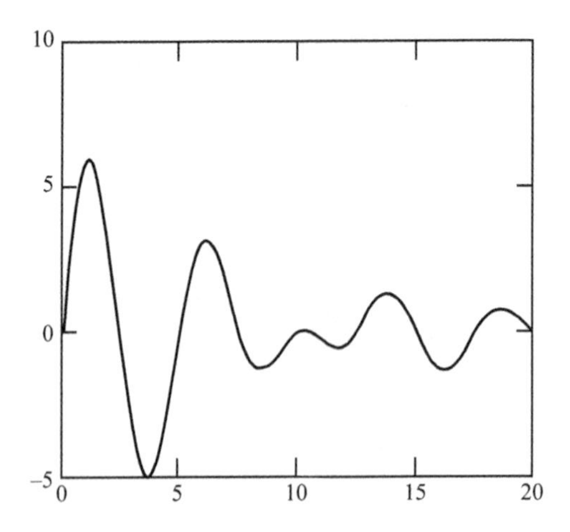

Figure C3. Partially coherent interference.

from the center. If there were more wavelengths, the curve would be even flatter after a bit. So there is coherent interference for a small distance. What is that distance? If the half maximum points of the sinc function are taken as the ends, then the uncertainty of time times the uncertainty of frequency is $1/4\pi$, that is $\Delta t\Delta v = 1/4\pi$. The **coherence** time is therefore $\Delta t = 1/4\pi\Delta\nu$, and the **coherence length** is $c\Delta t = c/4\pi\Delta\nu = \lambda^2/\Delta\lambda$ where the 4π has been dropped since this is an approximation and somewhat arbitrary, and the wavelength is the central one of the small bandwidth. These are the lengths and times over which there will be good interference. At longer lengths and longer times it will not be so good, if there is any interference at all.

C.3 Multiple beam interference

The transmission function for multiple-beam interference is $\tau = 1/(1 + F\sin^2\delta/2)$, where F is $4\rho/(1-\rho)^2$, ρ is the reflectivity, and δ is the phase function, which is equal to $(2\pi/\lambda)2nl$–$\cos\theta$, where λ is the wavelength, n is the refractive index, l is the length, and θ is the angle. As the reflectivity of the plates increases, two things happen. The overall transmission decreases, and the peaks become sharper. The first effect is because less light is transmitted. The second is because more and more waves interfere with each other. Figure C4 shows how the quantity F increases with increasing reflectivity. Figure C5 shows how the transmission decreases with increasing F. Figure C6 shows how the peaks get steeper with increasing F (from top to bottom $F = 1$, 3, 10, 20), and therefore reflectivity, and therefore more interference. They can be compared to a simple sinusoid for two-wave interference—almost like the curve for $F = 1$, the top curve of figure C6.

This is the sort of interference one gets in a Fabry–Pérot interferometer.

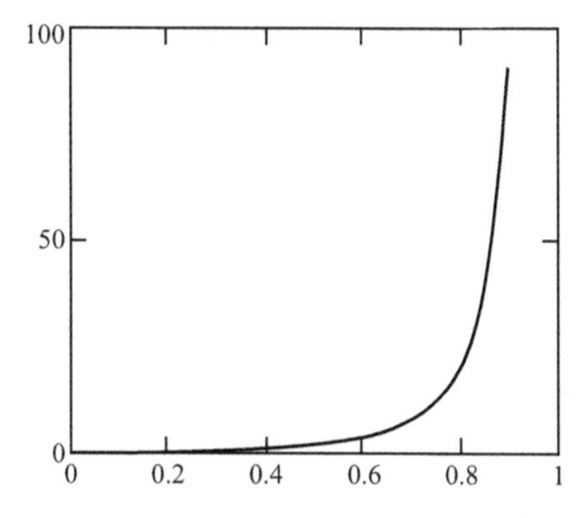

Figure C4. The function F.

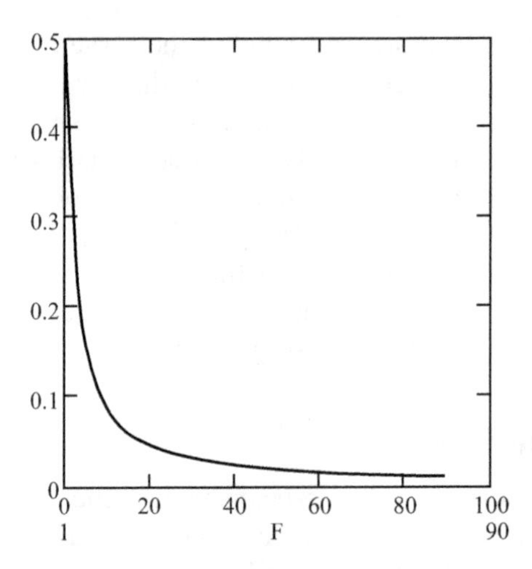

Figure C5. Transmission as a function of F.

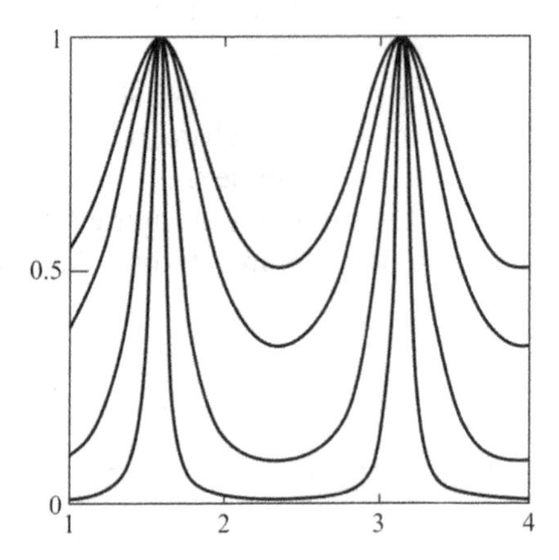

Figure C6. Multiple beam interference.

C.4 Young's double slit experiment

This experiment was critical in proving that light is a wave motion and not a bunch of corpuscles. A coherent light source is used to illuminate two slits, as shown in figure C7. The light from each of the slits expands behind them and combines on a screen, where they interfere. It was not so easy to do in Young's time, but we can use a laser today to do it. He had to use a source behind a single slit to get a small source and illuminate the other two with light from the same source. When the distances from the two slits to the screen are an equal number of wavelengths, there is constructive interference and a maximum in the pattern. That is when $y - x = n\lambda$, where n is an integer. Otherwise there is a certain amount of destructive interference and not a maximum. The pattern on the double slit arrangement screen is sinusoidal if the light is monochromatic (or nearly so).

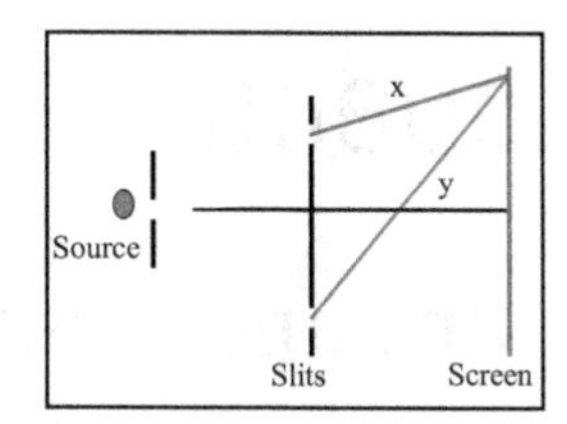

Figure C7. Double slit geometry.

IOP Publishing

Rays, Waves and Photons
A compendium of foundations and emerging technologies of pure and applied optics
William L Wolfe

Appendix D

Maxwell's equations and Planck's derivation

D.1 Maxwell's equations

The modern treatment for obtaining the Planck equation starts with the Maxwell expressions for the fields, derives the wave equation, separates and solves it, and calculates the density of states. It then calculates the energy per mode.

Maxwell's equations are

$$\nabla \cdot \mathbf{E} = \rho/\varepsilon \quad \nabla \cdot \mathbf{B} = 0 \quad \nabla X\mathbf{E} = -\partial B/\partial t \quad \nabla X\mathbf{B} = \mu\mathbf{J} + \sqrt{(\mu_0\varepsilon_0)}\partial\mathbf{E}/\partial t$$

where **E**, **B** and **J** are vectors, the electric field, magnetic field and current; μ_0 is the permeability of free space; ε_0 is the permittivity of free space; and ρ is the charge density. The del operator is defined as

$$\nabla = i\partial/\partial x + j\partial/\partial y + k\partial/\partial z$$

and the unit vectors are i, j, and k. In the absence of any charge or current, as in free space, the current and charge terms, J and ρ, are zero.

D.2 The wave equation

The first step is to take the curl of the third Maxwell equation.

$$\nabla X\nabla X\mathbf{E} = -\partial\nabla X\mathbf{B}/\partial t = -\mu_0\varepsilon_0\partial^2\mathbf{E}/\partial t^2$$

Then, applying the vector identity,

$$\nabla \mathrm{X} \nabla \mathrm{X}\mathbf{E} = \nabla(\nabla \cdot \mathbf{E}) - \nabla^2\mathbf{E}$$

one gets

$$(\nabla^2 - \mu_0\varepsilon_0\, \partial^2/\partial t^2)\mathbf{E} = 0$$

which is the vector wave equation.

D.3 Solution of the wave equation

The vector wave equation can be separated into its component parts, which is what those symbols stand for

$$i\partial_x^2/\partial x^2 + j\partial^2 E_y/\partial y^2 + k\partial^2 E_z\partial^2 z - i\,\mu_0\varepsilon_0\partial^2 E_x\partial t^2$$
$$-j\,\mu\varepsilon\partial^2 E_y/\partial t^2 - k\,\mu_0\varepsilon_0\partial^2 E_z/\partial t^2 = 0$$

Then this can be separated into its x, y, and z component equations, each of which has the form

$$(\partial^2/\partial x^2 - \mu_0\varepsilon_0\partial^2/\partial t^2)\mathbf{E} = 0 = \partial^2/\partial x^2 - 1/c^2(\partial^2/\partial t^2)$$

This is recognized as a wave equation (by those in the know) with the solution in the form of $f(x-ct)$. The spatial double derivative of f is f''; the temporal double derivative is $c^2 f''$. Their difference is zero if c is equal to the square root of $\mu_0\varepsilon_0$. This is another way of determining the speed of light—that is, $\sqrt{(\mu_0\varepsilon_0)}$. The solutions to this equation are sines and cosines, and the fact that the boundary conditions are zero on the walls dictate that the solutions are of the form sin $n\pi x/L$, where n is an integer for each spatial term and sin $2\pi ct/\lambda$ for the temporal term. The solutions plugged back into the equation reveal

$$(l\pi/L)^2 + (m\pi/L)^2 + (n\pi/L)^2 = (2\pi/\lambda)^2$$

which can be reduced to

$$l^2 + m^2 + n^2 = 4L^2/\lambda^2$$

D.4 The volume of modes, mode density

The volume of modes would then be (using the equation for the volume of a sphere)

$$(4\pi/3)(l^2 + m^2 + n^2)^{3/2}$$

But this has to be reduced for the number of redundant modes, the negative ones, eight of them, and increased for the two modes of polarization.

This determines the density of modes. Each mode must have an energy associated with it. If they are in steps, then each mode has an energy of $e^{nu/kT}$, where u is an energy unit to be determined, n is an integer, k is Boltzmann's constant, and T is the absolute temperature. The sum of an infinite series of these, since $n = 1, 2, 3$, etc, is 1 over the ratio minus one, or $1/(e^{u/kT} - 1)$. This is the result Planck obtained and associated the unit of energy with $h\nu$, although h was not yet called Planck's constant.

D.5 Variations on the theme

Planck's equation comes in many varieties depending upon whether energy or photons are being considered, whether it is in terms of frequency or wavelength, and whether it is for power, power per unit area, or power per unit area and unit solid angle. The most common form seems to be the power per unit area per wavelength:

$$M_\lambda = 2\pi c^2 h/\lambda^5(e^{hc/\lambda kT} - 1)\ \mathrm{Wm^{-3}}$$

Since blackbody radiation is Lambertian—that is, has the same radiance in all directions—the radiance is just the irradiance divided by π:

$$L_\lambda = 2c^2h/\lambda^5(e^{hc/\lambda kT} - 1)\ \mathrm{Wm^{-3}\,sr^{-1}}$$

Since the energy of a photon is hc/λ, the photon radiance (photons per second per unit area and per unit solid angle) is

$$L_\lambda = 2c/\lambda^4(e^{hc/\lambda kT} - 1)\ \mathrm{s^{-1}\,m^{-3}\,sr^{-1}}$$

Since $M_\lambda d\lambda = M_\nu, d\nu$ and $\nu = c/\lambda$, the three equivalent expressions in terms of frequency are:

$$M_\nu = 2\pi c^2h\nu^3(e^{hc/\lambda kT} - 1)\ \mathrm{s^{-1}\,m^{-2}}$$

The total integral of the distribution yields the **Stefan Boltzmann** law,

$$M = \sigma T^4 = 5.670 \times 10^{-8}\ \mathrm{Wm^{-2}}$$

The maximum, determined by taking the derivative and setting it to zero, is the **Wien displacement law**:

$$\lambda_{\mathrm{max}}\ T = 2998\quad \mu\mathrm{m\ K}$$

D.6 Examples

Two blackbody distributions of particular interest are those of the Sun, shown as a red line, and the Earth, shown as a blue line in figure D1. They are shown as normalized values and the irradiance (radiant incidance) of the Sun at the Earth compared to the radiant emittance (exitance) of the Earth. The Earth's radiant emittance is increased by three million to get it on the same level as the Sun's.

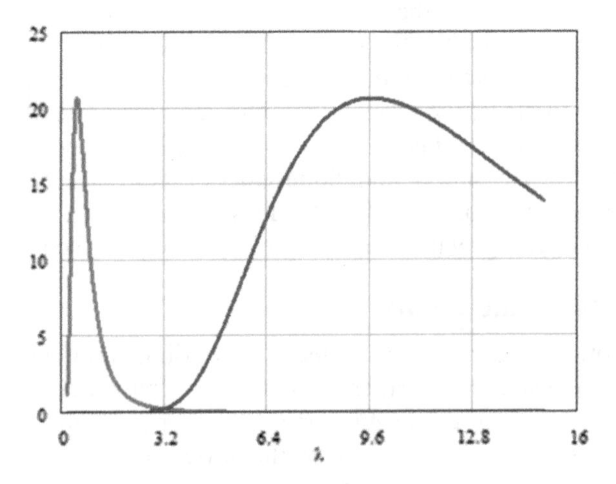

Figure D1. Normalized solar and Earth radiant emittances.

In figure D2, the spectral irradiance of the Sun at the Earth is shown and compared to the spectral emittance of the Earth at the Earth, both at the surface of the Earth but ignoring any atmospheric transmission. And with no adjustments.

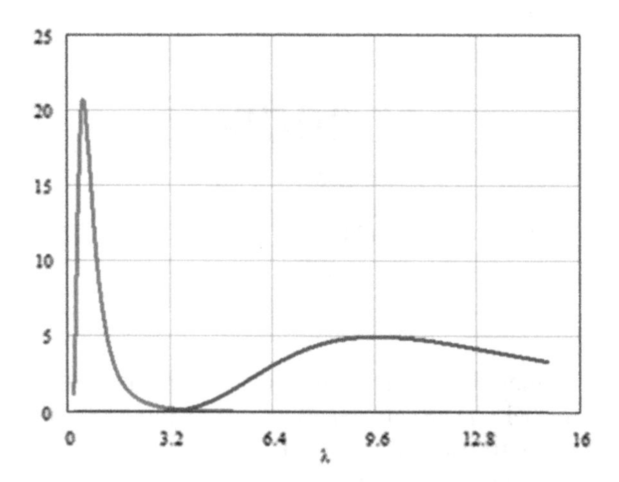

Figure D2. Solar irradiance and Earth emittance.

Appendix E

Fresnel's equations for reflection and refraction

I use the modern treatment based on electromagnetic theory, rather than the original development of Fresnel. There are three electromagnetic waves involved: the incident, reflected, and transmitted ones. The spatial components, denoted with subscripts i, r, and t, respectively, are

$$E_i \exp(\mathbf{k} \cdot \mathbf{r}) = E_i \exp(2\pi n_i x \sin \theta_i / \lambda)$$
$$E_r \exp(\mathbf{k} \cdot \mathbf{r}) = E_r \exp(2\pi n_i x \sin \theta_r / \lambda)$$
$$E_t \exp(\mathbf{k} \cdot \mathbf{r}) = E_t \exp(2\pi n_t x \sin \theta_t / \lambda)$$

It all happens at the same time, so the time variations need not be considered (and I didn't). The wave numbers, k, need to have the refractive indices, n, as part of them, since the wavelength changes in different media. The exponents can be equated. So for reflection for which the waves are both in the initial medium, one has

$$2\pi n_i x \sin \theta_i / \lambda = 2\pi n_i \; x \sin \theta_r / \lambda$$

Clearly, the two's are equal, the wavelengths are equal, the pi's are equal, and the x's are equal. So, as a result of all this equality, the two angles are equal; the angle of reflection equals the angle of incidence—that is, **the law** of **reflection.**

Similarly, for refraction involving the incident and second media

$$2\pi n_i x \sin \theta_i / \lambda = 2\pi n_i x \sin \theta_t / \lambda$$

Eliminating 2, π, λ, and x this leaves

$$n_i \sin \theta_i = n_t \sin \theta_t$$

Snell's law is obtained as easily as that!

These are the easy steps, and they came from boundary considerations of waves well before Maxwell enunciated his laws of electromagnetism. Fresnel probably knew the proper expressions for waves even if he did not have Maxwell's results to depend upon. The more difficult task is the derivation of the reflection and transmission coefficients for the s and p polarizations. We leave that derivation to many different textbooks, but quote and plot the results here. There are four such

doi:10.1088/978-0-7503-2612-4ch51

coefficients: two each for the amplitudes of the waves and two for the intensities of the waves. Two each for each of the polarizations that are indicated by the subscripts s and p. This is to avoid p for parallel and p for perpendicular. The German word *senkrecht* is used for perpendicular.

The amplitude reflection coefficient for a dielectric for s polarization is

$$r_s = -\sin(\theta - \theta')/\sin(\theta + \theta')$$

The transmission coefficient for s polarization is

$$t_2 = 2\sin(\theta')\cos(\theta)/\sin(\theta + \theta')$$

The reflection and transmission coefficients for p polarization are

$$r_p = \tan(\theta - \theta')/\tan(\theta + \theta') \text{ and } t_p = 2\sin(\theta')\cos(\theta)/\sin(\theta + \theta')\cos(\theta - \theta')$$

These are shown plotted in figure E1(a). The top two are the transmission coefficients and are almost identical. The bottom two are those for reflection. The solid curve represents perpendicular polarization, while the dashed one represents the parallel component. The intensity coefficients are the squares of the amplitude coefficients, shown in figure E1(b) with capitals.

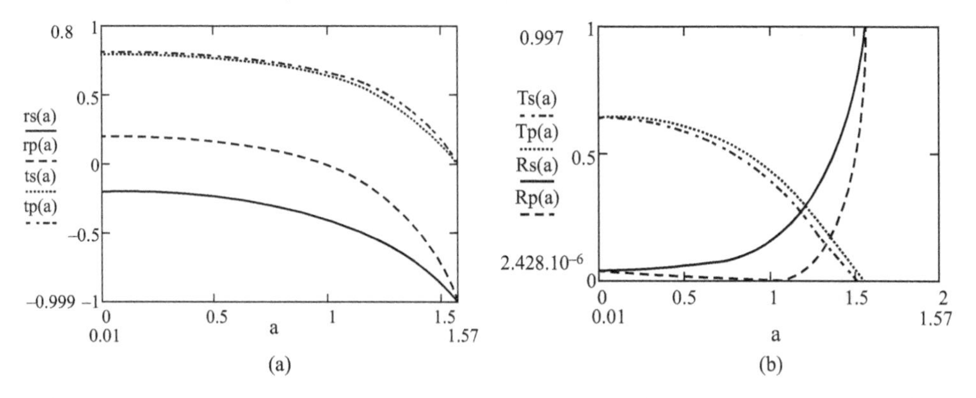

Figure E1. (a) Transmittance and reflectance amplitudes, (b) transmittance and reflectance intensities.

An easy way to derive the well-known law of reflection for normal incidence is as follows: use the expression for r_p and substitute sines for tangents, since the angles are small. Then expand the sine of the sum and difference of two angles. Then divide through by $\sin\theta'$.

$$r_p = \tan(\theta - \theta')/\tan(\theta + \theta') = \sin(\theta - \theta')/\sin(\theta + \theta)$$
$$= \sin\theta\cos\theta' - \cos\theta\sin\theta')/(\sin\theta\cos\theta' + \cos\theta\sin\theta'$$

Divide through by $\sin\theta'$ to get

$$r_p = (\sin\theta/\sin\theta')\cos\theta' - \cos\theta/(\sin\theta\sin\theta')\cos\theta' + \cos\theta$$

Now note that $\sin\theta/\sin\theta'$ is just the relative refractive index by Snell's law, and we have

$$r_p = (\text{ncos}\,\theta' - \cos\theta)/(\text{ncos}\,\theta' + \cos\theta).$$

Since the angle of incidence is 90°, the cosines are both 1, and the result is

$$r_p = (n - 1)/n + 1),$$

and its square is the well-known equation for the **intensity reflectivity of a dielectric at normal incidence:**

$$p = [(n - 1)/(n + 1)]^2$$

Glass has a refractive index of about 1.5 so that its normal reflectivity is

$$p = [(1.5 - 1)/(1.5 + 1)]^2 = (0.5/2.5)^2 = 0.2^2 = 0.04.$$

This is the well-known result that glass has a reflectivity of about four percent (at normal incidence).

Appendix F

Lorentz transformations

These transformations—those that adjust electromagnetics for moving media—can be very complicated, but a few simple cases are considered by way of example. They were developed by Lorentz and verified by Einstein's relativity.

The so-called simple *boost* is a translation of a frame of reference along the x-axis. Then the coordinates of the translated frame, indicated by primes, e.g. x', are

$$
\begin{aligned}
x' &= (x - vt)/\sqrt{(1 - \beta^2)} \\
y' &= y \\
z' &= z \\
t' &= (t - \beta x/c)/\sqrt{(1 - \beta^2)}
\end{aligned}
$$

This shows that physical dimensions in the direction of travel are shortened by a small amount, that the other physical dimensions are unchanged, and that time is shortened by a small amount. The term in parenthesis for x, that is $x - vt$, is just the translation due to the velocity v. The change in dimension is dictated by the denominator $\sqrt{1-\beta^2}$ where β is v/c, the ratio of the velocity of motion to the speed of light. Since that ratio is normally very small there are very small time and contraction effects. In an airplane flying 500 mph, β^2 is 5.6×10^{-13}. So $(1 - \beta^2)$ is 0.999 999 999 9999 and its square root is 0.999 9999. Any length moving at this velocity is shortened by this much. It may look like it is lengthened, but you have to keep track of which frame is moving with respect to which and which you are measuring in. The time will be changed by an even smaller factor.

doi:10.1088/978-0-7503-2612-4ch52

Appendix G

Units conversions and nomenclature

Throughout this text, various units in different systems of units are used. This is an attempt to make them easier to understand. I have also added some constants that may be of use.

G.1 Numbers

Throughout much of science, numbers are given in so-called scientific notation. I guess it is reasonable for scientists to use scientific notation. It works like this: A fairly large number like 124 000 is written as 1.24×10^5. It is the first number with a decimal after it and then ten raised to the number of zeros as there are digits after the first. Then 1.5 million, 1 500 000, is 1.5×10^6. (Note that the IOP convention of spaces rather than commas is used to delineate large numbers.) This has at least three advantages. The first is that one can point out the uncertainty in a number. The number 1.2345×10^6 has more significant digits, six, than 1.23×10^6, which has only three. We know its value better, although they are both about the same. The second advantage is in multiplication. The old dictum that in order to multiply numbers you add exponents applies here. The product of 8×10^5 and 9×10^4 is 8 times 9 and then times ten to the 5 plus 4, or 72×10^9, which we would then write as 7.2×10^{10}. That is much easier, I think, than multiplying 800 000 by 90 000. The third advantage is that these scientific numbers take up far less space, especially for really big ones.

Scientific notation applies to little numbers, too. 0.000 001 234 can be written as 1.234×10^{-5}. Just count the zeros in front of the number to get the value of the negative exponent. The comments above about significance and multiplication apply equally well here.

Although other **number bases** are not used in this text, I just had to add a little about them. We live in a digital age, and all computers know is up or down, yes or no, zero and one. They work on the number system of base two, the binary system. A bit is the portmanteau of binary digit. We use the decimal system or base ten. We have ten numerals, digits: 0, 1, 2, 3, 4, 5, 6, 7, 8, 9. The binary system has just two numerals, or bits: 0, 1. Thus, counting goes like this: 0 in the decimal system is 0 in the binary system;

doi:10.1088/978-0-7503-2612-4ch53

1 is 1; but 2 in the binary system is 10 because there is no such thing as a 2 in the binary system, only 1 and 0, so the two's row is used (like we use 10, putting a 1 in the ten's row). Then in the binary system 3 is 11, and 4 is 100, and so on. A bit is a binary digit. A byte is eight binary digits, like 100100100, and is the specification for most computers, which may have 4 gigs of memory, 4 gigabytes. Some people even call half a byte, four bits, a nibble! Metric prefixes are given below.

G.2 Temperature

There are four basic temperature scales: Kelvin, Centigrade or Celsius, which are metric in nature, and Fahrenheit and Rankine, which are more familiar to many of us Americanos, at least Fahrenheit. The Kelvin and Rankine scales, named after William Thomson, Lord Kelvin (1824–1907), and William John Macquorn Rankine (1820–1872), start at absolute zero so that when there is no molecular motion, absolute zero, the temperature in these scales is zero. The Centigrade scale assumes that the freezing point of water is at zero degrees and its boiling point is 100°. The freezing point of water on the Kelvin scale is 273.15, so these two scales are different by the simple additive factor of 273.15. For most purposes the decimal part can be neglected. The freezing point of water on the Fahrenheit scale is 32° and the boiling point is 212°, a 180° difference. The difference between the Fahrenheit and Rankine scales is 459.67 Fahrenheit degrees. So the freezing point of water on the Rankine scale is 492.67°, as if you cared.

The **Fahrenheit** scale was invented by—you guessed it—by Daniel Gabriel Fahrenheit (1686–1736) in 1724. The story of how he picked these values is a bit cloudy, but it is said that he worked with Ole Rømer (1644–1710) and multiplied his values by four to get rid of decimals. He used a brine solution as zero and about 90° as the temperature of the human body. It is amazing that we still use this system today! Lots of engineers use the **Rankine** system, a so-called absolute scale. He wanted an absolute scale and he wanted to use the Fahrenheit degree, so it came out that absolute zero is −459.67 Fahrenheit degrees.

The **Celsius** scale was originally upside down—water boiled at zero degrees and froze at 100. As you might guess, it was invented by Anders Celsius (1701–1744) in 1742. In the next year Jean-Pierre Christin (1683–1755) flipped it, although he was working independently. Since he lived in Lyon, it was called the Lyon thermometer, but today we honor Celsius, although many call it simply the **Centigrade** scale since there are 100° between freezing and boiling. Then the great Lord Kelvin developed his scale to account for the relatively new idea of absolute zero. Would you believe it is called the Kelvin scale? Note that degrees on this scale are often just called kelvins (without the capital), but there are no centigrades, fahrenheits or rankines.

The relations are

$$
\begin{aligned}
R &= F + 459.67 \\
K &= C + 273.15 \\
C &= 5/9(F - 32) = 0.55555(F - 32) \\
F &= 9/5C + 32 = 1.8C + 32
\end{aligned}
$$

Here is the curve that will allow you to change from Fahrenheit to Centigrade temperatures and the reverse. I don't think you need Kelvin or Rankine! Note that 32 °F, 72 °F, and 98 °F are 0 °C, 22.2 °C, and 36.6 °C (figures G1 and G2).

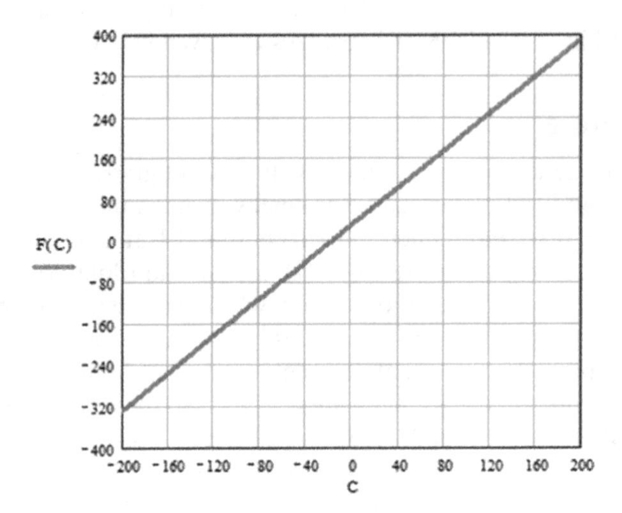

Figure G1. Conversion from *C* to *F*.

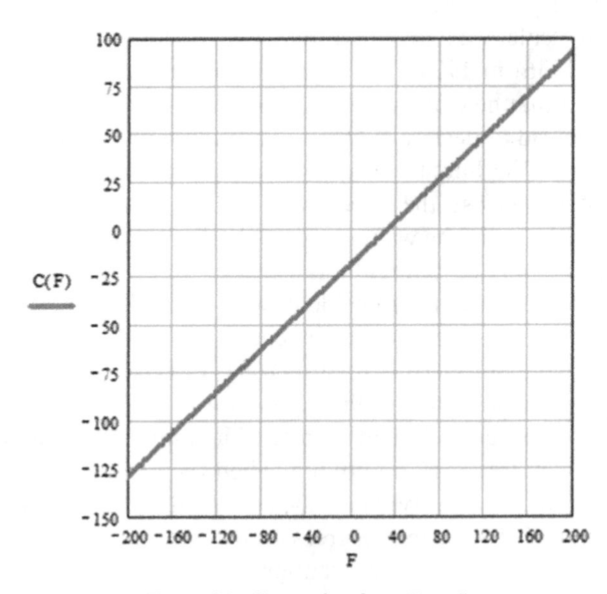

Figure G2. Conversion from *F* to *C*.

G.3 Radiometry and photometry nomenclature

The unit of power in radiometry is the watt. Incident power per unit area is called *irradiance* or *incidance* in watts per unit area. This power density exiting a surface is called *radiant emittance* or *exitance* in the same units. The power per unit area and solid angle is called *radiance* or *sterance.* In watts per unit area per steradian, a *steradian* is a measure of a solid angle. Radiant intensity or *pointance* is the power per unit solid angle in watts per steradian.

Photometry units are similar to radiometry units in geometry, but they are normalized to the response of the human eye. They are called luminous units. The unit of luminous power is the lumen. Luminous power incident per unit area is illuminance with units of lux or foot candle or lumen per square meter. That being emitted is luminous exitance, with the same units. The unit of luminance, which is equivalent to radiance, is the candela per square meter. The unit of luminous intensity is the candela, once called the candle. It is a lumen per steradian. In much of photometry, use is made of the fact that luminance and illuminance are related by π for an isotropic radiator, so many units have π associated with them and have different names, like stilb apostilb, and nit. But I will not go into that.

These units and names are confusing enough, but different fields of scientific endeavors use the same words in different ways. In astronomy, luminosity is the total power emitted by a star, not the luminous power. In astronomy, radiant intensity is watts per square meter per steradian; in radiometry, it is watts per steradian; and in electromagnetic theory, it is watts per square meter. Since Neville Woolf is an astronomer and Emil Wolf is an expert in electromagnetic theory, I declare that intensity is whatever Wo(o)lf(e) wants it to be!

G.4 The wavelengths and frequencies of light

There are a number of ways to specify light wavelengths as well as frequencies. The wavelength, with the universal symbol λ, can be specified in terms of micrometers µm, nanometers, or even ångströms. A micrometer is just a millionth of a meter, a nanometer a billionth of a meter, and an ångström is one ten-millionth of a meter. Therefore, 1 µm = 1000 µm = 10 000 Å. Thus, the yellow line in the middle of the visible spectrum is 0.5 µm = 500 nm = 5000 Å. The visible spectrum ranges from 0.4 µm to 0.8 µm, or 400 nm to 800 nm, or 4000 Å to 8000 Å.

The Ångström unit is in honor of Anders Ångström (1814–1874), a noted spectroscopist, but is not in much use today. As they say in nomenclature circles, it is to be deprecated. So is the term micron. That was in wide use until about 1980; instead use micrometer.

The frequency of light, usually denoted by ν, is inversely related to the wavelength, $\nu = c/\lambda$, where c is the speed of light and λ is the wavelength. So the frequency can be given in Hertz, in honor of Heinrich Hertz (1857–1894), the number of cycles per second, or cps (which is also now deprecated in favor of Hertz, Hz). But it can also be given in terms of the wave number, which is the number of waves per unit length, usually centimeters, $\sigma = 1/\lambda$. The Greek consonant σ should never be confused with the Stefan–Boltzmann constant, which is the same letter and named after Jozef Stefan (1835–1893) and Ludwig Boltzmann (1844–1906).

G.5 Lengths

The base unit of length in the **mks system**—meter kilogram second system—is the **meter**. It is about a yard in the English system, 39.375 inches. The length units that are larger are the decameter, **kilometer**, and megameter—ten meters, one thousand meters, and a million meters, abbreviated Dm, km, Mm. The smaller lengths are the decimeter, **centimeter**, **millimeter**, **micrometer**, **nanometer**, picometer, femtometer, and attometer, abbreviated dm, cm, mm, μm, nm, pm, fm, and am. These are, respectively, a tenth of a meter, one hundredth, one thousandth, one millionth, one billionth. A centimeter is about 2.54 inches, and a kilometer is about 0.6 miles. Really BIG lengths are a light year, the distance light travels in a year: 186 000 mile per second times 365 days per year times 24 h per day times 60 min per hour times 60 seconds per minute = 31 536 000 seconds or 5.8657×10^{12} = 5 865 700 000 000 miles. That's a long way. But a parsec is even longer; it is 3.28 light years. Get out your calculator if you wish.

The distance to the Sun is about 8 light minutes, 89 280 000 miles or about 9×10^7 miles. The moon is only 384 000 km, 384 Mm, 3.84×10^8 m. Earth's diameter is about 12 700 km, 12.7 Mm. Hardly anybody uses megameters, only me. The size of a human hair is about 20 μm.

G.6 Fundamental constants

These are cited in many places in the text and are therefore listed here:

Planck's constant: $h = 6.626\ 0755 \times 10^{34}$ J s
Speed of light: $c = 2.997\ 9248 \times 10^8$ m s^{-1}
Boltzmann's constant: $k = 1.380\ 658 \times 10^{-23}$ J K^{-1}
Stefan–Boltzmann constant: $8\pi^5 k^4/60h^3c^2 = 5.670\ 51 \times 10^{-8}$ W m^{-2} K^{-4}
Wien displacement constant: $b = h.c./k = 2.897\ 756 \times 10^{-3}$ m K
First radiation constant: $c_1 = 2\pi c^2 h = 3.714\ 77 \times 10^{-16}$ W m^{-2}
Second radiation constant: $c_2 = h.c./k = 1.438\ 7770 \times 10^{-2}$ m K

G.7 Metric prefixes

From very small to very large, the main ones are given below with the exponent of ten listed with them:

Prefix	Exponent	Symbol	Common name
Atto	−18	a	Quintillionth
Femto	−15	f	Quadrillionth
Pico	−12	p	Trillionth
Nano	−9	n	Billionth
Micro	−6	μ	Millionth
Milli	−3	m	Thousandth
Centi	−2	c	Hundredth
Deci	−1	d	Tenth

(*Continued*)

(*Continued*)

Prefix	Exponent	Symbol	Common name
Deca	1	da	Ten
Hector	2	h	Hundred
Kilo	3	k	Thousand
Mega	6	M	Million
Giga	9	G	Billion
Terra	12	T	Trillion
Peta	15	P	Quadrillion
Exa	18	E	Quintillion
Zetta	21	Z	Sextillion
Yotta	24	Y	Septillion

IOP Publishing

Appendix H

Symbols

c	speed of light = 299 792 458 m s^{-1}
c_1	first radiation constant = 3.741 771 53(17) × 10^{-16} W m^{-2}
c_2	second radiation constant = 1.438 7770(13) × 10^{-2} m K
E	electric field
H	magnetic field
h	Planck's constant = 6.626 069 57(29) × 10^{-34} J s
k	Boltzmann constant = 1.380 6488(13) × 10^{-23} J K^{-1}
m	meter
n	refractive index
n_c	refractive index at the C line
n_D	refractive index at the D line
n_F	refractive index at the F line
nm	nanometer
T	temperature
t	time
V	dispersion $(n_D - 1)/(n_F - n_C)$
α	angle
β	angle
μ	micro or millionth
ε	emissivity
λ	wavelength, usually in μm, but also in nm
v	frequency
Φ	flux
π	pi = 3.141 592 653 59
σ	Stefan Boltzmann constant = 5.670 373(21) × 10^{-8} W m^{-2} K^{-4}
θ	angle
ω	angular frequency = $2\pi v$
$\sum$	summation
$\sim$	approximately equal
∞	infinity
$\pm$	plus or minus
$\int$	integral
∇	del operator = $i\partial + j\partial + k\partial$

doi:10.1088/978-0-7503-2612-4ch54

i, **j**, **k**	unit vectors in the x, y and z directions
∂	partial derivative
d	derivative, as in dn/dT
$\nabla \mathbf{X} \mathbf{v}$	curl $\mathbf{v} = \mathbf{i}(\partial v_z/\partial y - \partial v_y/\partial z) + \mathbf{j}(\partial v_x/\partial z - \partial v_z/\partial x) + \mathbf{k}(\partial v_y/\partial x - \partial v_x/\partial y)$
$\nabla \mathbf{v}$	divergence of $\mathbf{v} = dv_x/dx + dv_y + dy + dv_z/dz$

The curl and divergence are for rectangular coordinates.

IOP Publishing

Rays, Waves and Photons

A compendium of foundations and emerging technologies of pure and applied optics

William L Wolfe

Appendix I

Journal names and abbreviations

Abh. Kgl. Sachs. Ges. Wiss. math-phys—Abhandlungen der Koniglich Sächscishe Gesellschaft der Wissenschaften mathematische-physik.
Am. J. Phys.—American Journal of Physics.
Am. J. Sci.—American Journal of Science.
Ann. Chim. Phys.—Annales de chimie et de phyisique.
Ann. Phys.—Annalen der Physik.
Ann. N. Y. Acad. Sci.—Annals of the New York Academy of Science.
Ann. Phys. Chem.—Annalen der Physik und Chemie.
Annu. Rev. Mater. Sci.—Annual Reviews of Material Science.
Appl. Opt.—Applied Optics.
Appl. Phys. Lett.—Applied Physics Letters.
Armada Intl.—Armada International.
ASAIO J.—American Society for Artificial Internal Organs Journal.
Bell Syst. Tech. J.—Bell System Technical Journal.
Camb. Dubl. Math. J.—Cambridge and Dublin Mathematical Journal.
Can. Med. Assoc. J.—Canadian Medical Association Journal.
C. R.—Comptes Rendus.
Electron. Lett.—Electronics Letters.
Eur. Phys. Lett.—European Physics Letters.
J. Electrochem. Soc.—Journal of the American Chemical Society.
J. Opt. Soc. Am.—Journal of the Optical Society of America.
J. Appl. Phys.—Journal of Applied Physics.
J. Phys. Chem. Solids—Journal of the Physics and Chemistry of Solids.
J. Sci. Instrum.—Journal of Scientific Instruments.
Med. Biol. Eng. Comput.—Medical and Biological Engineering and Computing.
Opt. Commun.—Optical Communications.
Philos. Mag.—Philosophical Magazine.
Philos. Mag. Suppl.—Philosophical Magazine Supplement.
Philos. Trans.—Philosophical Transactions.

doi:10.1088/978-0-7503-2612-4ch55

Philos. Trans. R. Soc. Lond.—Philosophical Transactions of the Royal Society of London.
Phys. Rev.—Physical Review.
Proc. Am. Philos. Soc.—Proceedings of the American Philosophical Society.
Proc. IEEE—Proceedings of the Institute of Electrical and Electronic Engineers.
Proc. Inst. Electron. Engrs.—Proceedings of the Institute of Electronic Engineers.
Proc. IRE—Proceedings of the Institute of Radio Engineers.
Proc. IRIS—Proceedings of the Infrared Information Symposia.
Proc. SPIE—Proceedings of the Society for Photographic Instrumentation Engineers.
Rev. Sci. Instrum.—Review of Scientific Instruments.
SMPTE—Society of the Motion Picture and Television Engineers.
Trans. Am. Phys. Soc.—Transaction of the American Physical Society.
Trans. Cambridge Philos. Soc.—Transaction of the Cambridge Philosophical Society.
Umschau Wiss.—Umschau Wissenschaften.
Verh. der Deutschen Phys. Gesellschaft—Verhandlungen der Deutschen Physikalischen Gesellschaft.
Z. Instrum.—Zeitschrift fur Instrumentkunde.
Zh. Exp. Phys. Tech. Pisma Red.—Zhurnal Experimental Technical Physics, Pisma Redaktor.
Zh. russ. fiz.-khim. obsh.—Zhurnal russkogo fiziko-khimicheskogo obshchestva.

IOP Publishing

Rays, Waves and Photons

A compendium of foundations and emerging technologies of pure and applied optics

William L Wolfe

Glossary

Abbe V number	A specification of the dispersion of a material, $V = (n_D - 1)/(n_F - n_c)$.
Aberrations	The lack of perfection in an optical system.
Absorption	The change of optical energy into some other form, like heat.
Absorptivity	The ratio of the absorption to that of a perfect absorber.
Accommodation	The adjustment of focus of the eye by altering the lens shape.
Absorptance	The ratio of absorption to that of a perfect absorber for a particular sample.
Achromatic lens	A lens corrected for two colors and spherical aberration.
Acousto-optics	The interaction of sound and light.
Afocal	Without a focus, pertaining to collimated beams.
Airy curve	The diffraction pattern of the image of a point source; also called Airy pattern.
Airy disc	The central disc of the two-dimensional Airy pattern.
Anamorphic optics	Optics that do not have circular symmetry.
Anastigmat	A lens corrected for spherical aberration, coma, and astigmatism.
Aperture stop	An optical stop that limits the collection area.
Aplanat	A lens corrected for spherical aberration and coma.
Apochromatic lens	A lens corrected for three colors and spherical aberration.
Arago spot	The bright spot behind an opaque disc caused by diffraction; same as Poisson spot.
Argument	In mathematics, the independent variable of a function, the x in sin (x).
Astigmatism	An off-axis aberration that images an object point into two lines.
Atmospheric sounding	Using several wavelengths to measure atmospheric properties at different altitudes.
Barrel distortion	An aberration that causes the image of a square to have bulging edges like a barrel.
Baud	One bit per second.
Biconvex lens	A lens with two convex surfaces.
Birefringence	The property in which a material has different refractive indices in different directions.
Blackbody	A perfect absorber and a perfect radiator.
Calorimetry	A measurement of the amount of heat.

Cerenkov radiation	Continuous spectrum radiation caused by a very fast charged particle.
Cervit	A special low-expansion glass formulation.
Chromatic aberration	Imperfect imagery due to the difference in refractive indices at different wavelengths.
Colorimeter	A device for performing colorimetry.
Colorimetry	The measurement of color, or the measurement of absorption at a given color.
Collimation	The act of generating a beam of parallel light.
Concave surface	A surface that bows inward like a bowl.
Contact lens	A correcting lens that is in contact with the eye.
Convex surface	A surface that bulges out like a balloon.
Coma	An off-axis aberration that images an object point into a comet shape.
Crown glass	Glass made from the crowns; that is, lumps from old doorways with $V > 50$.
Critical angle	The angle at which light from a dense to rarer medium is totally reflected.
Curvature of field	An aberration that results in a curved focal surface.
Decibel	A logarithmic representation of a ratio: 1 dB = 10 log (ratio); 3 dB = 0.5 = 50%.
Detectivity	The ratio of signal-to-noise to an input, usually power.
Dewar	A laboratory device to keep things cold, similar to a Thermos bottle.
Diffraction	The redistribution of light as a result of an obstacle.
Diffraction limit	The degree to which an optical element does not focus a point object perfectly to a point image due to diffraction.
Diffraction orders	The successive spectra that occur at regular intervals.
Direction cosines	The cosines of the angles between a ray and the coordinate axes.
Dispersion	The variation of refractive index with wavelength; indicated by V in the visible, $dn/d\lambda$ mathematically.
Double refraction	A single ray is split into two different rays in a material that has birefringence.
Electro-optics	The interaction of electricity and light.
Emission	The exiting of optical radiation from something.
Emissivity	The ratio of the emission of a given substance to that of a perfect radiator.
Emittance	The ratio of the emission from a particular surface to that of a perfect radiator.
Equatorial orbit	A satellite path that is over the equator.
Étendue	The same as throughput, which see.
Excimer laser	A laser emitting in the ultraviolet region from an excited dimer.
False color	The representation of light with wavelengths outside the visible by colors of visible light.
Fast	A fast lens is one that has a low F/number, ratio of focal length to aperture diameter.
Fellgett advantage	Same as multiplex advantage.
Field stop	An optical stop that defines the field of view.
First light	The first actual use of a telescope.
Flint glass	Glass made from flint with lead added with $V < 50$.

F number	Ratio of the focal length of an optical element to its aperture diameter; abbreviated F, F/, and f/.
Focal ratio	The same as F/number.
Geosynchronous satellite or orbit	One that orbits at the same rate that the Earth revolves, thereby staying over the same spot on the ground about 22 320 km high and 24 h orbit time.
Heliocentricity	A Sun at the center of a solar system.
Hertz	A unit of frequency, cycles per second.
Heterochromatic	Consisting of more than one color or wavelength.
Higher-order asphere	A shape (of a mirror or lens) that is not spherical or conic, but described by an equation that may have terms in the fourth, sixth or higher orders.
Hydrophobic	Hates water, repels water.
Hydrophilic	Loves water, interacts with water.
Hyperopia	Farsightedness.
Illuminance	Visible irradiance.
Immersion	See optical immersion.
Index of refraction	The ratio of the speed of light in vacuo to that in a material; $n = c/v$.
Irradiance	Incident power per unit area; also called radiant incidance.
Jacquinot advantage	Same as throughput advantage.
Jinc function	Bessel function of the first kind divided by its argument, $J^1(x)/x$.
Kirchhoff's law	Absorptivity equals emissivity (under certain conditions).
Kilometer	One thousand meters.
Kluge	A device that is clumsily put together, but works; from the German for smart?
Laser	Light amplification by stimulated emission of radiation; also LASER.
Lead telluride	An infrared detector material.
Lens	A transparent material with two surfaces that redirect light.
Lens speed	The focal length divided by the aperture diameter; see also relative aperture.
LIDAR	Light Detection And Ranging, or the portmanteau light radar.
Luminance	Visible power per unit projected area and solid angle, visible radiance.
Luminiferous ether	A substance by which light was thought to propagate; proved to be nonexistent.
Magneto-optics	The interaction of a magnetic field and light.
Maser	Microwave Amplification by Stimulated Emission of Radiation; also MASER.
Matrix	An array of numbers in columns and rows, 2 by 2, 4 by 4, 8 by 10, etc.
Maxwell's equation	Set of vector-differential equations about electric and magnetic fields.
MCT	See mercury cadmium telluride.
Meniscus lens	A lens consisting of two surfaces which both curve in the same direction.
Mercury cadmium telluride	A mixed crystal infrared detector.
Meter	A metric unit of length equal to about 39 inches.
Micrometer	A millionth of a meter, also called a micron; abbreviated μm.

Mie scattering	Light scattering by particles about as large or larger than the wavelength.
Mirror	A device that reflects light.
Modulation	The process of varying something with time or space.
Modulation transfer function	The ratio of the spatial spectrum of the image modulation to that of the object.
Multiplex advantage	Sensing all wavelengths at once, in a spectrometer.
Myopia	Nearsightedness.
Nanometer	One billionth of a meter, thousandth of a micrometer; abbreviated nm.
n_c	The refractive index at the C line, 656.3 nm.
n_D	The refractive index at the D line, 589.3 nm.
n_F	The refractive index at the F line, 486.1 nm.
Normal	Perpendicular; also opposite of unusual.
Numerical aperture	Abbreviated NA, the refractive index times the sine of the half angle of the cone of light; also the reciprocal of twice the F/number (with $n = 1$).
Optical immersion	A method in which the lens (of a microscope) is in a liquid that is on the sample to increase magnification; also used in infrared systems.
Optical stop	An element in an optical system that limits the range of the collection of rays.
Optical system	A collection of optical elements, lenses, mirrors, prisms, gratings, etc.
OSA	The Optical Society of America, now just the Optical Society.
Panchromatic	Covering the entire visible spectral band.
Petzval sum	The total curvature in an optical (lens or mirror) system.
Phonon	A quantum of sound energy.
Photon	A quantum of light energy.
Photoconductor	A material that conducts electricity when illuminated but doesn't when not.
Photodetector	One that converts photons into electrons; also photon detector.
Photoemissive surface	One that emits electrons when photons of sufficient energy are incident upon it.
Photosphere	The outer portions of the Sun, the shell.
Pincushion distortion	An aberration in which the image of a square has indented sides like a pincushion.
Pixel	Picture element; often an individual detector in an array of detectors or LED in illuminators.
Planck's law	The equation for the spectral distribution of blackbody radiation.
Plano-concave lens	A lens with one plane and one concave surface.
Plano-convex lens	A lens with one plane and one convex surface.
Plasma	Roughly, a gas of free positive and negative charges that is essentially neutral.
Point spread function	The distribution of energy in the image of a point object.
Poisson spot	The bright spot behind an opaque object caused by diffraction; same as Arago spot.
Polar orbit	Satellite path that goes over the poles.
Presbyopia	Lack of accommodation of the eye.

Push-broom scan	A scan in which an array of detectors is perpendicular to the direction of motion of an aircraft or satellite.
Radiance	Power per unit projected area and solid angle; also called sterance.
Radiant emittance	Power per unit area into a hemisphere; also called radiant exitance.
Radiant incidence	Incident power per unit area; also called irradiance.
Radiative cooler	A device that is cooled by viewing the cold of outer space.
Radiometer	A device for measuring the amount of radiation.
Rayleigh resolution criterion	A specification of optical resolution in which two point sources produce Airy images where the first maximum of one falls on the first minimum of the other.
Rayleigh scattering	Light scattering by particles smaller than the wavelength.
Reflection	The return of light back from a material.
Reflectivity	The ratio of reflected light to incident light.
Reflectance	Reflectivity for a particular material.
Refraction	The bending of light as it goes between two surfaces of different refractive index.
Refractive index	The ratio of the speed of light in vacuo with that in a material; $n = c/v$.
Relative aperture	The focal length divided by the diameter of the aperture.
Responsivity	The ratio of an output signal to an input, e.g. volts per watt.
Relative aperture	The focal length divided by the aperture diameter; same as focal ratio.
Scattering	The redirection of light by particles and/or irregularities.
Scophony	Early mechanical television systems.
Seidel aberrations	Lens imperfections described by the Seidel polynomial.
Sinc function	Sine of an argument divided by the argument, $\sin(x)/x$.
Sodium D-lines	Two bright, closely spaced yellow lines characteristic of sodium.
Sparrow resolution criterion	Two points are resolved when there is no dip between the Airy curves of the images of two points.
Specific detectivity	Detectivity normalized to area and bandwidth.
Spectral resolving power	A measure of spectral purity, $\lambda/d\lambda$.
Spectrum	A plot of the intensity of light against the wavelength (or frequency).
Speculum metal	An alloy of tin and brass.
Speed (of a lens)	The F/number, ratio of the focal length to the aperture diameter.
Speed of light	Approximately 186 000 miles per second, or 300 000 km s^{-1}.
Spherical aberration	The aberration caused by the use of spherical surfaces.
SPIE	Society of Photo-Optical Instrumentation Engineers.
Stefan–Boltzmann law	Describing total radiation of a blackbody, $M = \sigma T^4$.
Sterance	Power per unit projected area and solid angle; also called radiance.
Stop	Optical stop; an element in an optical system that limits the range of rays.
Sun synchronous satellite or orbit	A polar orbit that is always at the same local time, 600–800 km high, and 95–100 min orbit time.
Telephoto lens	A lens arrangement in which the focal length is greater than the physical length.
Thermal detector	One that turns infrared energy into an electrical signal by increasing its temperature.

Throughput	Generally how much light gets through a system; technically the $A\Omega$, area-solid angle product.
Throughput advantage	In a spectrometer having larger throughput.
Transmission	The passage of light through a medium.
Transmissivity	The ratio of transmitted light to incident light.
Transmittance	Transmissivity for a particular sample.
UHF	Ultra high frequency, from about 300 MHz to 1 GHz.
V number	A measure of dispersion in the visible, $V = (n_D - 1)/(n_F - n_c)$; also called dispersion.
Voxel	A volume pixel, a three-dimensional pixel.
Wafer	As used in the semiconductor industry, the slice of silicon on which the circuit is made.
Wollaston prism	A special prism that polarizes light.
Whisk-broom scan	A scan with an array of detectors that is parallel to the direction of vehicle motion and is moved from side to side perpendicular to that direction.
Wien's displacement law	Describing the maximum of the blackbody curve, $\lambda_m T = 2889$ μmK.
WYSIWYG	What you see is what you get.
Zernike polynomial	A method of describing optical aberrations.
Zerodur	A special low-expansion glass formulation.
Zoom lens	An optical system wherein the image stays in focus as the focal length changes.

CPSIA information can be obtained
at www.ICGtesting.com
Printed in the USA
BVHW010440020421
603972BV00004B/54